This book provides a comprehensive and detailed introduction to the principles of particle detectors used in physics, biology and medicine.

Introductory chapters review the interactions of particles and radiation with matter, introduce the principles of detector operation and describe and define different types of measurement and their units. The main body of the book encompasses all currently used detectors and counters. Each description covers basic principles, potential uses and limitations. The scope of the book includes detectors for ionization and track measurement, methods for time, energy and momentum measurement, and for particle identification. Two chapters are dedicated to electronics (readout methods, monitoring, data acquisition) and data analysis. A final chapter gives examples of detector systems. The book concludes with a detailed glossary of terms, tables of units and physical constants and a detailed reference list.

The book is written as a reference guide for researchers, graduate students and users of particle detectors in physics, medicine and biology.

CAMBRIDGE MONOGRAPHS ON PARTICLE PHYSICS, NUCLEAR PHYSICS AND COSMOLOGY

CAMBRIDGE MONOGRAPHS ON
PARTICLE PHYSICS,
NUCLEAR PHYSICS AND COSMOLOGY
5

General Editors: T. Ericson, P. V. Landshoff

PARTICLE DETECTORS

The computer reconstruction of a three-jet event
from the Z^0-decay recorded in the ALEPH experiment at the
Large Electron-Positron storage ring LEP at CERN.

29 September 1992

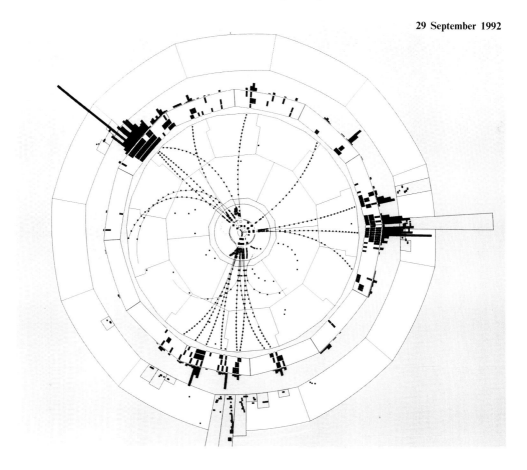

PARTICLE DETECTORS

CLAUS GRUPEN

University of Siegen

with the cooperation of
ARMIN BÖHRER and LUDĚK SMOLÍK

CAMBRIDGE
UNIVERSITY PRESS

Published by the Press Syndicate of the University of Cambridge
The Pitt Building, Trumpington Street, Cambridge CB2 1RP
40 West 20th Street, New York, NY 10011-4211, USA
10 Stamford Road, Oakleigh, Melbourne 3166, Australia

First published 1996

Printed and bound in Great Britain by
Biddles Ltd, Guildford and King's Lynn

A catalogue record for this book is available from the British Library

Library of Congress cataloguing in publication data

Grupen, Claus.
Particle detectors / Claus Grupen.
p.   cm. – (Cambridge monographs on particle physics,
nuclear physics, and cosmology; 5)
Includes bibliographical references and index.
ISBN 0 521 55216 8 (hc)
1. Nuclear counters.   I. Title   II. Series.
QC787.G6G78   1996
539.7′7–dc20   95-18455 CIP

ISBN 0 521 55216 8 hardback

Contents

Preface

The basic motive which drives the scientist to new discoveries and understanding of nature is curiosity. Progress is achieved by carefully directed questions to nature, by experiments. To be able to analyze these experiments, results must be recorded. The most simple instruments are the human senses, but for modern questions, these natural detection devices are not sufficiently sensitive or they have a range which is too limited. This becomes obvious if one considers the human eye. To have a visual impression of light, the eye requires approximately 20 photons. A photomultiplier, however, is able to 'see' single photons. The dynamical range of the human eye comprises half a frequency decade (wavelengths from 400 to 800 nm), while the spectrum of electromagnetic waves from domestic current over radiowaves, microwaves, infrared radiation, visible light, ultraviolet light, X-rays, and gamma rays, covers 23 frequency decades!

Therefore, for many questions to nature, precise measurement devices or detectors had to be developed to deliver objective results over a large dynamical range. In this way, the human being has sharpened his 'senses' and has developed new ones. For many experiments, new and special detectors are required and these involve in most cases not only just one sort of measurement. However, a multifunctional detector which allows one to determine all parameters at the same time does not exist yet.

To peer into the world of the microcosm, one needs microscopes. Structures can only be resolved to the size of the wavelength used to observe them; for visible light this is about $0.5\,\mu$m. The microscopes of elementary particle physicists are the present day accelerators with their detectors. Because of the inverse proportionality between wavelengths and momentum (de Broglie relation), particles with high momentum allow small structures to be investigated. At the moment, resolutions of the order of 10^{-17} cm can be reached, which is an improvement compared to the optical microscope of a factor of 10^{13}.

To investigate the macrocosm, the structure of the universe, energies in the ranges between some one hundred microelectron volts (μeV, cosmic microwave background radiation) up to 10^{20} eV (high energy cosmic rays) must be recorded. To master all these problems, particle detectors are required which can measure parameters like time, energy, momentum, velocity and the spatial coordinates of particles and radiation. Furthermore, the nature of particles must be identified. This can be achieved by a combination of a number of different techniques.

In this book, particle detectors are described which are in use in elementary particle physics, in cosmic ray studies, in high energy astrophysics, nuclear physics, and in the fields of radiation protection, biology and medicine. Apart from the description of the working principles and characteristic properties of particle detectors, fields of application of these devices are also given.

This book originated from lectures which I have given over the past 20 years. In most cases these lectures were titled 'Particle Detectors'. However, also in other lectures like 'Introduction to Radiation Protection', 'Elementary Particle Processes in Cosmic Rays', 'Gamma Ray Astronomy' and 'Neutrino Astronomy' special aspects of particle detectors were described. This book is an attempt to present the different aspects of radiation and particle detection in a comprehensive manner. The application of particle detectors for experiments in elementary particle physics and cosmic rays is, however, one of the main aspects.

I would like to mention that excellent books on particle detectors do already exist. In particular I want to emphasize the four editions of the book of Kleinknecht [1] and the slightly out-of-date book of Allkofer [2]. But also other presentations of the subject deserve attention [3]-[25].

Without the active support of many colleagues and students, the completion of this book would have been impossible. I thank Dr U. Schäfer and Dipl. Phys. S. Schmidt for many suggestions and proposals for improvement. Mr R. Pfitzner and Mr J. Dick have carefully done the proof reading of the manuscript. Dr G. Cowan and Dr H. Seywerd have significantly improved my translation of the book into English. I thank Mrs U. Bender, Mrs C. Tamarozzi and Mrs R. Sentker for the production of a ready-for-press manuscript and Mr M. Euteneuer, Mrs C. Tamarozzi as well as Mrs T. Stöcker for the production of the many drawings. I also acknowledge the help of Mr J. Dick, Dipl. Phys.-Ing. K. Reinsch, Dipl. Phys. T. Stroh, Mr R. Pfitzner, Dipl. Phys. G. Gillessen and Mr Cornelius Grupen for their help with the computer layout of the text and the figures.

Siegen, January 1996   Claus Grupen

Introduction

The development of particle detectors practically starts with the discovery of radioactivity by Henri Becquerel in the year 1896. He noticed that the radiation emanating from uranium salts could blacken photosensitive paper. Almost at the same time X-rays, which originated from materials after the bombardment by energetic electrons, were discovered by Wilhelm Conrad Röntgen.

The first nuclear particle detectors (X-ray films) were thus extremely simple. Also the zinc-sulfide scintillators in use at the beginning of the century were very primitive. Studies of scattering processes — e.g., of α-particles — required tedious and tiresome optical registration of scintillation light with the human eye.

In the course of time the measurement methods have been greatly refined. Today, it is generally insufficient only to detect particles and radiation. One wants to identify their nature, i.e., one would like to know whether one is dealing, for example, with electrons, muons, pions or energetic γ-rays. On top of that, an accurate energy and momentum measurement is often required. For the majority of applications an exact knowledge of the spatial coordinates of particle trajectories is of interest. From this information particle tracks can be reconstructed by means of optical (for example in spark chambers, streamer chambers, bubble and cloud chambers) or electronic (in multiwire proportional or drift chambers) detection.

The trend of particle detection has shifted in the course of time from optical measurement to purely electronic means [26]. Therefore, the electronic processing of signals from particle detectors plays an increasingly important rôle. Also the storage of data on magnetic disks or tapes and computer aided preselection of data is already an integral part of complex detection systems. This means of data acquisition allows also very high detection rates. While optical registration, for example with cloud chambers, allows only the recording of one event per minute, it is possible today, for example with organic scintillation counters, to process data rates of the order of 10^9 Hertz.

In this book the detectors are ordered according to the object of measurement. This ordering scheme has been introduced consistently by Kleinknecht [1]. In individual cases it is not always easy to decide which sort of measurement a certain detector type can be associated with. Each detector determines to some degree the spatial coordinates of particles because it only registers these particles if they went through the limited volume of the detector. The detectors, therefore, are ordered according to their primary purpose. This is not necessarily obvious or unique because solid state detectors, for example in nuclear physics, are used to make very precise energy measurements, but as solid state strip detectors in elementary particle physics they are used for accurate track reconstruction.

In this book the application of particle detectors in nuclear physics, elementary particle physics, in the physics of cosmic rays, astronomy and astrophysics as well as in biology and medicine are weighted in a different manner. The main object of this book is the application of particle detectors in elementary particle physics. This also includes astrophysical applications and techniques from the field of cosmic rays because these fields are very close to particle physics.

1
Interactions of particles and radiation with matter

Particles and radiation cannot be detected directly, but rather only through their interactions with matter. There are specific interactions for charged particles which are different from those of neutral particles, e.g., of photons. One can say that every interaction process can be used as a basis for a detector concept. The variety of interactions is quite rich and, as a consequence, a large number of detection devices for particles and radiation exist. In addition, for one and the same particle, different interaction processes at different energies may be relevant.

In this chapter the main interaction mechanisms will be presented in a comprehensive fashion. Special effects will be dealt with when the individual detectors are being presented. The interaction processes and their cross sections will not be derived from basic principles but are presented only in their results, as they are used for particle detectors.

The main interactions of charged particles with matter are ionization and excitation. For relativistic particles, bremsstrahlung energy losses must also be considered. Neutral particles must produce charged particles in an interaction that are then detected via their characteristic interaction processes. In the case of photons these processes are the photoelectric effect, Compton scattering, and pair production of electrons. The electrons produced in these photon interactions can be observed through their ionization in the sensitive volume of the detector.

1.1 Interactions of charged particles

Charged particles passing through matter lose kinetic energy by excitation of bound electrons and by ionization. Excitation processes like

$$
\begin{aligned}
e^- + \text{atom} \quad &\longrightarrow \quad \text{atom}^* + e^- \\
&\hookrightarrow \text{atom} + \gamma
\end{aligned}
\tag{1.1}
$$

lead to low energy photons and are therefore useful for particle detectors which can record this luminescence. Of greater importance are pure scattering processes in which incident particles transfer a certain amount of their energy to atomic electrons so that they are liberated from the atom.

The maximum transferable kinetic energy to an electron depends on the mass m and the momentum of the incident particle. Given the momentum of the incident particle

$$p = mv = \gamma m_0 \beta c \, , \tag{1.2}$$

where γ is the Lorentz factor $(= E/m_0 c^2)$, $\beta c = v$ the velocity and m_0 the rest mass, the maximum energy that may be transferred to an electron (mass m_e) is given by [27]

$$E_{\text{kin}}^{\text{max}} = \frac{2 m_e c^2 \beta^2 \gamma^2}{1 + 2\gamma m_e/m_0 + (m_e/m_0)^2} = \frac{2 m_e p^2}{m_0^2 + m_e^2 + 2 m_e E/c^2} \, . \tag{1.3}$$

In this case it makes sense to give the kinetic energy, rather than total energy, for the electron since the electron is already there and does not have to be produced. The kinetic energy E_{kin} is related to the total energy E according to

$$E_{\text{kin}} = E - m_0 c^2 = c\sqrt{p^2 + m_0^2 c^2} - m_0 c^2 \, . \tag{1.4}$$

For low energies

$$2\gamma m_e/m_0 \ll 1 \tag{1.5}$$

and under the assumption that the incident particles are heavier than electrons ($m_e < m_0$) equation (1.3) can be approximated by

$$E_{\text{kin}}^{\text{max}} = 2 m_e c^2 \beta^2 \gamma^2 \, . \tag{1.6}$$

A particle (e.g., a muon, $m_\mu c^2 = 106 \, \text{MeV}$) with a Lorentz factor of $\gamma = E/m_0 c^2 = 10$ corresponding to $E = 1.06 \, \text{GeV}$ can transfer approximately $100 \, \text{MeV}$ to an electron (mass $m_e c^2 = 0.511 \, \text{MeV}$).

If one neglects the quadratic term in the denominator of equation (1.3) $((m_e/m_0)^2 \ll 1)$, which is a good assumption for all incident particles except for electrons, it follows that

$$E_{\text{kin}}^{\text{max}} = \frac{p^2}{\gamma m_0 + m_0^2/2 m_e} \, . \tag{1.7}$$

For relativistic particles $E_{\text{kin}} \approx E$ and $pc \approx E$ holds. Consequently, the maximum transferable energy is

$$E^{\text{max}} = \frac{E^2}{E + m_0^2 c^2/2 m_e} \tag{1.8}$$

which for muons gives

$$E^{\text{max}} = \frac{E^2}{E + 11} \quad ; \quad E \text{ in GeV} . \tag{1.9}$$

In the extreme relativistic case $(E \gg m_0^2 c^2 / 2m_e)$ the total energy can be transferred to the electron.

If the incident particle is an electron, these approximations are no longer valid. In this case, one gets (compare equation (1.3))

$$E_{\text{kin}}^{\text{max}} = \frac{p^2}{m_e + E/c^2} = \frac{E^2 - m_e^2 c^4}{E + m_e c^2} \tag{1.10}$$

$$E_{\text{kin}}^{\text{max}} = E - m_e c^2 \tag{1.11}$$

which is also expected in classical non-relativistic kinematics for particles of equal mass for a central collision.

1.1.1 Energy loss by ionization and excitation

The treatment of the maximum transferable energy has already shown that incident electrons, in contrast to heavy particles $(m_0 \gg m_e)$, play a special rôle. Therefore, to begin with, we give the energy loss for 'heavy' particles. Following Bethe and Bloch [5, 6],[28]-[32]* the average energy loss dE per length dx is given by

$$-\frac{dE}{dx} = 4\pi N_A r_e^2 m_e c^2 z^2 \frac{Z}{A} \frac{1}{\beta^2} \left[\ln \frac{2m_e c^2 \gamma^2 \beta^2}{I} - \beta^2 - \frac{\delta}{2} \right] , \tag{1.12}$$

where

z – charge of the incident particle in units of the elementary charge

Z, A – atomic number and atomic weight of the absorber

m_e – electron mass

r_e – classical electron radius $\left(r_e = \frac{1}{4\pi\varepsilon_0} \cdot \frac{e^2}{m_e c^2} \right.$ with ε_0 - permittivity of free space)

N_A – Avogadro number (= number of atoms per gram atom) $= 6.022 \cdot 10^{23} \text{ mol}^{-1}$

I – ionization constant, characteristic of the absorber material, which can be approximated by

$$I = 16 \, Z^{0.9} \, \text{eV} \quad \text{for } Z > 1 .$$

To a certain extent, I also depends on the molecular state of the absorber atoms, e.g., $I = 15 \, \text{eV}$ for atomic and $19.2 \, \text{eV}$ for molecular hydrogen. For liquid hydrogen I is $21.8 \, \text{eV}$.

---

* For the following considerations and formulae, not only the original literature but also secondary literature was used, mainly [27]-[35] and references therein.

δ – is a parameter which describes how much the extended transverse electric field of incident relativistic particles is screened by the charge density of the atomic electrons. In this way, the energy loss is reduced ('density effect', 'Fermi plateau' of the energy loss). As already indicated by the name, this density effect is important in dense absorber materials, such as lead or iron. For gases under normal pressure and for not too high energies it can be neglected.

For energetic particles, δ can be approximated by

$$\delta = 2\ln\gamma + \zeta \, ,$$

where ζ is a material-dependent constant.

Various approximations for δ and material dependences for parameters, which describe the density effect, are discussed extensively in the literature [33].

A useful constant appearing in equation (1.12) is

$$4\pi N_A r_e^2 m_e c^2 = 0.3071 \, \frac{\text{MeV}}{\text{g/cm}^2} \, .$$

In the logarithmic term of equation (1.12), the quantity $2m_e c^2 \gamma^2 \beta^2$ occurs in the numerator, which, according to equation (1.6), is identical to the maximum transferable energy. The average energy of electrons produced in the ionization process in gases equals approximately the ionization energy [5, 6].

If one uses the approximation for the maximum transferable energy (equation (1.6)) and the shorthand

$$\kappa = 2\pi N_A r_e^2 m_e c^2 z^2 \cdot \frac{Z}{A} \cdot \frac{1}{\beta^2} \tag{1.13}$$

the Bethe-Bloch formula can be written as

$$-\frac{\mathrm{d}E}{\mathrm{d}x} = 2\kappa \left[\ln \frac{E_{\text{kin}}^{\text{max}}}{I} - \beta^2 - \frac{\delta}{2} \right] \, . \tag{1.14}$$

The energy loss $-\frac{\mathrm{d}E}{\mathrm{d}x}$ is usually given in units of $\frac{\text{MeV}}{\text{g/cm}^2}$. The length unit $\mathrm{d}x$ (in g/cm^2) is commonly used, because the energy loss per surface mass density

$$\mathrm{d}x = \varrho \cdot \mathrm{d}s \tag{1.15}$$

with ϱ density (in g/cm^3) and ds length (in cm) is largely independent of the properties of the material. This length unit $\mathrm{d}x$ consequently gives the surface mass density of the material.

Equation (1.12) represents only an approximation for the energy loss of charged particles by ionization and excitation in matter which is, however, precise at the level of a few percent up to energies of several hundred GeV. However, equation (1.12) cannot be used for slow particles, that is, for particles which move with velocities which are comparable to those of atomic electrons or slower. For these velocities ($\alpha z \gg \beta \geq 10^{-3}$, $\alpha = \frac{e^2}{4\pi\varepsilon_0\hbar c}$; fine structure constant) the energy loss is proportional to β. The energy loss of slow protons, e.g., in silicon, can be described by [34, 35]

$$- \frac{\mathrm{d}E}{\mathrm{d}x} = 61.2 \; \beta \quad \frac{\mathrm{GeV}}{\mathrm{g/cm^2}} \quad , \quad \beta < 5 \cdot 10^{-3} \; . \qquad (1.16)$$

Equation (1.12) is valid for all velocities

$$\beta \gg \alpha z \; .$$

In the low energy domain the energy loss decreases like $1/\beta^2$ and reaches a broad minimum of ionization near $\beta\gamma \approx 4$. Relativistic particles ($\beta \approx 1$), which have an energy loss corresponding to this minimum are called 'minimum-ionizing particles'. In light absorber materials, where the ratio $Z/A \approx 0.5$, the energy loss of minimum-ionizing particles can be roughly represented by

$$- \frac{\mathrm{d}E}{\mathrm{d}x}\bigg|_{\mathrm{min}} \approx 2 \; \frac{\mathrm{MeV}}{\mathrm{g/cm^2}} \; . \qquad (1.17)$$

In table 1.1 the energy losses of minimum-ionizing particles in different materials are given; for further values see [34, 35].

The energy loss increases again for $\gamma > 4$ ('logarithmic rise') because of the logarithmic term in the bracket of equation (1.12). The increase follows approximately a dependence like $2 \ln \gamma$.

The decrease of the energy loss at the ionization minimum with increasing atomic number of the absorber originates mainly from the Z/A term in equation (1.12). A large fraction of the logarithmic rise relates to large energy transfers to few electrons in the medium (δ or knock-on electrons). Because of the density effect the logarithmic rise of the energy loss saturates at high energies.

The energy loss according to equation (1.12) describes only energy losses due to ionization and excitation. At high energies, radiation losses become more and more important (see section 1.1.4). The energy loss due to bremsstrahlung is proportional to the energy of the particle. It dominates the total energy loss, even for heavy particles, at very high energies ($> 1 \, \mathrm{TeV}$).

The energy loss by ionization and excitation in iron for muons is shown in figure 1.1 [34, 35, 36].

Figure 1.2 shows the energy loss for electrons, muons, pions, protons, deuterons, and α-particles in air [37].

Table 1.1. *Average energy loss of minimum-ionizing particles in various materials [34, 35]; gases under standard pressure and temperature*

absorber	$\frac{dE}{dx}\big|_{min}\left[\frac{MeV}{g/cm^2}\right]$	$\frac{dE}{dx}\big|_{min}\left[\frac{MeV}{cm}\right]$
Hydrogen (H_2)	4.12	$0.37 \cdot 10^{-3}$
Helium	1.94	$0.35 \cdot 10^{-3}$
Lithium	1.58	0.84
Beryllium	1.61	2.98
Carbon (Graphite)	1.78	4.03
Nitrogen	1.82	$2.28 \cdot 10^{-3}$
Oxygen	1.82	$2.60 \cdot 10^{-3}$
Air	1.82	$2.35 \cdot 10^{-3}$
Carbon dioxide	1.82	$3.60 \cdot 10^{-3}$
Neon	1.73	$1.56 \cdot 10^{-3}$
Aluminum	1.62	4.37
Silicon	1.66	3.87
Argon	1.51	$2.69 \cdot 10^{-3}$
Titanium	1.51	6.86
Iron	1.48	11.65
Copper	1.44	12.90
Germanium	1.40	7.45
Tin	1.26	9.21
Xenon	1.24	$7.30 \cdot 10^{-3}$
Tungsten	1.16	22.39
Platinum	1.15	24.67
Lead	1.13	12.83
Uranium	1.09	20.66
Water	2.03	2.03
Lucite	1.95	2.30
Shielding concrete	1.70	4.25
Quartz (SiO_2)	1.72	4.54

Equation (1.12) gives only the average energy loss of charged particles by ionization and excitation. For thin absorbers (in the sense of equation (1.15)), in particular, strong fluctuations around the average energy loss exist. The energy loss distribution for thin absorbers, i.e., mainly in gases, is strongly asymmetric [5, 6]. This distribution can be parametrized by a Landau distribution. The Landau distribution is described by the inverse Laplace transform of the function s^s [38]-[41]. A reasonable approxima-

a)

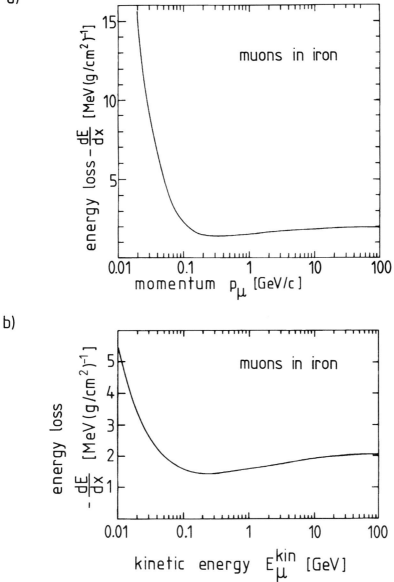

Fig. 1.1. Energy loss by ionization and excitation for muons in iron and its dependence on the muon momentum (a) and on the muon energy (b) [34, 35, 36].

tion of the Landau distribution is given by [42, 43, 44]

$$L(\lambda) = \frac{1}{\sqrt{2\pi}} \cdot \exp\left\{-\frac{1}{2}(\lambda + e^{-\lambda})\right\} , \qquad (1.18)$$

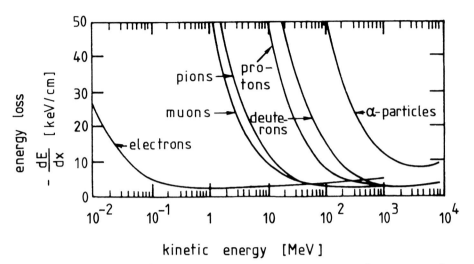

Fig. 1.2.   Energy loss for electrons, muons, pions, protons, deuterons and α-particles in air [37].

where λ is the deviation from the most probable energy loss

$$\lambda = \frac{\Delta E - \Delta E^{\mathrm{W}}}{\xi} \tag{1.19}$$

ΔE – actual energy loss in a layer of thickness x
ΔE^{W} – most probable energy loss in a layer of thickness x

$$\xi = 2\pi N_{\mathrm{A}} r_e^2 m_e c^2 z^2 \frac{Z}{A} \cdot \frac{1}{\beta^2} \varrho x = \kappa \varrho x \tag{1.20}$$

(ϱ – density in $\mathrm{g/cm^3}$; x – absorber thickness in cm).

For argon and electrons of energies up to $3.54\,\mathrm{MeV}$ from a $^{106}\mathrm{Rh}$-source the most probable energy loss is [42]

$$\Delta E^{\mathrm{W}} = \xi \left\{ \ln \left[\frac{2 m_e c^2 \gamma^2 \beta^2}{I^2} \xi \right] - \beta^2 + 0.423 \right\} \ . \tag{1.21}$$

As an example let us consider the ionization energy loss of $250\,\mathrm{MeV}$ electrons in an argon layer of $1\,\mathrm{cm}$ thickness. The most probable energy loss, that is, the energy loss corresponding to the maximum of the energy loss distribution, is, if we take equation (1.21) as an approximation,

$$\Delta E^{\mathrm{W}} = 2.4 \,\mathrm{keV} \ .$$

This most probable energy loss, of course, is smaller than the average energy loss in an argon layer of $1\,\mathrm{cm}$ thickness.

Using

$$\xi = 0.125 \, \text{keV}$$

from equation (1.20), the energy loss distribution under these circumstances according to equation (1.18) is sketched in figure 1.3 in linear and semilogarithmic scale. Experimentally, one finds that the actual energy loss distribution is frequently broader than that represented by the Landau distribution. [5, 6, 42].

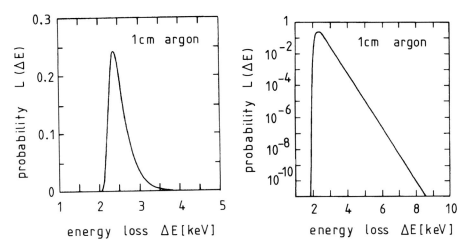

Fig. 1.3. Landau distribution for the ionization energy loss of 250 MeV electrons in gaseous argon of 1 cm thickness in linear and semilogarithmic scale.

Correspondingly, the most probable energy loss for minimum-ionizing particles ($\beta\gamma = 4$) in 1 cm argon is $\Delta E^{\text{W}} = 1.2 \, \text{keV}$, which is significantly smaller than the average energy loss (2.69 keV, see table 1.1) [5, 6, 42, 45].

For thick absorber layers, the tail of the Landau distribution originating from high energy transfers, however, is reduced [46]. For very thick absorbers $\left(\frac{\mathrm{d}E}{\mathrm{d}x} \cdot x \gg 2m_e c^2 \beta^2 \gamma^2 \right)$ the energy-loss distribution can be approximated by a Gaussian distribution.

The energy loss $\mathrm{d}E/\mathrm{d}x$ in a compound of various elements i is given by

$$\frac{\mathrm{d}E}{\mathrm{d}x} \approx \sum_i f_i \left. \frac{\mathrm{d}E}{\mathrm{d}x} \right|_i , \qquad (1.22)$$

where f_i is the mass fraction of the i-th element and $\left. \frac{\mathrm{d}E}{\mathrm{d}x} \right|_i$ the average energy loss in this element. Corrections to this relation because of the dependence of the ionization constant on the molecular structure can be safely neglected.

The Landau fluctuations of the energy loss are related to a large extent to very high energy transfers to atomic electrons. The energy transfers to these ionization electrons can be so large that these electrons can cause further ionization. These electrons are called δ or knock-on electrons. The energy spectrum of knock-on electrons is given by [27, 34, 35, 47]

$$\frac{dN}{dE_{\mathrm{kin}}} = \xi \cdot \frac{F}{E_{\mathrm{kin}}^2} \qquad (1.23)$$

for $I \ll E_{\mathrm{kin}} \leq E_{\mathrm{kin}}^{\mathrm{max}}$.

F is a spin-dependent factor of order unity, if $E_{\mathrm{kin}} \ll E_{\mathrm{kin}}^{\mathrm{max}}$. Of course, the energy spectrum of knock-on electrons falls to zero if the maximum transferable energy is reached. This kinematic limit also constrains the factor F [27, 47]. The spin dependence of the spectrum of the knock-on electrons only manifests itself close to the maximum transferable energy [27, 47].

The strong fluctuations of the energy loss in thin absorber layers are quite frequently not observed by a detector. Detectors only measure the energy which is actually deposited in their sensitive volume, and this energy may not be the same as the energy lost by the particle. For example, the energy which is transferred to knock-on electrons may only be partially deposited in the detector because the knock-on electrons can leave the sensitive volume of the detector.

Therefore, quite frequently it is of practical interest to consider only that part of the energy loss with energy transfers E smaller than a given cut value E_{cut}. This truncated energy loss is given by [34, 35, 48]

$$-\left.\frac{dE}{dx}\right|_{\leq E_{\mathrm{cut}}} = \kappa \left[\ln \frac{2 m_e c^2 \beta^2 \gamma^2 E_{\mathrm{cut}}}{I^2} - \beta^2 - \delta \right] , \qquad (1.24)$$

where κ is defined by equation (1.13). Equation (1.24) is similar, but not identical, to equation (1.12). Distributions of the truncated energy loss do not show a pronounced Landau tail as the distributions of equations (1.12) and (1.18). Because of the density effect — expressed by δ in equations (1.12) or (1.24), respectively — the truncated energy loss approaches a constant at high energies, which is called the Fermi plateau.

So far, the energy loss by ionization and excitation has been described for heavy particles. Electrons as incident particles, however, play a special rôle in the treatment of the energy loss. On the one hand, the total energy loss of electrons even at low energies (MeV-range) is influenced by bremsstrahlung processes. On the other hand, the ionization loss requires special treatment because the mass of the incident particle and the target electron is the same.

In this case, one can no longer distinguish between the primary and secondary electron after the collision. Therefore, the energy transfer probability must be interpreted in a different manner. One electron after the collision receives the energy E_{kin} and the other electron the energy $E - m_e c^2 - E_{\text{kin}}$ (E is the total energy of the incident particle). All possible cases are considered if one allows the energy transfer to vary between 0 and $\frac{1}{2}(E - m_e c^2)$ and not up to $E - m_e c^2$.

This effect can be most clearly seen if in equation (1.12) the maximum energy transfer $E_{\text{kin}}^{\text{max}}$ of equation (1.6) is replaced by the corresponding expression for electrons. For relativistic particles the term $\frac{1}{2}(E - m_e c^2)$ can be approximated by $E/2 = \frac{1}{2}\gamma m_e c^2$. Using $z = 1$, the ionization loss of electrons then can be approximated by

$$-\frac{\mathrm{d}E}{\mathrm{d}x} = 4\pi N_A r_e^2 m_e c^2 \frac{Z}{A} \cdot \frac{1}{\beta^2} \left[\ln \frac{\gamma m_e c^2}{2I} - \beta^2 - \frac{\delta^*}{2} \right] , \qquad (1.25)$$

where δ^* takes a somewhat different value for electrons compared to the parameter δ appearing in equation (1.12). A more precise calculation considering the specific differences between incident heavy particles and electrons yields a more exact formula for the energy loss of electrons due to ionization and excitation [49]

$$\begin{aligned}
-\frac{\mathrm{d}E}{\mathrm{d}x} &= 4\pi N_A r_e^2 m_e c^2 \frac{Z}{A} \cdot \frac{1}{\beta^2} \left\{ \ln \frac{\gamma m_e c^2 \beta \sqrt{\gamma-1}}{\sqrt{2} I} \right. \\
&\left. + \frac{1}{2}(1 - \beta^2) - \frac{2\gamma - 1}{2\gamma^2} \ln 2 + \frac{1}{16} \left(\frac{\gamma - 1}{\gamma} \right)^2 \right\} . \qquad (1.26)
\end{aligned}$$

This equation takes into account the kinematics of electron-electron collisions and also screening effects.

The treatment of the ionization loss of positrons is similar to that of electrons if one considers that these particles are of equal mass, but not identical particles.

For completeness, we also give the ionization loss of positrons [50]:

$$\begin{aligned}
-\frac{\mathrm{d}E}{\mathrm{d}x} &= 4\pi N_A r_e^2 m_e c^2 \frac{Z}{A} \frac{1}{\beta^2} \left\{ \ln \frac{\gamma m_e c^2 \beta \sqrt{\gamma-1}}{\sqrt{2} \cdot I} \right. \\
&\left. - \frac{\beta^2}{24} \left(23 + \frac{14}{\gamma + 1} + \frac{10}{(\gamma + 1)^2} + \frac{4}{(\gamma + 1)^3} \right) \right\} . \qquad (1.27)
\end{aligned}$$

Since positrons are antiparticles of electrons, there is, however, an additional consideration; if positrons come to rest, they will annihilate with an electron normally into two photons which are emitted anticollinearly. Both photons have energies of $511\,\text{keV}$ in the center of mass system corresponding to the rest mass of the electrons. The cross section for annihilation in flight is given by [50]

$$\sigma(Z, E) = \frac{Z\pi r_{\mathrm{e}}^2}{\gamma + 1} \left\{ \frac{\gamma^2 + 4\gamma + 1}{\gamma^2 - 1} \ln(\gamma + \sqrt{\gamma^2 - 1}) - \frac{\gamma + 3}{\sqrt{\gamma^2 - 1}} \right\}. \quad (1.28)$$

More details about the ionization process of elementary particles, in particular its spin dependence can be taken from the books of Rossi and Sitar *et al.* [5, 6, 27].

1.1.2   Ionization yield

The average energy loss by ionization and excitation can be transformed into the number of electron-ion pairs produced along the track of a charged particle. One must distinguish between primary ionization, that is, the number of primary produced electron-ion pairs, and the total ionization. A sufficiently large amount of energy can be transferred to some primary produced electrons so that they also can ionize (knock-on electrons). This secondary ionization together with the primary ionization forms the total ionization.

The average energy required to form an electron-ion pair (W-value) exceeds the ionization potential of the gas because inner shells of the gas atoms can also be involved in the ionization process, and a fraction of the energy of the incident particle can be dissipated by excitation processes which do not lead to free electrons. The W-value of a material is constant for relativistic particles and increases only slightly for low velocities of incident particles.

For gases the W-values are around 30 eV. They can, however, strongly depend on impurities in the gas. Table 1.2 shows the W-values for some gases together with the number of primary (n_{p}) and total (n_{T}) produced electron-ion pairs for minimum-ionizing particles (see table 1.1) [1, 34, 35, 51, 52].

The numerical values for n_{p} are somewhat uncertain because experimentally it is very difficult to distinguish between primary and secondary ionization. The total ionization (n_{T}) can be computed from the total energy loss ΔE in the detector according to

$$n_{\mathrm{T}} = \frac{\Delta E}{W} . \quad (1.29)$$

This is only true if the transferred energy is completely deposited in the sensitive volume of the detector.

In solid state detectors charged particles produce electron-hole pairs. For the production of an electron-hole pair on the average 3.6 eV in silicon and 2.85 eV in germanium are required. This means that the number of charge carriers produced in solid state detectors is much larger compared to the production rate of electron-ion pairs in gases. Therefore, the statistical fluctuations in the number of produced charge carriers for a

Table 1.2. *Compilation of some properties of gases. Given is the average effective ionization potential per electron I_0, the average energy loss W per produced ion pair, the number of primary (n_p) and total (n_T) produced electron-ion pairs per cm at standard pressure and temperature for minimum-ionizing particles [1, 34, 35, 51, 52]*

gas	density $\varrho[g/cm^3]$	$I_0[eV]$	$W[eV]$	$n_p[cm^{-1}]$	$n_T[cm^{-1}]$
H_2	$8.99 \cdot 10^{-5}$	15.4	37	5.2	9.2
He	$1.78 \cdot 10^{-4}$	24.6	41	5.9	7.8
N_2	$1.25 \cdot 10^{-3}$	15.5	35	10	56
O_2	$1.43 \cdot 10^{-3}$	12.2	31	22	73
Ne	$9.00 \cdot 10^{-4}$	21.6	36	12	39
Ar	$1.78 \cdot 10^{-3}$	15.8	26	29	94
Kr	$3.74 \cdot 10^{-3}$	14.0	24	22	192
Xe	$5.89 \cdot 10^{-3}$	12.1	22	44	307
CO_2	$1.98 \cdot 10^{-3}$	13.7	33	34	91
CH_4	$7.17 \cdot 10^{-4}$	13.1	28	16	53
C_4H_{10}	$2.67 \cdot 10^{-3}$	10.8	23	46	195

given energy loss is much smaller in solid state detectors than in gaseous detectors.

The production of pairs of charge carriers for a given energy loss is a statistical process. If, on average, N charge carrier pairs are produced one would naïvely expect this number to fluctuate according to Poisson statistics with an error of $\sqrt{N}$. Actually the fluctuation around the average value is smaller by a factor $\sqrt{F}$ depending on the material; this was demonstrated for the first time by Fano [53]. If one considers the situation in detail, the origin of the Fano factor is clear. For a given energy deposit, the number of produced charge carriers is limited by energy conservation.

In the following, a formal justification for the Fano factor will be given [53, 54]. Let $E = E_{total}$ be the fixed energy deposited in a detector, e.g., by an X-ray photon or a stopping α-particle. This energy is transferred in p steps to the detector medium, in general, in unequal portions E_p in each individual ionization process. For each interaction step m_p electron-ion pairs are produced. After N steps the total energy is completely absorbed (see figure 1.4).

Let

$m_p^{(e)} = \frac{E_p}{W}$ be the expected number of ionizations in the step p, and

$\overline{n}^{(e)} = \frac{E}{W}$ be the average expected number of the totally produced electron-ion pairs.

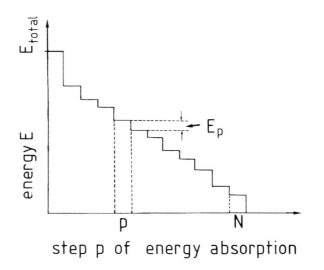

Fig. 1.4.   Energy loss in N discrete steps with energy transfer E_p in the p-th step [54].

The quantity which will finally describe the energy resolution, is

$$\sigma^2 = \overline{(n - \overline{n})^2} \, , \tag{1.30}$$

where $\overline{n}$ is the average value over many experiments for fixed energy absorption.

$$\sigma^2 = \frac{1}{L} \sum_{k=1}^{L} (n_k - \overline{n})^2 \, . \tag{1.31}$$

That is, we perform L gedankenexperiments where in experiment k a total number n_k electron-ion pairs is produced. In experiment k the energy is transferred to the detector medium in N_k steps, where in the p-th interval the number of produced electron-ion pairs is m_{pk}.

$$
\begin{aligned}
n_k - \overline{n} &= \sum_{p=1}^{N_k} m_{pk} - \frac{E}{W} \\
&= \sum_{p=1}^{N_k} m_{pk} - \frac{1}{W} \sum_{p=1}^{N_k} E_{pk} \, .
\end{aligned} \tag{1.32}
$$

The second term in the sum constrains the statistical character of the charge carrier production rate through energy conservation. Therefore, one would expect that the fluctuations are smaller compared to an unconstrained accidental energy-loss process.

The energy E is subdivided consequently into N_k discrete steps each with energy portion E_{pk}. If we introduce

$$\nu_{pk} = m_{pk} - \frac{E_{pk}}{W} \; , \tag{1.33}$$

it follows that

$$n_k - \overline{n} = \sum_{p=1}^{N_k} \nu_{pk} \; . \tag{1.34}$$

The variance for L experiments is given by

$$\sigma^2(n) = \frac{1}{L} \cdot \underbrace{\sum_{k=1}^{L}}_{L \text{ experiments}} \underbrace{\left(\sum_{p=1}^{N_k} \nu_{pk} \right)^2}_{\text{per experiment}} \tag{1.35}$$

$$\sigma^2(n) = \frac{1}{L} \left(\sum_{k=1}^{L} \sum_{p=1}^{N_k} \nu_{pk}^2 + \sum_{k=1}^{L} \sum_{i \neq j}^{N_k} \nu_{ik}\nu_{jk} \right) \; . \tag{1.36}$$

Let us consider the mixed term at first

$$\frac{1}{L} \sum_{k=1}^{L} \sum_{i \neq j}^{N_k} \nu_{ik}\nu_{jk} = \frac{1}{L} \sum_{k=1}^{L} \sum_{i=1}^{N_k} \nu_{ik} \left(\sum_{j=1}^{N_k} \nu_{jk} - \nu_{ik} \right) \; . \tag{1.37}$$

The last term in the bracket of equation (1.37) originates from the suppression of the product $\nu_{ik}\nu_{jk}$ for $i = j$, which is already contained in the quadratic terms.

For a given event k the average value

$$\overline{\nu}_k = \frac{1}{N_k} \sum_{j=1}^{N_k} \nu_{jk} \tag{1.38}$$

can be introduced. Using this quantity, one gets

$$\frac{1}{L} \sum_{k=1}^{L} \sum_{i \neq j}^{N_k} \nu_{ik}\nu_{jk} = \frac{1}{L} \sum_{k=1}^{L} N_k \overline{\nu}_k (N_k \overline{\nu}_k - \overline{\nu}_k) \; . \tag{1.39}$$

In this equation the last term ν_{ik} has been approximated by the average value $\overline{\nu}_k$. Under these conditions one obtains

$$\frac{1}{L} \sum_{k=1}^{L} \sum_{i \neq j}^{N_k} \nu_{ik}\nu_{jk} = \frac{1}{L} \sum_{k=1}^{L} N_k (N_k - 1)\overline{\nu}_k^2 = (\overline{N^2} - \overline{N})\overline{\nu}^2 \; , \tag{1.40}$$

if one assumes that N_k and $\overline{\nu}_k$ are uncorrelated, and $\overline{\nu}_k = \overline{\nu}$, if N_k is sufficiently large.

The average value of ν, however, vanishes according to equation (1.33), consequently the second term in equation (1.36) does not contribute. The remaining first term gives

$$\sigma^2(n) = \frac{1}{L}\sum_{k=1}^{L}\sum_{p=1}^{N_k}\nu_{pk}^2 = \frac{1}{L}\sum_{k=1}^{L}N_k\overline{\nu_k^2} = \overline{N\nu^2}$$
$$= \overline{N}\cdot\overline{(m_p - E_p/W)^2}\,. \tag{1.41}$$

In this case m_p is the actually measured number of electron-ion pairs in the energy absorption step p with energy deposit E_p.

Remembering that $\overline{N} = \frac{\overline{n}}{\overline{m_p}}$, leads to

$$\sigma^2(n) = \frac{\overline{(m_p - E_p/W)^2}}{\overline{m_p}}\,\overline{n}\,. \tag{1.42}$$

The variance of n consequently is

$$\sigma^2(n) = F\cdot\overline{n} \tag{1.43}$$

with the Fano factor

$$F = \frac{\overline{(m_p - E_p/W)^2}}{\overline{m_p}}\,. \tag{1.44}$$

As a consequence, the energy resolution is improved by the factor $\sqrt{F}$ compared to Poisson fluctuations. However, it must be remembered that one has to distinguish between the occasional very large fluctuations of the energy loss (Landau fluctuations) in thin absorber layers and the fluctuation of the number of produced electron-ion pairs for a given fixed well-defined energy loss. This last case is true for all particles which deposit their total energy in the sensitive volume of the detector.

Table 1.3 lists some Fano factors for various substances [54]. The improvement on the energy resolution can be quite substantial.

Table 1.3. *Fano factors for typical detector materials [54]*

source	energy	absorber	F
X-rays	5.9 keV	Ar + 10 % CH$_4$	0.21
"	2.6 keV	"	0.31
α	5.03 MeV	"	0.18
α	5.68 MeV	Ar + 0.8 % CH$_4$	0.19
p	1...4.5 MeV	Si	0.16

1.1.3 Multiple scattering

A charged particle traversing matter will be scattered by the Coulomb potentials of nuclei and electrons. This leads to a large number of scattering processes with very low deviations from the original path (see figure 8.4). The distribution of scattering angles due to Coulomb scattering is described by Molière's theory [34, 35, 55]. For small scattering angles it is normally distributed around the average scattering angle $\Theta = 0$. Larger scattering angles caused by collisions of charged particles with nuclei are, however, more frequent than expected from a Gaussian distribution.

The root mean square of the projected scattering angle distribution is given by [34, 35]

$$\Theta_{\text{rms}}^{\text{proj.}} = \sqrt{\langle\Theta^2\rangle} = \frac{13.6\,\text{MeV}}{\beta c p}\, z \sqrt{\frac{x}{X_0}}\, [1 + 0.038\,\ln(x/X_0)]\,, \qquad (1.45)$$

where p (in MeV/c) is the momentum, βc the velocity and z the charge of the scattered particle. x/X_0 is the thickness of the scattering medium, measured in units of the radiation length (see section 1.1.4) [27, 56, 57]

$$X_0 = \frac{A}{4\alpha N_A Z^2 r_e^2\,\ln(183\,Z^{-1/3})}\,, \qquad (1.46)$$

where Z and A are the atomic number and the atomic weight of the absorber, respectively.

Equation (1.45) is an approximation. For most practical applications equation (1.45) can be further approximated for particles with $z = 1$ by

$$\Theta_{\text{rms}}^{\text{proj.}} = \sqrt{\langle\Theta^2\rangle} = \frac{13.6}{\beta c p}\sqrt{\frac{x}{X_0}}\,. \qquad (1.47)$$

Equations (1.45) or (1.47) give the root mean square of the projected distribution of the scattering angles. Such a projected distribution is, for example, of interest for detectors, like layers of flash tubes, which provide only a two-dimensional view of an event. The corresponding root mean square deviation for non-projected scattering angles is increased by factor $\sqrt{2}$, so that we have

$$\Theta_{\text{rms}}^{\text{space}} = \frac{19.2}{\beta c p}\sqrt{\frac{x}{X_0}}\,. \qquad (1.48)$$

1.1.4 Bremsstrahlung

Fast charged particles lose, in addition to their ionization loss, energy by interactions with the Coulomb field of the nuclei of the traversed medium. If the charged particles are decelerated in the Coulomb field of the nucleus, a fraction of their kinetic energy will be emitted in form of photons (bremsstrahlung).

The energy loss by bremsstrahlung for high energies can be described by [27]

$$-\frac{\mathrm{d}E}{\mathrm{d}x} = 4\alpha \cdot N_A \cdot \frac{Z^2}{A} \cdot z^2 \left(\frac{1}{4\pi\varepsilon_0} \cdot \frac{e^2}{mc^2}\right)^2 \cdot E \, \ln \frac{183}{Z^{1/3}} \, . \tag{1.49}$$

In this equation

Z, A – are the atomic number and atomic weight of the medium,
z, m, E – are the charge, mass and energy of the incident particle.

The bremsstrahlung energy loss of electrons is given correspondingly by

$$-\frac{\mathrm{d}E}{\mathrm{d}x} = 4\alpha N_A \cdot \frac{Z^2}{A} r_e^2 \cdot E \, \ln \frac{183}{Z^{1/3}} \tag{1.50}$$

if $E \gg m_e c^2/\alpha Z^{1/3}$.

It should be pointed out that, in contrast to the ionization energy loss (equation (1.12)), the energy loss by bremsstrahlung is proportional to the energy of the particle and inversely proportional to the mass squared of the incident particles.

Because of the small size of the electron mass, the bremsstrahlung energy loss plays an especially important rôle for electrons. For electrons ($z = 1$, $m = m_e$) equation (1.49), or equation (1.50), respectively, can be written in the following fashion:

$$-\frac{\mathrm{d}E}{\mathrm{d}x} = \frac{E}{X_0} \, . \tag{1.51}$$

This equation defines the radiation length X_0. An approximation for X_0 has already been given by equation (1.46).

The proportionality

$$X_0^{-1} \propto Z^2 \tag{1.52}$$

in equation (1.46) originates from the interaction of the incident particle with the Coulomb field of the target nucleus.

Bremsstrahlung, however, is also emitted in interactions of incident particles with the electrons of the target material. The cross section for this process follows closely the calculation of the bremsstrahlung energy loss on the target nucleus, the only difference being that for atomic target electrons the charge is always equal to unity, and therefore one obtains an additional contribution to the cross section, which is proportional to the number of target electrons, that is $\propto Z$. The cross section for bremsstrahlung must be extended by this term [33]. Therefore, the factor Z^2 in equation (1.46) must be replaced by $Z^2 + Z = Z(Z + 1)$, which leads to a better description of the radiation length, accordingly

$$X_0 = \frac{A}{4\alpha N_A Z(Z+1)r_e^2 \ln(183\ Z^{-1/3})}\ [\text{g/cm}^2]\,. \tag{1.53}$$

In addition, one has to consider that the atomic electrons will screen the Coulomb field of the nucleus to a certain extent. If screening effects are taken into account, the radiation length can be approximated, using recent calculations [34, 35], by

$$X_0 = \frac{716.4 \cdot A}{Z(Z+1)\ln(287/\sqrt{Z})}\ [\text{g/cm}^2]\,. \tag{1.54}$$

The numerical results for the radiation length based on equation (1.54) deviate from those of equation (1.46) by a few percent.

Usually, radiation length is defined for electrons as incident particles. Consequently, it depends only on the properties of the material. Because of the proportionality

$$X_0 \propto r_e^{-2} \tag{1.55}$$

and the relation

$$r_e = \frac{1}{4\pi\varepsilon_0} \cdot \frac{e^2}{m_e c^2} \tag{1.56}$$

the radiation length, however, also has a dependence on the mass of the incident particle

$$X_0 \propto m^2\,. \tag{1.57}$$

The radiation lengths given in the literature, however, are always meant for electrons.

Integrating equations (1.49) or (1.51), respectively, leads to

$$E = E_0 e^{-x/X_0}\,. \tag{1.58}$$

This function describes the exponential attenuation of the *energy* of charged particles by radiation losses. Note the distinction from the exponential attenuation of the *intensity* of a photon beam passing through matter (see section 1.2, equation (1.76)).

The radiation length of a mixture of elements or a compound can be approximated by

$$X_0 = \frac{1}{\sum_{i=1}^N f_i/X_0^i}\,, \tag{1.59}$$

where f_i are the mass fractions of the components with the radiation length X_0^i.

Energy losses due to bremsstrahlung are proportional to the energy while ionization energy losses beyond the minimum of ionization are proportional to the logarithm of the energy. The energy, where these two interaction processes for electrons lead to equal energy losses, is called the critical energy E_c.

$$-\frac{dE}{dx}(E_c)\bigg|_{\text{ionization}} = -\frac{dE}{dx}(E_c)\bigg|_{\text{bremsstrahlung}} . \qquad (1.60)$$

In principle, the critical energy can be calculated from the equations (1.12) and (1.49) using equation (1.60). Numerical values for the critical energy of electrons are given in the literature [33, 34, 35]. For heavy elements ($Z \geq 13$) the equation

$$E_c = \frac{550 \, \text{MeV}}{Z} \qquad (1.61)$$

describes the critical energies quite satisfactorily [58].

Table 1.4 lists the radiation lengths and critical energies for some materials [33, 34, 35]. The critical energy — as well as the radiation length — scales as the square of the mass of the incident particles. For muons ($m_\mu = 106 \, \text{MeV}/c^2$) in iron one obtains:

$$E_c^\mu \approx E_c^e \cdot \left(\frac{m_\mu}{m_e}\right)^2 = 890 \, \text{GeV} . \qquad (1.62)$$

1.1.5   Direct electron-pair production

Apart from bremsstrahlung losses, additional energy-loss mechanisms come into play, particularly at high energies. Electron-positron pairs can be produced by virtual photons in the Coulomb field of the nuclei. For high energy muons this energy-loss mechanism is even more important than bremsstrahlung losses. The energy loss by 'trident production' (e.g., like $\mu + \text{nucleus} \rightarrow \mu + e^+ + e^- + \text{nucleus}$) is also proportional to the energy and can be parametrized by

$$-\frac{dE}{dx}\bigg|_{\text{pair pr.}} = b_{\text{pair}}(Z, A, E) \cdot E ; \qquad (1.63)$$

the $b(Z, A, E)$ parameter varies only slowly with energy for high energies. For 100 GeV muons in iron the energy loss due to direct electron-pair production can be described by [47, 59, 60]

$$-\frac{dE}{dx}\bigg|_{\text{pair pr.}} = 3 \cdot 10^{-6} \cdot E \quad [\frac{\text{MeV}}{\text{g/cm}^2}] , \qquad (1.64)$$

$$\text{i.e.,} \quad -\frac{dE}{dx}\bigg|_{\text{pair pr.}} = 0.3 \, \frac{\text{MeV}}{\text{g/cm}^2} .$$

Table 1.4. *Radiation lengths and critical energies for some absorber materials [33, 34, 35]. The values for the radiation lengths agree with equation (1.54) within a few percent. Only the experimental value for helium shows a some-what larger deviation. The numerical results for the critical energies of electrons scatter quite significantly in the literature. The effective values for Z and A of mixtures and compounds can be calculated for A by $A_{eff} = \sum_{i=1}^{N} f_i A_i$, where f_i are the mass fractions of the components with atomic weight A_i. Correspond-ingly one obtains the effective atomic numbers using equations (1.54) and (1.59). Neglecting the logarithmic Z-dependence in equation (1.54) Z_{eff} can be calculated from $Z_{eff} \cdot (Z_{eff} + 1) = \sum_{i=1}^{N} f_i Z_i (Z_i + 1)$, where f_i are the mass fractions of the components with charge numbers Z_i. For the practical calculation of an effec-tive radiation length of a compound one determines first the radiation length of the contributing components and then determines the effective radiation length according to equation (1.59)*

material	Z	A	$X_0[\text{g/cm}^2]$	X_0/ϱ [cm]	$E_c[\text{MeV}]$
Hydrogen	1	1.01	63	700 000	350
Helium	2	4.00	94	530 000	250
Lithium	3	6.94	83	156	180
Carbon	6	12.01	43	18.8	90
Nitrogen	7	14.01	38	30 500	85
Oxygen	8	16.00	34	24 000	75
Aluminum	13	26.98	24	8.9	40
Silicon	14	28.09	22	9.4	39
Iron	26	55.85	13.9	1.76	20.7
Copper	29	63.55	12.9	1.43	18.8
Silver	47	109.9	9.3	0.89	11.9
Tungsten	74	183.9	6.8	0.35	8.0
Lead	82	207.2	6.4	0.56	7.40
Air	7.3	14.4	37	30 000	84
SiO_2	11.2	21.7	27	12	57
Water	7.5	14.2	36	36	83

The energy spectrum of directly produced electron-positron pairs at high energy transfers is steeper than the spectrum of bremsstrahlung photons. High fractional energy transfers are therefore dominated by bremsstrahlung processes.

1.1.6   Energy loss by photonuclear interactions

Charged particles can interact inelastically via virtual gauge particles (in this case, photons) with nuclei of the absorber material, thereby losing energy (nuclear interactions).

In the same way as for energy losses through bremsstrahlung or direct electron-pair production, the energy loss by photonuclear interactions is proportional to the particle's energy.

$$-\frac{dE}{dx}\Big|_{\text{photonucl.}} = b_{\text{nucl.}}(Z, A, E) \cdot E \ . \tag{1.65}$$

For 100 GeV muons in iron the energy-loss parameter b is given by $b_{\text{nucl.}} = 0.4 \cdot 10^{-6} \, \text{g}^{-1}\text{cm}^2$ [47], that is,

$$-\frac{dE}{dx}\Big|_{\text{photonucl.}} = 0.04 \, \frac{\text{MeV}}{\text{g/cm}^2} \ . \tag{1.66}$$

1.1.7   Total energy loss

In contrast to energy losses due to ionization those by bremsstrahlung, direct electron-pair production, and photonuclear interactions are characterized by large energy transfers with correspondingly large fluctuations. Therefore, it is somewhat problematic to speak of an average energy loss for these processes because extremely large fluctuations around this average value can occur [61, 62].

Nevertheless, the total energy loss of charged particles by the above mentioned processes can be parametrized by

$$-\frac{dE}{dx}\Big|_{\text{total}} = -\frac{dE}{dx}\Big|_{\text{ionization}} -\frac{dE}{dx}\Big|_{\text{brems.}} -\frac{dE}{dx}\Big|_{\text{pair pr.}} -\frac{dE}{dx}\Big|_{\text{photonucl.}}$$

$$= a(Z, A, E) + b(Z, A, E) \cdot E \ , \tag{1.67}$$

where $a(Z, A, E)$ describes the energy loss according to equation (1.12) and $b(Z, A, E)$ the sum over the energy losses due to bremsstrahlung, direct electron-pair production and photonuclear interactions. The parameters a and b, and their energy dependence for various particles and materials are given in the literature.

Figure 1.5 shows the b parameters and figure 1.6 the various energy-loss mechanisms for muons in iron in their dependence on the muon energy [59].

Up to energies of several hundred GeV the energy loss in iron due to ionization and excitation is dominant. For energies in excess of several TeV direct electron-pair production and bremsstrahlung represent the main energy-loss processes. Photonuclear interactions contribute only at

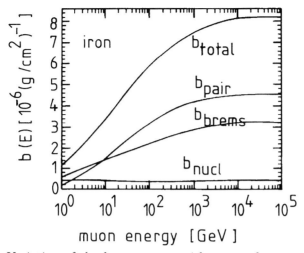

Fig. 1.5. Variation of the *b* parameters with energy for muons in iron. Plotted are the fractional energy losses by direct electron-pair production (b_{pair}), bremsstrahlung (b_{brems}), and photonuclear interactions ($b_{\text{nucl.}}$) as well as their sum (b_{total}) [59].

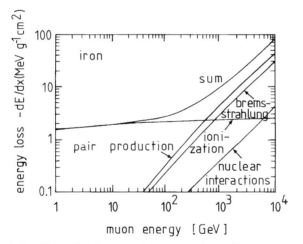

Fig. 1.6. Contributions to the energy loss of muons in iron [59].

the 10 % level. Since the energy loss due to these processes is proportional to the muon's energy, this opens up the possibility of muon calorimetry by means of energy-loss sampling.

The dominance of the energy proportional interaction processes over ionization and excitation depends, of course, on the target material. For uranium this transition starts around several $100\,\text{GeV}$, while in hydrogen bremsstrahlung and direct electron-pair production only prevail at energies in excess of $10\,\text{TeV}$.

1.1.8 *Energy-range relations for charged particles*

Because of the different energy-loss mechanisms it is nearly impossible to give a simple representation of the range of charged particles in matter. The definition of a range is in any case complicated because of the fluctuations of the energy loss by catastrophic energy loss processes, i.e., by interactions with high energy transfers, and because of the multiple Coulomb scattering in the material, all of which lead to substantial range straggling. In the following, therefore, some empirical formulae are given, which are valid for certain particle species in fixed energy ranges.

Generally speaking, the range can be calculated from:

$$R = \int_0^E \frac{\mathrm{d}E}{\mathrm{d}E/\mathrm{d}x} \ . \tag{1.68}$$

However, since the energy loss is a complicated function of the energy, in most cases approximations of this integral are used. For the determination of the range of low energy particles, in particular, the difference between the total energy E and the kinetic energy E_{kin} must be taken into account, because only the kinetic energy can be transferred to the material.

For α-particles with energies between $2.5\,\text{MeV} \leq E_{\text{kin}} \leq 20\,\text{MeV}$ the range in air $(15\,^\circ\text{C},\ 760\,\text{Torr})$ can be described by

$$R_\alpha = 0.31\, E_{\text{kin}}^{3/2}\,[\text{cm}] \tag{1.69}$$

(E_{kin} in MeV, R_α in cm) [63]. For rough estimates of the range of α-particles in other materials one can use

$$R_\alpha = 3.2 \cdot 10^{-4} \frac{\sqrt{A}}{\varrho} \cdot R_{\text{air}}[\text{cm}] \tag{1.70}$$

(ϱ in g/cm^3; A atomic weight) [63]. The range of α-particles in air is shown in figure 1.7.

For protons with energies between $0.6\,\text{MeV} \leq E_{\text{kin}} \leq 20\,\text{MeV}$ the range in air [63] can be approximated by

$$R_{\text{p}} = 100 \cdot \left(\frac{E_{\text{kin}}}{9.3}\right)^{1.8}\,[\text{cm}] \tag{1.71}$$

(E_{kin} in MeV, R_{p} in cm).

The range of low energy electrons $(0.5\,\text{MeV} \leq E_{\text{kin}} \leq 5\,\text{MeV})$ in aluminum is described [63] by

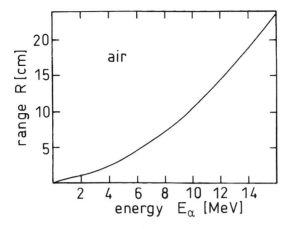

Fig. 1.7. Range of α-particles in air [63].

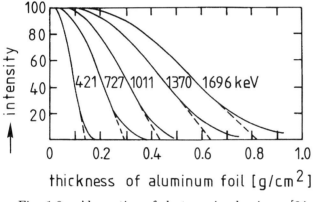

Fig. 1.8. Absorption of electrons in aluminum [64, 65].

$$R_e \;=\; 0.526\,E_{\text{kin}} - 0.094\ [\text{g/cm}^2] \qquad (1.72)$$
$$(E_{\text{kin}} \text{ in MeV},\ R_e \text{ in g/cm}^2)\,.$$

Figure 1.8 shows the absorption of electrons in aluminum [64, 65]. Plotted is the fraction of electrons (with the energy E_{kin}), which penetrate through a certain absorber thickness.

This figure shows the difficulty in the definition of a range of a particle due to the pronounced range straggling. The extrapolation of the linear part of the curves shown in figure 1.8 to the intersection with the abscissa defines the practical range [65]. The range of electrons defined in this way is shown in figure 1.9 for various absorbers [65].

For higher energies the range of electrons, muons, pions, protons and α-particles in nuclear emulsions (see section 4.16) can be taken from fig-

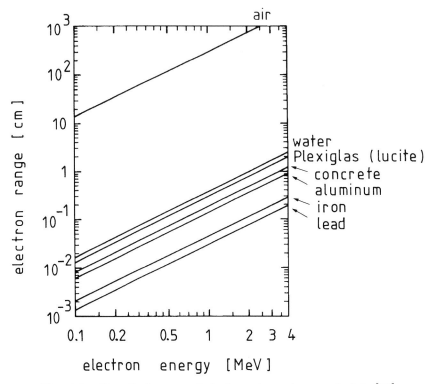

Fig. 1.9.   Practical range of electrons in various materials [65].

ure 1.10 [63, 66]. The curve for electrons is dashed for $\geq 1\,\text{MeV}$ because, for higher electron energies, bremsstrahlung energy losses become important. If such losses occur, equation (1.68) in principle gives the range; however, because bremsstrahlung energy losses are partly characterized by high energy transfers, the range straggling is much more pronounced in this case compared to energy loss processes by ionization and excitation, so that a definition of a range under these circumstances is somewhat questionable. This situation becomes even more problematic when high energy electrons initiate electromagnetic showers (see section 7.2).

The range of high energy muons can be obtained by integrating equation (1.68), using equations (1.67) and (1.12), and neglecting the logarithmic term in equation (1.12). This leads to

$$R_\mu(E_\mu) = \frac{1}{b}\ln(1 + \frac{b}{a}E_\mu)\,. \tag{1.73}$$

For $1\,\text{TeV}$ muons in iron equation (1.73) yields

$$R_\mu(1\,\text{TeV}) = 265\,\text{m}\,. \tag{1.74}$$

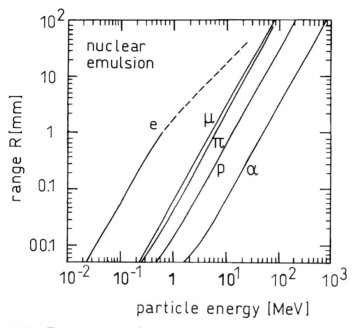

Fig. 1.10. Energy-range relation in nuclear emulsions for electrons, muons, pions, protons and α-particles [63, 66]. The curve for electrons is dashed for energies in excess of 1 MeV, because for higher energies bremsstrahlung processes dominate and cause problems in the definition of the practical range (see section 7.2).

A numerical integration for the range of muons in rock (standard rock with $Z = 11$, $A = 22$) yields for $E_\mu > 10 \, \text{GeV}$ [67]

$$R_\mu(E_\mu) = \left[\frac{1}{b} \ln(1 + \frac{b}{a} E_\mu)\right] \left[0.96 \frac{\ln E_\mu - 7.894}{\ln E_\mu - 8.074}\right] \qquad (1.75)$$

with $a = 2.2 \, \frac{\text{MeV}}{\text{g/cm}^2}$; $b = 4.4 \cdot 10^{-6} \text{g}^{-1} \text{cm}^2$ and E_μ in MeV. This energy-range dependence of muons in rock is shown in figure 1.11.

1.2   Interactions of photons

Photons are detected indirectly via interactions in the medium of the detector. In these processes, charged particles are produced which are recorded through their subsequent ionization in the sensitive volume of the detector. Interactions of photons are fundamentally different from ionization processes of charged particles because in every photon interaction, the photon is either completely absorbed (photoelectric effect, pair production) or scattered through a relatively large angle (Compton effect).

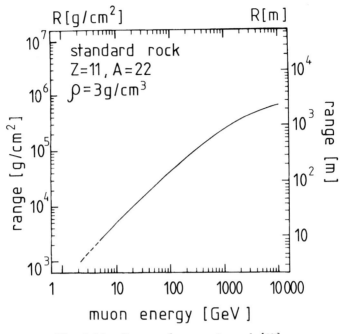

Fig. 1.11.   Range of muons in rock [67].

Since the absorption or scattering is a statistical process, it is impossible to define a range for γ-rays. A photon beam is attenuated exponentially in matter according to

$$I = I_0\, e^{-\mu x}\ .\qquad(1.76)$$

The mass absorption coefficient μ is related to the cross sections for the various interaction processes of photons according to

$$\mu = \frac{N_{\mathrm{A}}}{A} \sum_i \sigma_i\ ,\qquad(1.77)$$

where σ_i is the atomic cross section for the process i, A the atomic weight and N_{A} the Avogadro number.

The mass absorption coefficient (according to equation (1.77) given per g/cm^2) depends strongly on the photon energy. For low energies (100 keV $\geq E_\gamma \geq$ ionization energy) the photoelectric effect dominates

$$\gamma + \text{atom} \rightarrow \text{atom}^+ + e^-\ .\qquad(1.78)$$

In the range of medium energies ($E_\gamma \approx 1\,\text{MeV}$) the Compton effect dominates, which is the scattering of photons off quasi-free atomic electrons

$$\gamma + e^- \rightarrow \gamma + e^-,\qquad(1.79)$$

and at high energies $(E_\gamma \gg 1\,\text{MeV})$ the cross section for pair production is largest

$$\gamma + \text{nucleus} \rightarrow e^+ + e^- + \text{nucleus} \,. \tag{1.80}$$

The length x in equation (1.76) is a surface mass density with the dimension g/cm^2. If the length is measured in cm, the mass absorption coefficient μ must be multiplied by the density ϱ of the material.

1.2.1 Photoelectric effect

Atomic electrons can absorb the energy of a photon completely, while — because of momentum conservation — this is not possible for free electrons. The absorption of a photon by an atomic electron requires a third collision partner which in this case is the atomic nucleus. The cross section for absorption of a photon of energy E_γ in the K-shell is particularly large ($\approx 80\,\%$ of the total cross section), because of the proximity of the third collision partner, the atomic nucleus, which takes the recoil momentum. The total photoelectric cross section in the non-relativistic range away from the absorption edges is given in the non-relativistic Born approximation by [68]

$$\sigma_{\text{photo}}^{\text{K}} = \left(\frac{32}{\varepsilon^7}\right)^{1/2} \alpha^4 \cdot Z^5 \cdot \sigma_{\text{Th}}^e \qquad [\text{cm}^2/\text{atom}] \,, \tag{1.81}$$

where $\varepsilon = E_\gamma / m_e c^2$ is the reduced photon energy and $\sigma_{\text{Th}}^e = \frac{8}{3}\pi r_e^2 = 6.65 \cdot 10^{-25}\,\text{cm}^2$ is the Thomson cross section for elastic scattering of photons on electrons. Close to the absorption edges, the energy dependence of the cross section is modified by a function $f(E_\gamma, E_\gamma^{\text{edge}})$. For high energies ($\varepsilon \gg 1$) the energy dependence of the cross section for the photoelectric effect is much less pronounced

$$\sigma_{\text{photo}}^{\text{K}} = 4\pi r_e^2 Z^5 \alpha^4 \cdot \frac{1}{\varepsilon} \,. \tag{1.82}$$

In equations (1.81) and (1.82) the Z-dependence of the cross section is approximated by Z^5. This indicates that the photon does not interact with an isolated atomic electron. Z-dependent corrections, however, cause σ_{photo} to be a more complicated function of Z. In the energy range between $0.1\,\text{MeV} \leq E_\gamma \leq 5\,\text{MeV}$ the exponent of Z varies between 4 and 5.

As a consequence of the photoelectric effect in an inner shell (e.g., of the K-shell) the following secondary effects may occur. If the free place, e.g., in the K-shell is filled by an electron from a higher shell, the energy difference between those two shells can be liberated in the form of X-rays of characteristic energy. However, this energy difference can also be transferred to an electron of the *same* atom. If this energy is larger than the binding energy of the shell in question, a further electron can leave the atom (Auger effect, Auger electron). The energy of these Auger electrons is necessarily small compared to the energy of the primary photoelectrons.

If the photoionization occurs in the K-shell (binding energy B_K), and if the hole in the K-shell is filled up by an electron from the L-shell (binding energy B_L), the excitation energy of the atom $(B_K - B_L)$ can be transferred to an L-electron. If $B_K - B_L > B_L$, the L-electron can leave the atomic shell with an energy $B_K - 2B_L$ as an Auger electron.

1.2.2   Compton effect

The Compton effect describes the scattering of photons off quasi-free atomic electrons. In the treatment of this interaction process, the binding energy of the atomic electrons is neglected. The total cross section for Compton scattering per electron is given by the Klein-Nishina formula [69]

$$
\sigma_c^e = 2\pi r_e^2 \left[\left(\frac{1+\varepsilon}{\varepsilon^2} \right) \left\{ \frac{2(1+\varepsilon)}{1+2\varepsilon} - \frac{1}{\varepsilon} \ln(1+2\varepsilon) \right\} + \frac{1}{2\varepsilon} \ln(1+2\varepsilon) \right.
$$
$$
\left. - \frac{1+3\varepsilon}{(1+2\varepsilon)^2} \right] \; [\text{cm}^2/\text{electron}] . \tag{1.83}
$$

For Compton scattering from atoms the cross section is increased by the factor Z, because there are exactly Z electrons as possible scattering partners in an atom; consequently $\sigma_c^{\text{atomic}} = Z \cdot \sigma_c^e$.

At high energies the energy dependence of the Compton scattering cross section can be approximated by [70]

$$
\sigma_c^e \propto \frac{\ln \varepsilon}{\varepsilon} . \tag{1.84}
$$

The ratio of scattered (E'_γ) to incident photon energy is given by

$$
\frac{E'_\gamma}{E_\gamma} = \frac{1}{1 + \varepsilon(1 - \cos\theta_\gamma)} , \tag{1.85}
$$

where θ_γ is the scattering angle of the photon in the laboratory system (see figure 1.12).

For backscattering $(\theta_\gamma = \pi)$ the energy transfer to the electron reaches a maximum value, leading to a ratio of scattered to incident photon energy of

$$
\frac{E'_\gamma}{E_\gamma} = \frac{1}{1 + 2\varepsilon} . \tag{1.86}
$$

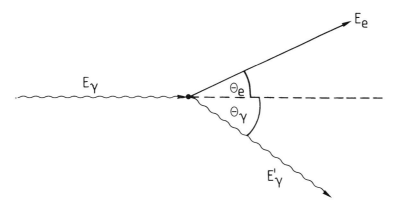

Fig. 1.12. Definition of kinematic variables in Compton scattering.

The scattering angle of the electron with respect to the direction of the incident photon can be obtained from

$$\cot \theta_e = (1 + \varepsilon) \tan \frac{\theta_\gamma}{2} \; . \tag{1.87}$$

Because of momentum conservation the scattering angle of the electron θ_e can never exceed $\pi/2$.

In Compton-scattering processes only a fraction of the photon energy is transferred to the electron. Therefore, one defines an energy scattering cross section

$$\sigma_{cs} = \frac{E'_\gamma}{E_\gamma} \cdot \sigma_c^e \tag{1.88}$$

and subsequently an energy absorption cross section

$$\sigma_{ca} = \sigma_c^e - \sigma_{cs} \; . \tag{1.89}$$

The latter is relevant for absorption processes and is related to the probability that an energy $E_{kin} = E_\gamma - E'_\gamma$ is transferred to the target electron.

In passing, it should be mentioned that in addition to the normal Compton scattering of photons on target electrons at rest, inverse Compton scattering also exists. In this case, an energetic electron collides with a low energy photon and transfers a fraction of its kinetic energy to the photon which is blue-shifted to higher frequencies. This inverse Compton-scattering process plays an important rôle, e.g., in astrophysics. Starlight photons (eV-range) can be shifted in this way by collisions with energetic electrons into the X-ray (keV) or gamma (MeV) range.

Naturally, Compton scattering does not only occur with electrons, but also for other charged particles. For the measurement of photons in particle detectors, however, Compton scattering off atomic electrons is of special importance.

1.2.3   Pair production

The production of electron-positron pairs in the Coulomb field of a nucleus is only possible if the photon energy exceeds a certain threshold. This threshold energy is given by the rest masses of two electrons plus the recoil energy which is transferred to the nucleus. From energy and momentum conservation, this threshold energy can be calculated to be

$$E_\gamma \geq 2m_{\mathrm{e}}c^2 + 2\frac{m_{\mathrm{e}}^2}{m_{\mathrm{nucleus}}}c^2 \; . \tag{1.90}$$

Since $m_{\mathrm{nucleus}} \gg m_{\mathrm{e}}$, the effective threshold can be approximated by

$$E_\gamma \geq 2m_{\mathrm{e}}c^2 \; . \tag{1.91}$$

If, however, the electron-positron pair production proceeds in the Coulomb field of an electron, the threshold energy is:

$$E_\gamma \geq 4m_{\mathrm{e}}c^2 \; . \tag{1.92}$$

Electron-positron pair production in the Coulomb field of an electron is, however, strongly suppressed compared to pair production in the Coulomb field of the nucleus.

In the case that the nuclear charge is not screened by atomic electrons, (for low energies the photon must come relatively close to the nucleus to make pair production probable, which means that the photon sees only the 'naked' nucleus)

$$1 \ll \varepsilon < \frac{1}{\alpha Z^{1/3}} \tag{1.93}$$

the pair production cross section is given by [27]

$$\sigma_{\mathrm{pair}} = 4\alpha r_{\mathrm{e}}^2 Z^2 \left(\frac{7}{9}\ln 2\varepsilon - \frac{109}{54}\right) \qquad [\mathrm{cm}^2/\mathrm{atom}] \; ; \tag{1.94}$$

for complete screening of the nuclear charge, however, $\left(\varepsilon \gg \frac{1}{\alpha Z^{1/3}}\right)$ [27]

$$\sigma_{\mathrm{pair}} = 4\alpha r_{\mathrm{e}}^2 Z^2 \left(\frac{7}{9}\ln \frac{183}{Z^{1/3}} - \frac{1}{54}\right) \qquad [\mathrm{cm}^2/\mathrm{atom}] \; . \tag{1.95}$$

(At high energies pair production can also proceed at relatively large impact parameters of the photon with a respect to the nucleus. But in this case, the screening of the nuclear charge by the atomic electrons must be taken into account.)

For large photon energies, the pair production cross section approaches an energy independent value which is given by equation (1.95). Neglecting the small term $\frac{1}{54}$ in the bracket of this equation, this asymptotic value is given by

$$
\begin{aligned}
\sigma_{\text{pair}} &\approx \frac{7}{9}\, 4\alpha\, r_{\text{e}}^2 Z^2 \ln \frac{183}{Z^{1/3}} \\
&\approx \frac{7}{9} \cdot \frac{A}{N_{\text{A}}} \cdot \frac{1}{X_0}
\end{aligned} \tag{1.96}
$$

(see equation (1.46)).

The partition of the energy between the produced electrons and positrons is symmetric at low and medium energies and becomes strongly asymmetric at large energies. The differential cross section for an energy transfer E_+ to the positron is given by

$$
\frac{\mathrm{d}\sigma_{\text{pair}}}{\mathrm{d}E_+} = \frac{\alpha r_{\text{e}}^2}{E_\gamma - 2m_{\text{e}}c^2} \cdot Z^2 \cdot f(\varepsilon, Z) \qquad [\text{cm}^2/\text{MeV} \cdot \text{atom}] . \tag{1.97}
$$

$f(\varepsilon, Z)$ is a dimensionless, non-trivial function of ε and Z. The trivial Z^2-dependence of the cross section is, of course, already considered in a factor separated from $f(\varepsilon, Z)$. Therefore, $f(\varepsilon, Z)$ depends only weakly (logarithmically) on the atomic number of the absorber (see equation (1.95)). $f(\varepsilon, Z)$ varies with Z only by few percent [37]. The dependence of this function on the energy-partition parameter

$$
x = \frac{E_+ - m_{\text{e}}c^2}{E_\gamma - 2m_{\text{e}}c^2} = \frac{E_+^{\text{kin}}}{E_{\text{pair}}^{\text{kin}}} \tag{1.98}
$$

for average Z values is shown in figure 1.13 for various parameters ε [37, 71, 72]. The curves shown in figure 1.13 do not just include the pair production on the nucleus, but also the pair-production probability on atomic electrons ($\propto Z$), so that the Z^2-dependence of the pair production cross section (equation (1.97)) is modified to $Z(Z+1)$ in a similar way as was argued when the electron bremsstrahlung process was presented (see equation (1.53)).

1.2.4 Total photon absorption cross section

The total mass absorption coefficient, which is related to the cross sections according to equation (1.77) is shown in figure 1.14 for the absorbers water, air, aluminum, and lead [63, 73, 74, 75].

Since Compton scattering plays a special rôle for photon interactions, because only part of the photon energy is transferred to the target electron, one has to distinguish between the mass attenuation coefficient and the mass absorption coefficient. The mass attenuation coefficient μ_{cs} is related to the Compton energy scattering cross section σ_{cs} (see equation (1.88)) according to equation (1.77). Correspondingly the mass absorption coefficient μ_{ca} is calculated from the energy absorption cross sec-

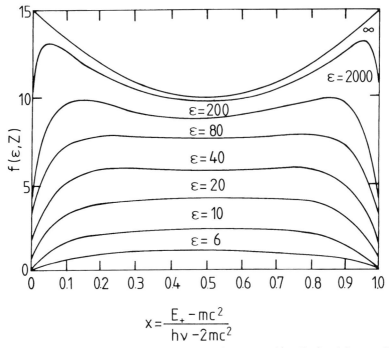

$$x = \frac{E_+ - mc^2}{h\nu - 2mc^2}$$

Fig. 1.13.   Form of the energy-partition function $f(\varepsilon, Z, x)$ with $\varepsilon = E_\gamma/m_e c^2$ as parameter. The total pair production cross section is given by the area under the corresponding curve in units of $Z(Z+1)\alpha r_e^2$ [37, 71, 72].

tion σ_{ca} (equation (1.89) and equation (1.77)). For various absorbers the Compton-scattering cross sections, or absorption coefficients shown in figure 1.14, have been multiplied by the atomic number of the absorber, since the Compton-scattering cross section (equation (1.83)) given by the Klein-Nishina formula is valid per electron, but in this case, the atomic cross sections are required.

Ranges in which the individual photon interaction processes dominate, are plotted in figure 1.15 as a function of the photon energy and the atomic number of the absorber [37, 65, 68].

Further interactions of photons (photonuclear interactions, photon-photon scattering, etc.) are governed by extremely low cross sections. Therefore, these processes are of little importance for the detection of photons. However, these processes are of large interest in elementary particle physics and astrophysics.

a)

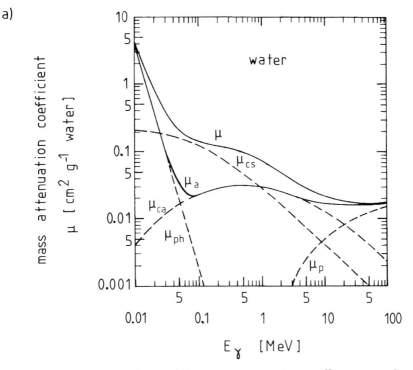

Fig. 1.14 a.   Energy dependence of the mass attenuation coefficient μ and mass absorption coefficient μ_a for photons in water [63, 73, 74, 75]. μ_{ph} describes the photoelectric effect, μ_{cs} the Compton scattering, μ_{ca} the Compton absorption, and μ_p the pair production. μ_a is the total mass absorption coefficient ($\mu_a = \mu_{ph} + \mu_p + \mu_{ca}$) and μ is the total mass attenuation coefficient ($\mu = \mu_{ph} + \mu_p + \mu_c$, where $\mu_c = \mu_{cs} + \mu_{ca}$).

1.3   Strong interactions of hadrons

Apart from the electromagnetic interactions of charged particles strong interactions may also play a rôle for particle detection. In the following we will sketch the strong interactions of hadrons.

In this case, we are dealing mostly with inelastic processes, where secondary strongly interacting particles are produced in the collision. The total cross section for proton-proton scattering approaches a constant value of 50 mb ($1\,\text{mb} = 10^{-27}\,\text{cm}^2$) for high energies ($>$ several GeV). Both the elastic and inelastic part of the cross section show a rather strong energy dependence at low energies [76]

$$\sigma_{\text{total}} = \sigma_{\text{elastic}} + \sigma_{\text{inelastic}} \; . \tag{1.99}$$

The specific quantity that characterizes the inelastic processes is the av-

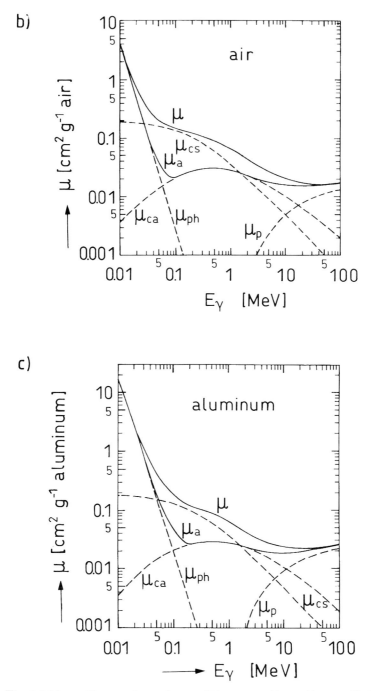

Fig. 1.14 b,c.   Energy dependence of the mass attenuation coefficient μ and mass absorption coefficient μ_a for photons in air b) and aluminum c) [63, 73, 74, 75].

d)

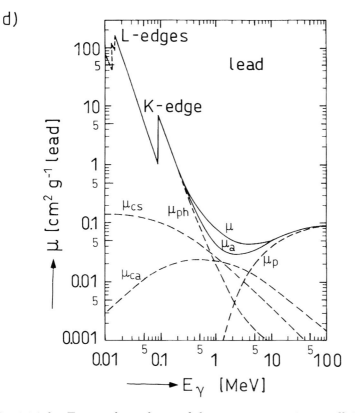

Fig. 1.14 d. Energy dependence of the mass attenuation coefficient μ and mass absorption coefficient μ_a for photons in lead [63, 73, 74, 75].

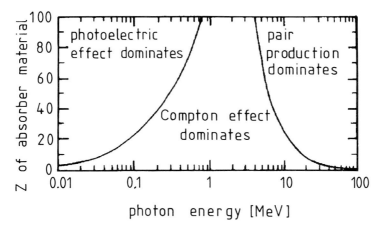

Fig. 1.15. Ranges, in which the photoelectric effect, Compton effect, and pair production dominate as a function of the photon energy and the target charge Z [37, 65, 68].

erage absorption length λ_a, which describes the absorption of hadrons in matter according to

$$N = N_0 e^{-x/\lambda_a} .\tag{1.100}$$

λ_a can be calculated from the inelastic part of the hadronic cross section as follows:

$$\lambda_a = \frac{A}{N_A \cdot \varrho \cdot \sigma_{\text{inel}}} .\tag{1.101}$$

If A is given in g/mol, N_A in mol^{-1}, ϱ in g/cm^3 and the cross section in cm^2, then λ_a has the unit cm. The surface mass density corresponding to λ_a [cm] would be $\lambda_a \cdot \varrho$ [g/cm^2]. The nuclear interaction length λ_w is related to the total cross section σ_{total} according to

$$\lambda_w = \frac{A}{N_A \cdot \varrho \cdot \sigma_{\text{total}}} .\tag{1.102}$$

Since $\sigma_{\text{total}} > \sigma_{\text{inel}}$, it follows that $\lambda_w < \lambda_a$.

The absorption and interaction lengths for various materials are given in table 1.5 [34, 35].

Table 1.5.   *Total and inelastic cross sections as well as interaction and absorption lengths for various materials derived from the corresponding cross sections [34, 35]*

material	Z	A	σ_{total} [barn]	σ_{inel} [barn]	$\lambda_w \cdot \varrho$ [g/cm^2]	$\lambda_a \cdot \varrho$ [g/cm^2]
Water	1	1.01	0.0387	0.033	43.3	50.8
Helium	2	4.0	0.133	0.102	49.9	65.1
Beryllium	4	9.01	0.268	0.199	55.8	75.2
Carbon	6	12.01	0.331	0.231	60.2	86.3
Nitrogen	7	14.01	0.379	0.265	61.4	87.8
Oxygen	8	16.0	0.420	0.292	63.2	91.0
Aluminum	13	26.98	0.634	0.421	70.6	106.4
Silicon	14	28.09	0.660	0.440	70.6	106.0
Iron	26	55.85	1.120	0.703	82.8	131.9
Copper	29	63.55	1.232	0.782	85.6	134.9
Tungsten	74	183.85	2.767	1.65	110.3	185
Lead	82	207.19	2.960	1.77	116.2	194
Uranium	92	238.03	3.378	1.98	117.0	199

Strictly speaking, the hadronic cross sections depend on the energy and vary somewhat for different strongly interacting particles. For the calculation of the interaction and absorption lengths, however, the cross sections σ_{total} and σ_{inel} have been assumed to be energy independent and independent of the particle species (protons, pions, kaons, etc.).

For target materials with $Z \geq 6$ the interaction and absorption lengths, respectively, are much larger than the radiation lengths X_0 (compare table 1.4).

The definitions for λ_a and λ_w are not uniform in the literature. Equations (1.101) and (1.102), however, define these characteristic quantities in a clear fashion.

The cross sections can be used to calculate the probabilities for interactions in a simple manner. If σ_N is the nuclear interaction cross section (i.e., per nucleon), the corresponding probability for an interaction per g/cm^2 is calculated to be

$$\phi[\text{g}^{-1}\text{cm}^2] = \sigma_N \cdot N_A \ , \tag{1.103}$$

where N_A is Avogadro's number. In the case that the atomic cross section σ_A is given, it follows that

$$\phi[\text{g}^{-1}\text{cm}^2] = \sigma_A \cdot \frac{N_A}{A} \ , \tag{1.104}$$

where A is the atomic weight.

1.4   Drift and diffusion in gases[†]

Electrons and ions, produced in an ionization process, quickly lose their energy by multiple collisions with atoms and molecules of a gas. They approach the thermal energy distribution, corresponding to the temperature of the gas.

Their average energy at room temperature is

$$\varepsilon = \frac{3}{2}kT = 40\,\text{meV} \ , \tag{1.105}$$

where k is the Boltzmann constant and T the temperature in Kelvin. They follow a Maxwell-Boltzmann distribution of energies like

$$F(\varepsilon) = \text{const} \cdot \sqrt{\varepsilon} \cdot e^{-\varepsilon/kT} \ . \tag{1.106}$$

The locally produced ionization diffuses by multiple collisions correspond-

---

[†] Extensive literature to these processes is given in [1, 4, 5, 6, 51], [77]-[80].

ing to a Gaussian distribution

$$\frac{\mathrm{d}N}{N} = \frac{1}{\sqrt{4\pi Dt}} \exp\left(-\frac{x^2}{4Dt}\right) \mathrm{d}x \ , \tag{1.107}$$

where $\frac{\mathrm{d}N}{N}$ is the fraction of the charge which is found in the length element $\mathrm{d}x$ at a distance x after a time t. D is the diffusion coefficient. For linear or volume diffusion, respectively, one obtains

$$\begin{aligned} \sigma_x &= \sqrt{2Dt} \\ \sigma_{\mathrm{vol}} &= \sqrt{3} \cdot \sigma_x = \sqrt{6Dt} \ . \end{aligned} \tag{1.108}$$

The average mean free path in the diffusion process is

$$\lambda = \frac{1}{N\sigma(\varepsilon)} \ , \tag{1.109}$$

where $\sigma(\varepsilon)$ is the energy dependent collision cross section, and $N = \frac{N_A}{A}\varrho$ the number of molecules per unit volume. For noble gases one has $N = 2.69 \cdot 10^{19}$ molecules/cm^3 at standard pressure and temperature.

If the charge carriers are exposed to an electric field, an ordered drift along the field will be superimposed over the statistically disordered diffusion. A drift velocity can be defined according to

$$\vec{v}_{\mathrm{drift}} = \mu(E) \cdot \vec{E} \cdot \frac{p_0}{p} \ , \tag{1.110}$$

where

$\mu(E)$ – energy dependent charge-carrier mobility,
$\vec{E}$ – electric field strength, and
p/p_0 – pressure normalized to standard pressure.

The statistically disordered transverse diffusion, however, is not influenced by the electric field.

The drift of free charge carriers in an electric field requires, however, that electrons and ions do not recombine and that they are also not attached to atoms or molecules of the medium in which the drift proceeds.

Table 1.6 contains numerical values for the average mean free path, the diffusion constant and the mobilities of ions [51, 81].

The corresponding quantity for electrons strongly depends on the energy of the electrons and thereby on the field strength. The mobilities of electrons in gases exceed those of ions by approximately three orders of magnitude.

Figure 1.16 shows the root mean square deviation of an originally localized electron cloud for a drift of 1 cm [51, 82]. The width of the electron cloud $\sigma_x = \sqrt{2Dt}$ per 1 cm drift varies significantly with the field strength and shows characteristic dependencies on the gas. For a gas mixture of

Table 1.6. *Average mean free path* $\lambda_{\rm ion}$, *diffusion constant* $D_{\rm ion}$ *and mobilities* $\mu_{\rm ion}$ *of ions in some gases for standard pressure and temperature [51, 81]*

gas	$\lambda_{\rm ion}[\rm cm]$	$D_{\rm ion}[\rm cm^2/s]$	$\mu_{\rm ion}\left[\frac{\rm cm/s}{\rm V/cm}\right]$
H_2	$1.8 \cdot 10^{-5}$	0.34	13.0
He	$2.8 \cdot 10^{-5}$	0.26	10.2
Ar	$1.0 \cdot 10^{-5}$	0.04	1.7
O_2	$1.0 \cdot 10^{-5}$	0.06	2.2

argon (75 %) and isobutane (25 %) values around $\sigma_x \approx 200\,\mu$m are measured, which limit the spatial resolution of drift chambers. In principle, one has to distinguish between the longitudinal diffusion in the direction of the field and a transverse diffusion perpendicular to the electric field. The spatial resolution of drift chambers, however, is limited primarily by the longitudinal diffusion.

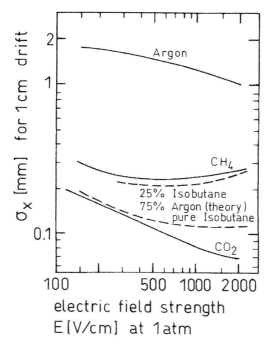

Fig. 1.16. Dependence of the root mean square deviation of an originally localized electron cloud after a drift of 1 cm in various gases [51, 82].

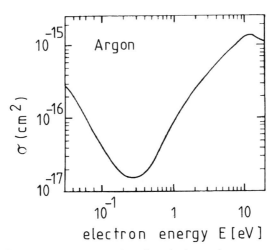

Fig. 1.17. Ramsauer cross section for electrons in argon as a function of the electron energy [84]-[89].

In a simple theory [83] the drift velocity can be expressed by

$$\vec{v}_{\text{drift}} = \frac{e}{m}\vec{E}\ \tau(\vec{E}, \varepsilon) ,\qquad(1.111)$$

where $\vec{E}$ is the field strength and τ the time between two collisions, which in itself depends on $\vec{E}$. The collision cross section, and as a consequence also τ, depends strongly on the electron energy ε and passes through pronounced maxima and minima (Ramsauer effect). These phenomena are caused by interference effects, if the electron wavelength $\lambda = h/p$ (h – Planck's constant, p – electron momentum) approaches molecular dimensions. Of course, the electron energy and electric field strength are correlated. Figure 1.17 shows the Ramsauer cross section for electrons in argon as a function of the electron energy [84]-[89].

Even small contaminations of a gas can drastically modify the drift velocity (figure 1.18, [51, 84, 90, 91]).

Figure 1.19 shows the drift velocities for electrons in argon-methane mixtures [51, 92, 93, 94] and figure 1.20 those in argon-isobutane mixtures [51, 93, 95, 96, 97].

As an approximate value for high field strengths in argon-isobutane mixtures a typical value for the drift velocity of

$$v_{\text{drift}} = 5\,\text{cm}/\mu\text{s}\qquad(1.112)$$

is observed. The dependence of the drift velocity on the field strength, however, may vary considerably for different gases [79, 93, 98]. Under comparable conditions the ions in a gas are slower by three orders of magnitude compared to electrons.

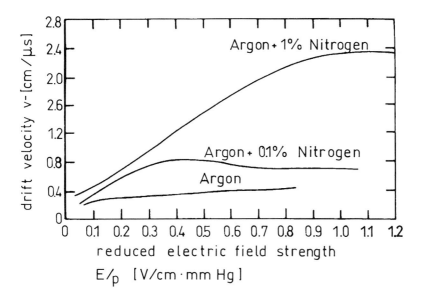

Fig. 1.18. Drift velocities of electrons in pure argon and in argon with minor additions of nitrogen [51, 84, 90, 91].

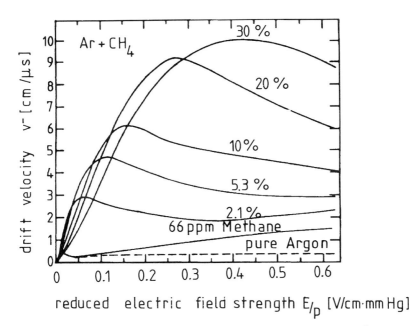

Fig. 1.19. Drift velocities for electrons in argon-methane mixtures [51, 92, 93, 94].

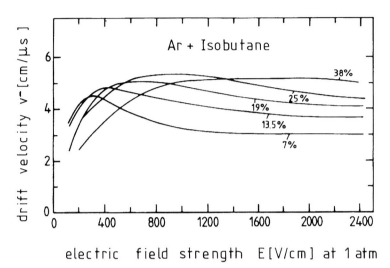

Fig. 1.20.   Drift velocities for electrons in argon-isobutane mixtures [51, 93, 95, 96, 97].

The drift velocity and the drift properties of electrons in gases are strongly modified in the presence of a magnetic field. In addition to the electric force, now the Lorentz force also acts on the charge carriers and forces the charge carriers into circular or spiral orbits.

The equation of motion for the free charge carriers reads

$$m\ddot{\vec{x}} = q\vec{E} + q \cdot \vec{v} \times \vec{B} + m\vec{A}(t) \,, \qquad (1.113)$$

where $m\vec{A}(t)$ is a time dependent stochastic force, which has its origin in collisions with gas molecules. If one assumes that the time average of the product $m \cdot \vec{A}(t)$ can be represented by a velocity proportional friction force $-m\vec{v}/\tau$, where τ is the average time between two collisions, the drift velocity can be derived from equation (1.113) [1] to be

$$\vec{v}_{\mathrm{drift}} = \frac{\mu}{1 + \omega^2\tau^2}\left(\vec{E} + \frac{\vec{E} \times \vec{B}}{B}\omega\tau + \frac{(\vec{E} \cdot \vec{B}) \cdot \vec{B}}{B^2}\omega^2\tau^2\right) \,, \qquad (1.114)$$

if one assumes that for a constant electric field a drift with constant velocity is approached, that is, $\dot{\vec{v}}_{\mathrm{drift}} = 0$.

In equation (1.114)

$\mu = e \cdot \tau/m$ is the mobility of the charge carriers, and
$\omega = e \cdot B/m$ is the cyclotron frequency (from $mr\omega^2 = evB$).

In the presence of electric and magnetic fields the drift velocity has components in the direction of $\vec{E}$, of $\vec{B}$ and perpendicular to $\vec{E}$ and $\vec{B}$ [99] (see also equation (1.114)). If $\vec{E} \perp \vec{B}$, the drift velocity $\vec{v}_{\mathrm{drift}}$ along a line

forming an angle α with the electric field can be derived from equation
(1.114) to be

$$|\vec{v}_{\text{drift}}| = \frac{\mu E}{\sqrt{1 + \omega^2 \tau^2}} \; .$$ (1.115)

The angle between the drift velocity $\vec{v}_{\text{drift}}$ and $\vec{E}$ (Lorentz angle) can be
calculated from equation (1.114) under the assumption of $\vec{E} \perp \vec{B}$

$$\tan \alpha = \omega \tau \; ;$$ (1.116)

if τ is taken from equation (1.111), it follows that

$$\tan \alpha = v_{\text{drift}} \cdot \frac{B}{E} \; .$$ (1.117)

This result may also be derived if the ratio of the acting Lorentz force
$e \vec{v} \times \vec{B}$ (with $\vec{v} \perp \vec{B}$) to the electric force $e\vec{E}$ is considered.

For $E = 500 \, \text{V/cm}$ and a drift velocity in the electric field of $v_{\text{drift}} = 3.5 \, \text{cm}/\mu\text{s}$, a drift velocity in a combined electric and magnetic field ($\vec{E} \perp \vec{B}$) is obtained from equation (1.115) for $B = 1.5$ Tesla on the basis of these simple considerations to be

$$v(E = 500 \, \text{V/cm}, B = 1.5 \, \text{T}) = 2.4 \, \text{cm}/\mu\text{s} \; ;$$

correspondingly the Lorentz angle is calculated from equation (1.117) to be

$$\alpha = 46° \; ,$$

which is approximately consistent with the experimental findings and the results of a more exact calculation (figure 1.21) [51, 95].

Small admixtures of electronegative gases (e.g., oxygen) considerably modify the drift behaviour due to electron attachment. For a 1% fraction of oxygen in argon at a drift field of $1 \, \text{kV/cm}$ the average mean free path of electrons for attachment is of the order $5 \, \text{cm}$. Small admixtures of electronegative gases will reduce the charge signal and in case of strong electronegative gases (such as chlorine) operation of a drift chamber may be even impossible.

Because of the high density the effect of impurities is even more pronounced for liquefied gases. For liquid noble gas chambers the oxygen concentration must stay below the ppm ($\equiv 10^{-6}$) level. 'Warm' liquids, like tetramethylsilane (TMS) even require to reduce the concentration of electronegative impurities to below ppb ($\equiv 10^{-9}$) (see section 4.5).

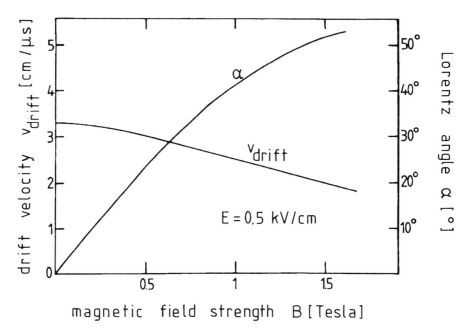

Fig. 1.21. Dependence of the electron drift velocity $\vec{v}_{\mathrm{drift}}$ and the Lorentz angle α on the magnetic field for low electric field strengths (500 V/cm) in a gas mixture of argon (67.2 %), isobutane (30.3 %) and methylal (2.5 %) [51, 95].

2

Characteristic
properties of detectors

The criterion by which to judge the quality of a detector is its resolution for the quantity to be measured (energy, time, spatial coordinates, etc.). If a quantity with true value z_0 is given (e.g., the monoenergetic γ-radiation of energy E_0), the measured results of a detector form a distribution function $D(z)$; the expectation value for this quantity is

$$\langle z \rangle = \int z \cdot D(z)\,\mathrm{d}z \Big/ \int D(z)\,\mathrm{d}z \ , \qquad (2.1)$$

where the integral in the denominator normalizes the distribution function.

The variance of the measured quantity is

$$\sigma_z^2 = \int (z - \langle z \rangle)^2 D(z)\,\mathrm{d}z \Big/ \int D(z)\,\mathrm{d}z \ . \qquad (2.2)$$

The integrals extend over the full range of possible values of the distribution function.

As an example, the expectation value and the variance for a rectangular distribution will be calculated. In a multiwire proportional chamber (see section 4.6) with wire spacing δz, the coordinates of charged particles passing through the chamber are to be determined. There is no drift time measurement on the wires. How accurate can this measurement be? Let us assume that one particular wire has fired. The distribution function $D(z)$ is constant $= 1$ from $-\delta z/2$ up to $+\delta z/2$ around the wire which has fired, and outside this interval the distribution function is zero (see figure 2.1).

The expectation value for z is evidently zero ($\hat{=}$ position of the fired wire):

$$\langle z \rangle = \int_{-\delta z/2}^{+\delta z/2} z \cdot 1\,\mathrm{d}z \Big/ \int_{-\delta z/2}^{+\delta z/2} \mathrm{d}z = \left.\frac{z^2}{2}\right|_{-\delta z/2}^{+\delta z/2} \Big/ \left. z \right|_{-\delta z/2}^{+\delta z/2} = 0 \ ; \qquad (2.3)$$

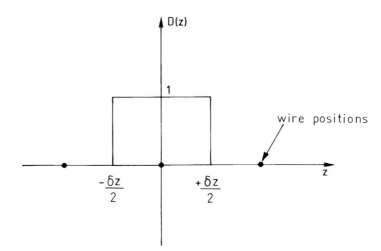

Fig. 2.1.   Schematic drawing for the determination of the variance of a rectangular distribution.

correspondingly, the variance is calculated to be

$$\sigma_z^2 = \int_{-\delta z/2}^{+\delta z/2} (z-0)^2 \cdot 1 \, \mathrm{d}z \bigg/ \delta z = \frac{1}{\delta z} \int_{-\delta z/2}^{+\delta z/2} z^2 \mathrm{d}z$$

$$= \frac{1}{\delta z} \frac{z^3}{3} \bigg|_{-\delta z/2}^{+\delta z/2} = \frac{1}{3\delta z} \left(\frac{(\delta z)^3}{8} + \frac{(\delta z)^3}{8} \right) = \frac{(\delta z)^2}{12} , \qquad (2.4)$$

which means

$$\sigma_z = \frac{\delta z}{\sqrt{12}} . \qquad (2.5)$$

The quantities δz and σ_z have dimensions. The relative values $\delta z/z$ or σ_z/z, respectively, are dimensionless.

   In many cases experimental results are normally distributed, corresponding to a distribution function (figure 2.2)

$$D(z) = \frac{1}{\sigma_z \sqrt{2\pi}} e^{-(z-z_0)^2/2\sigma_z^2} . \qquad (2.6)$$

The variance determined according to equation (2.2) for this Gaussian distribution implies that exactly 68.27 % of all experimental results lie between $z_0 - \sigma_z$ and $z_0 + \sigma_z$. Within $2\sigma_z$ there are 95.45 % and within $3\sigma_z$ there are 99.73 % of all experimental results.

   Frequently it is useful to define the confidence level for the measured quantity and the related probability that the true value lies within a given interval. For this purpose, we plot the normalized distribution function in

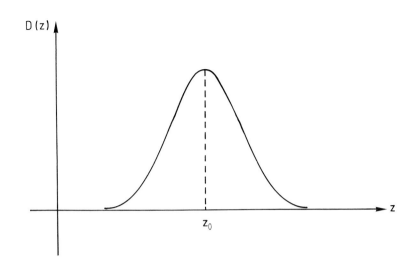

Fig. 2.2. Normal distribution (Gaussian distribution around the average value z_0).

its dependence on $z - \langle z \rangle$ (figure 2.3). For a normalized probability distribution with an expectation value $\langle z \rangle$ and root mean squared deviation σ_z

$$1 - \alpha = \int_{\langle z \rangle - \delta}^{\langle z \rangle + \delta} D(z) \mathrm{d}z \tag{2.7}$$

is the probability that the true value z_0 lies in the interval $\pm \delta$ around the measured quantity z, or equivalently: $100 \cdot (1 - \alpha)\,\%$ of all measured values lie in an interval $\pm \delta$, centered on the average value $\langle z \rangle$.

The choice of $\delta = \sigma_z$ for a Gaussian distribution leads to a confidence interval, which is called the standard error, and whose probability is $1 - \alpha = 0.6827$ (corresponding to $68.27\,\%$). On the other hand, if a confidence level is given, the related width of the measurement interval can be calculated. For a confidence level of $1 - \alpha \cong 95\,\%$ one gets an interval width of $\delta = \pm 1.96\,\sigma_z$; $1 - \alpha \cong 99.9\,\%$ yields a width of $\delta = \pm 3.29\,\sigma_z$ [34, 35].

A frequently used quantity for a resolution is the half width of a distribution which can easily be seen or calculated. The half width of a distribution is the full width at half maximum (fwhm). For normal distributions one gets

$$\Delta z(\mathrm{fwhm}) = 2\sqrt{2\ln 2}\,\sigma_z = 2.355\,\sigma_z \ . \tag{2.8}$$

The Gaussian distribution is a continuous distribution function. If one observes particles in detectors the events frequently follow a Poisson distribution. This distribution is asymmetric (negative values do not occur), and discrete.

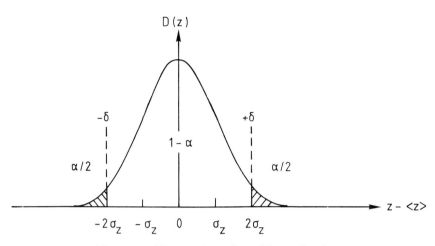

Fig. 2.3.   Illustration of confidence levels.

For a mean value μ the individual results n are distributed according to

$$f(n,\mu) = \frac{\mu^n e^{-\mu}}{n!} , \quad n = 0, 1, 2, \ldots \quad . \tag{2.9}$$

The expectation value for this distribution is equal to the mean value μ with a variance of $\sigma^2 = \mu$.

Let us assume that after many event counting experiments the average value is three events. The probability to find, in an individual experiment, e.g., no event, is $f(0,3) = e^{-3} = 0.05$, or equivalently, if one finds no event in a single experiment, then the true value is smaller than or equal to 3 with a confidence level of 95 %. For large values of n the Poisson distribution approaches the Gaussian.

The determination of the efficiency of a detector represents a random experiment with only two possible results: either the detector was efficient with probability p or not with probability $1 - p = q$. The probability that the detector was efficient exactly r times in n experiments is given by the binomial distribution (Bernoulli distribution)

$$f(n, r, p) = \binom{n}{r} p^r q^{n-r} = \frac{n!}{r!(n-r)!} p^r q^{n-r} . \tag{2.10}$$

The expectation value of this distribution is $\langle r \rangle = n \cdot p$ and the variance is $\sigma^2 = n \cdot p \cdot q$.

Let the efficiency of a detector be $p = 95 \%$ for 100 triggers (95 particles were observed, 5 not). In this example the standard deviation (σ of the expectation value $\langle r \rangle$) is given by

$$\sigma = \sqrt{n \cdot p \cdot q} = \sqrt{100 \cdot 0.95 \cdot 0.05} = 2.18 \qquad (2.11)$$

resulting in

$$p = (95 \pm 2.18)\,\% \ .$$

Note that with this error calculation the efficiency cannot exceed $100\,\%$, as is correct. Using a Poissonian error $(\pm\sqrt{95})$ would lead to a wrong result.

The methods for the statistical treatment of experimental results presented so far include only the most important distributions. For low event rates Poisson-like errors lead to inaccurate limits. If, e.g., one genuine event of a certain type has been found in a given time interval, the experimental value which is obtained from the Poisson distribution $n \pm \sqrt{n}$, in this case 1 ± 1, cannot be correct. Because, if one has found a genuine event, the experimental value can never be compatible with zero, also not within the error.

The statistics of small numbers therefore has to be modified, leading to the Regener statistics [100]. In table 2.1 the $\pm 1\sigma$ limits for the quoted event numbers are given. For comparison the normal error which is the square root of the event rate is also shown.

Table 2.1. *Statistics of low numbers. Quoted are the $\pm 1\sigma$ errors on the basis of the Regener statistics [100] and the $\pm 1\sigma$ square root errors of the Poisson statistics*

lower limit		number of events	upper limit	
square root error	statistics of low numbers		statistics of low numbers	square root error
0	0	0	1.84	0
0	0.17	1	3.3	2
0.59	0.71	2	4.64	3.41
1.27	1.37	3	5.92	4.73
6.84	6.89	10	14.26	13.16
42.93	42.95	50	58.11	57.07

The determination of errors or confidence levels is even more complicated if one considers counting statistics with low event numbers in the presence of background processes which are detected along with searched-for events. The corresponding formulae for such processes are given in the literature [34, 35, 101, 102, 103].

A general word of caution, however, is in order in the statistical treatment of experimental results. The definition of statistical characteristics in the literature is not always consistent.

In the case of determination of resolutions or experimental errors, one is frequently only interested in relative quantities, that is, $\delta z / \langle z \rangle$ or $\sigma_z / \langle z \rangle$; one has to bear in mind that the average result of a number of experiments $\langle z \rangle$ must not necessarily be equal to the true value z_0. To obtain the relation between the experimental answer $\langle z \rangle$ and the true value z_0, the detectors must be calibrated. Not all detectors are linear, like

$$\langle z \rangle = c \cdot z_0 + d \,, \tag{2.12}$$

where c, d are constants. Nonlinearities such as

$$\langle z \rangle = c(z_0) z_0 + d$$

may, however, be particularly awkward and require an exact knowledge of the calibration function. In many cases the calibration parameters are also time dependent.

In the following some characteristic quantities of detectors will be discussed.

Energy resolutions, spatial resolutions and time resolutions are calculated as discussed above. Apart from the time resolution there are in addition a number of further characteristic times [104].

The *dead time* τ_D is the time which has to pass between the registration of one set of incident particles and being sensitive to another set. The dead time, in which no further particles can be detected, is followed by a phase where particles can again be measured; however, the detector may not respond to the particle with full sensitivity. After a further time, the *recovery time* τ_R, the detector can again supply a signal of normal amplitude.

Let us illustrate this behaviour using the example of a Geiger-Müller counter (see section 4.3) (figure 2.4). After the passage of the first particle the counter is completely insensitive for further particles for a certain time τ_D. Slowly, the field in the Geiger-Müller counter recovers so that for times $t > \tau_D$ a signal can again be recorded, although not at full amplitude. After a further time τ_R the counter has recovered so that again the initial conditions are established.

The *sensitive time* τ_S is of importance for pulsed detectors. It is the time interval in which particles can be recorded, independent of whether these are correlated with the triggered event or not. If, for example, in an accelerator experiment the detector is triggered by a beam interaction (i.e., is made sensitive), usually a time window of defined length (τ_S) is opened, in which the event is recorded. If by chance in this time interval τ_S a cosmic-ray muon passes through the detector, it will also be recorded because the detector having been made sensitive once cannot distinguish between particles of interest and particles which just happen to pass through the detector in this time window.

In th
tion.
   Th
the
and
   In
cles
of in
abili
tane
secti
with
very
secti
hit'
(up

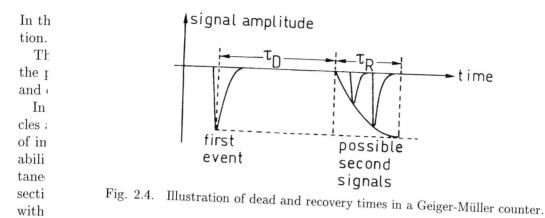

Fig. 2.4.   Illustration of dead and recovery times in a Geiger-Müller counter.

   The *readout time* is the time that is required to read the event, possibly into an electronic memory. For other than electronic registering (e.g., film) the readout time can be considerably long. Closely related to the readout time is the *repetition time*, which describes the minimum time which must pass between two subsequent events, so that they can be distinguished. The length of the repetition time is determined by the slowest element in the chain detector, readout and registering.

   The *memory time* of a detector is the maximum allowed time delay between particle passage and trigger signal, which still yields a 50 % efficiency.

   The previously mentioned time resolution characterizes the minimum time difference where two events can still be separated. This time resolution is very similar to the repetition time, the only difference being that the time resolution refers, in general, to an individual component of the whole detection system (e.g., only the front-end detector), while the repetition time includes all components. For example, the time resolution of a detector can be extremely short, but the whole speed can be lost by a slow readout.

   The term time resolution is also used for the precision with which the arrival time of a particle in a detector can be recorded. The time resolution for individual events defined in this way is determined by the fluctuation of the rise time of the detector signal.

   A very important quantity of each detector is its efficiency, that is, the probability that a particle which passes through the detector is also seen by it. This efficiency ε can vary considerably depending on the type of detector and radiation. For example, γ-rays are measured in gas counters with probabilities of the order of a percent, whereas charged particles in scintillation counters or gas detectors are seen with a probability of 100 %. Neutrinos can only be recorded with extremely low probabilities ($\sim 10^{-18}$ for MeV-neutrinos in a massive detector).

spectroscopy of particles of higher charge because in this case the deposited energies are in general larger compared to those of singly charged minimum ionizing particles.

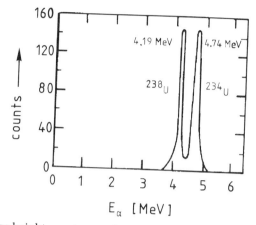

Fig. 4.2.   Pulse height spectrum of α-particles emitted from a ^{234}U/^{238}U isotope mixture, recorded with a Frisch grid ionization chamber [2].

Apart from planar ionization chambers, cylindrical ionization chambers are also in use. Because of the cylindrical arrangement of the electrodes the electric field in this case is no longer constant. The potential distribution in the cylindrical ionization chamber can be derived from the Laplace equation

$$\Delta V = 0 \ . \tag{4.12}$$

In cylinder coordinates, which are appropriate to this problem, the Laplace equation reads

$$\frac{1}{r}\frac{\partial}{\partial r}\left(r\frac{\partial V}{\partial r}\right) + \frac{\partial^2 V}{\partial z^2} + \frac{1}{r^2}\frac{\partial^2 V}{\partial \varphi^2} = 0 \ ; \tag{4.13}$$

z is the coordinate along the anode wire and φ the azimuthal angle. If we neglect edge effects at the end of the counting wire, equation (4.13) simplifies because of the azimuthal symmetry to

$$\frac{1}{r}\frac{\partial}{\partial r}\left(r\frac{\partial V}{\partial r}\right) = 0 \ . \tag{4.14}$$

Integrating this equation and considering the constraints $V = 0$ for $r = r_\mathrm{a}$ and $V = U_0$ for $r = r_\mathrm{i}$ (r_a radius of cylindrical cathode, r_i anode wire radius; see figure 4.3) leads to

$$V = \frac{U_0 \ln r/r_\mathrm{a}}{\ln r_\mathrm{i}/r_\mathrm{a}} \ . \tag{4.15}$$

In these extreme cases Poisson-like errors can be used as an approximation.

The efficiency of a detector normally also depends on the point where the particle has passed through the detector (homogeneity, uniformity) and on the angle of incidence (isotropy).

In many applications of detectors it is necessary to record many particles at the same time. For this reason, the multiparticle efficiency is also of importance. The multiparticle efficiency can be defined as the probability that exactly N particles are registered if N particles have simultaneously passed through the detector. For normal spark chambers (see section 4.15) the multitrack efficiency defined this way decreases rapidly with increasing N, while for scintillation counters it will probably vary very little with N. The multiparticle efficiency for drift chambers (see section 4.7) can also be affected by the way the readout is done ('single hit' where only one track is recorded or 'multiple hit' where many tracks (up to a preselected maximum) can be analyzed).

3

Units of radiation measurement

Many measurements and tests with detectors are made with radioactive sources. A basic knowledge of the units of radiation measurement and the biological effects of radiation are useful [65], [105]-[119].

Let us assume that there are initially N_0 nuclei of a certain radioactive element. The number will decrease in the course of time t due to decay according to

$$N = N_0 e^{-t/\tau} \; , \qquad (3.1)$$

where τ is the lifetime of the radioisotope. One has to distinguish between the lifetime and the half-life $T_{1/2}$. The half-life can be calculated from equation (3.1) as

$$N(t = T_{1/2}) = \frac{N_0}{2} = N_0 e^{-T_{1/2}/\tau} \; , \qquad (3.2)$$

$$T_{1/2} = \tau \cdot \ln 2 \; . \qquad (3.3)$$

The decay constant of the radioactive element is

$$\lambda = \frac{1}{\tau} = \frac{\ln 2}{T_{1/2}} \; . \qquad (3.4)$$

The activity of a source gives the number of decays per second

$$A = -\frac{dN}{dt} = \frac{1}{\tau} N = \lambda N \; . \qquad (3.5)$$

The unit of the activity is Becquerel (Bq). 1 Bq means 1 decay per second. (In passing it should be mentioned that the physical quantity with the dimension s^{-1} already has a name: Hertz! However, this unit Hz is mostly used for periodic phenomena, while Bq is used for statistically distributed events.) The unit Bq supersedes the old unit Curie (Ci). Historically 1 Ci was the activity of 1 gram of radium

$$1 \, \text{Ci} = 3.7 \cdot 10^{10} \, \text{Bq} \qquad (3.6)$$

56

or

$$1\,\mathrm{Bq} = 27 \cdot 10^{-12}\,\mathrm{Ci} = 27\,\mathrm{pCi}\,. \tag{3.7}$$

1 Bq is a very small unit of the activity.

The activity in Bq does not say very much about possible biological effects. These are related to the energy which is deposited per unit mass by a radioactive source.

The energy dose D (absorbed energy per mass unit)

$$D = \frac{1}{\varrho}\frac{\mathrm{d}W}{\mathrm{d}V} \tag{3.8}$$

($\mathrm{d}W$ – absorbed energy; ϱ – density; $\mathrm{d}V$ – unit of volume) is measured in Grays (1 Gray = 1 J/kg). The old cgs-unit rad (**r**öntgen **a**bsorbed **d**ose, 1 rad = 100 erg/g) is related to Gray according to

$$1\,\mathrm{Gy} = 100\,\mathrm{rad}\,. \tag{3.9}$$

Gray and rad describe only physical energy absorption, and do not take into account any biological effect. Since, however, α, β, γ, and neutron emitting sources have different biological effects for the same energy absorption, a **r**elative **b**iological **e**fficiency (RBE) is defined. The energy dose D_γ obtained from the exposure to γ or X-rays serves as reference. The energy dose of an arbitrary radiation which yields the same biological effect as D_γ, leads to the definition of the relative biological efficiency as

$$D_\gamma = RBE \cdot D\,. \tag{3.10}$$

The energy dose multiplied by the *RBE*-factor is called equivalent dose H and is measured in a different unit — although the *RBE*-factor has no dimension. The unit of the equivalent dose is 1 Sievert (Sv).

$$H[\mathrm{Sv}] = RBE \cdot D[\mathrm{Gy}]\,. \tag{3.11}$$

The old cgs-unit ($H[\mathrm{rem}] = RBE \cdot D[\mathrm{rad}]$, rem = **r**öntgen **e**quivalent **m**an) is related to Sievert according to

$$1\,\mathrm{Sv} = 100\,\mathrm{rem}\,. \tag{3.12}$$

Some *RBE*-factors are listed in table 3.1.

It should be mentioned that the biological effect of radiation is also influenced, for example, by the time sequence of absorption (e.g., fractionated irradiation), the energy spectrum of radiation, or the question whether the irradiated person has been sensitized by a pharmaceutical drug. In this way the *RBE*-factors are modified to the more adequate quality factors q which have to be multiplied to the dose D to give the equivalent dose H.

Table 3.1. *RBE-factors of various types of radiation*

radiation	RBE-factor
α	20
β	1
γ	1
X-rays	1
fast neutrons	10
thermal neutrons	3
protons	10
heavy recoil nuclei	20

Apart from these units, there is still another one describing the quantity of produced charge, which is the Röntgen (R). One Röntgen is the radiation dose for X-ray or γ-radiation which produces under normal conditions one electrostatic charge unit (esu) of electrons and ions in $1\,\mathrm{cm}^3$ of air.

The charge of an electron is $1.6 \cdot 10^{-19}$ C or $4.8 \cdot 10^{-10}$ esu. (The esu is a cgs-unit with $1\,\mathrm{esu} = \frac{1}{3 \cdot 10^9}$ C.) If one electrostatic charge unit is produced, the number of generated electrons per cm^3 is given by

$$N = \frac{1}{4.8 \cdot 10^{-10}} = 2.08 \cdot 10^9 \ . \tag{3.13}$$

If the unit Röntgen is transformed into an ion dose in C/kg, it gives

$$1\,\mathrm{R} = \frac{N \cdot q_e[\mathrm{C}]}{m_{\mathrm{air}}(1\,\mathrm{cm}^3)[\mathrm{kg}]} = \frac{1\,\mathrm{esu}}{m_{\mathrm{air}}(1\,\mathrm{cm}^3)[\mathrm{kg}]} \ , \tag{3.14}$$

where q_e is the electron charge in Coulomb, $m_{\mathrm{air}}(1\,\mathrm{cm}^3)$ is the mass of $1\,\mathrm{cm}^3$ air; consequently

$$1\,\mathrm{R} = 2.59 \cdot 10^{-4}\,\mathrm{C}/\,\mathrm{kg}_{\mathrm{air}} \ . \tag{3.15}$$

If Röntgen has to be converted to an energy dose, one has to consider that the production of an electron-ion pair in air requires an energy of about $W = 34\,\mathrm{eV}$.

$$\begin{aligned} 1\,\mathrm{R} &= N \cdot \frac{W}{m_{\mathrm{air}}} \\ &= 0.88\,\mathrm{rad} \ . \end{aligned} \tag{3.16}$$

To obtain a feeling for these abstract units, it is quite useful to establish a natural scale by considering the radiation load from the environment.

The radioactivity of the human body amounts to $\approx 7500\,\text{Bq}$, mainly caused by the radioisotope ^{14}C and the potassium isotope ^{40}K. The average radioactive load (at sea level) by cosmic radiation ($\sim 0.3\,\text{mSv/yr}$), by terrestrial radiation ($\sim 0.5\,\text{mSv/yr}$) and by incorporation of radioisotopes (inhalation $\sim 1.1\,\text{mSv/yr}$, ingestion $\sim 0.3\,\text{mSv/yr}$) are all of approximately the same order of magnitude, just as the radiation load caused by civilization ($\sim 0.8\,\text{mSv/yr}$), which is mainly caused by X-ray diagnostics and treatment and by exposures in nuclear medicine. The total annual per-capita dose consequently is about $3\,\text{mSv}$.

The natural radiation load, of course, depends on the place where one lives; it has a typical fluctuation corresponding to a factor of two. The radiation load caused by civilization naturally has a much larger fluctuation. The average value in this case results from relatively high doses obtained by few persons.

The maximum permissible whole-body dose for persons who work in an area of controlled access amounts to $50\,\text{mSv/yr}$ ($= 5\,\text{rem/yr}$). The lethal whole-body dose (50 % mortality in 30 days without medical treatment) is $4\,\text{Sv}$ ($= 400\,\text{rem}$).

In table 3.2 some α, β and γ-ray emitters, which are found to be quite useful for detector tests, are listed [34, 35]. (For β-ray emitters the maximum energies of the continuous energy spectra are given; EC means electron capture from the K-shell.)

If gaseous detectors are to be tested, an ^{55}Fe source is very convenient. The ^{55}Fe nucleus captures an electron from the K-shell leading to the emission of characteristic X-rays of manganese of $5.89\,\text{keV}$. X-rays or γ-rays do not provide a trigger. If one wants to test gaseous detectors with triggered signals, one should look for electron emitters with an electron energy as high as possible. Energetic electrons have a high range making it possible to penetrate the detector and also a trigger counter. ^{90}Y produced in the course of ^{90}Sr-decay has a maximum energy of $2.28\,\text{MeV}$ (corresponding to $\approx 4\,\text{mm}$ aluminum). A Sr/Y-radioactive source has the convenient property that almost no γ-rays, which are hard to shield, are emitted. If one wants to achieve even higher electron energies, one can use a ^{106}Rh source, it being a daughter element of ^{106}Ru. The electrons of this source with a maximum energy of $3.54\,\text{MeV}$ have a range of $\sim 6.5\,\text{mm}$ in aluminum. The EC emitter ^{207}Bi emits monoenergetic conversion electrons, and is therefore particularly well suited for an energy calibration and a study of the energy resolution of detectors.

Conversion electrons are produced if the nucleus is in an excited state after an electron capture from the K-shell ($p + e^- \rightarrow n + \nu_e$). This discrete excitation energy of the nucleus can be directly transferred to an electron of the atomic shell ('conversion electron').

Table 3.2. *A compilation of useful radioactive sources along with their charac-teristic properties [34, 35, 120, 121, 122]*

radio-isotope	decay mode/ branching fraction	$T_{1/2}$	energy of radiation	
			β, α	γ
$^{22}_{11}$Na	β^+ (89 %) EC (11 %)	2.6 a	β^+_1 1.83 MeV (0.05 %) β^+_2 0.54 MeV (90 %)	1.28 MeV
$^{55}_{26}$Fe	EC	2.7 a		Mn X-rays 5.89 keV (24 %) 6.49 keV (2.9 %)
$^{57}_{27}$Co	EC	267 d		14 keV (10 %) 122 keV (86 %) 136 keV (11 %)
$^{60}_{27}$Co	β^-	5.27 a	β^- 0.316 MeV (100 %)	1.173 MeV (100 %) 1.333 MeV (100 %)
$^{90}_{38}$Sr $\rightarrow {}^{90}_{39}$Y	β^- β^-	28.5 a 64.8 h	β^- 0.546 MeV (100 %) β^- 2.283 MeV (100 %)	
$^{106}_{44}$Ru $\rightarrow {}^{106}_{45}$Rh	β^- β^-	1.0 a 30 s	β^- 0.039 MeV (100 %) β^-_1 3.54 MeV (79 %) β^-_2 2.41 MeV (10 %) β^-_3 3.05 MeV (8 %)	0.512 MeV (21 %) 0.62 MeV (11 %)
$^{109}_{48}$Cd	EC	1.27 a	monoenergetic conversion electrons 63 keV (41 %) 84 keV (45 %)	88 keV (3.6 %) Ag X-rays
$^{137}_{55}$Cs	β^-	30 a	β^-_1 0.514 MeV (94 %) β^-_2 1.176 MeV (6 %)	0.662 MeV (85 %)
$^{207}_{83}$Bi	EC	32.2 a	monoenergetic conversion electrons 0.482 MeV (2 %) 0.554 MeV (1 %) 0.976 MeV (7 %) 1.048 MeV (2 %)	0.570 MeV (98 %) 1.063 MeV (75 %) 1.770 MeV (7 %)
$^{241}_{95}$Am	α	433 a	α 5.443 MeV (13 %) α 5.486 MeV (85 %)	60 keV (36 %) Np X-rays

If higher energies are required, or more penetrating radiation, one can take advantage of test beams at accelerators or use muons from cosmic radiation.

The flux of cosmic ray muons through a horizontal area amounts to approximately $1/(\text{cm}^2 \cdot \text{min})$ at sea level. The muon flux per solid angle from near vertical directions through a horizontal area is $8 \cdot 10^{-3} \, \text{cm}^{-2} \, \text{s}^{-1} \, \text{sr}^{-1}$ [123].

The angular distribution of muons roughly follows a $\cos^2 \theta$-law, where θ is the zenith angle measured with respect to the vertical direction. Muons account for 80% of all charged cosmic ray particles at sea level.

4

Detectors for ionization
and track measurements

A particular type of detector does not necessarily make only one sort of measurement. For example, a segmented calorimeter can be used to determine particle tracks; however, the primary aim of such a detector is to measure the energy. For semiconductor detectors the classification is somewhat more complicated. Lithium-drifted germanium or silicon counters are used almost exclusively for the energy measurement in the MeV-range. Silicon strip counters on the other hand are certainly excellent track detectors [124].

This chapter on detectors for ionization and track measurements mainly describes gaseous detectors ('chambers'). In most cases these chambers serve a dual purpose, namely of ionization and track measurements [5, 6], where the gas detectors with only one anode preferentially measure the ionization while those with many anode wires [125], in addition, also measure the spatial coordinates of particles.*

4.1 Ionization chambers

An ionization chamber is a gaseous detector which measures the ionization loss of a charged particle or the energy loss of a photon. This is done by separating the produced charge carrier pairs in an electric field and guiding them to the anode or cathode, respectively, where the corresponding signals can be recorded. If a particle is totally absorbed in an ionization chamber, this detector type measures its energy [126, 127].

In the most simple case an ionization chamber consists of a system of parallel electrodes. A voltage applied across the electrodes produces a homogeneous electric field. The electrode pair is mounted in a gas-tight

---

* The presentation of detectors according to their aim of measurement has been consistently introduced by K. Kleinknecht [1].

container that is filled with a gas mixture which allows electron and ion drift. That means, it does not contain any, or only very low amounts of, electronegative gases.

In principle the counting gas can also be a liquid or even a solid (solid state ionization chamber). The essential properties of ionization chambers are not changed by the state of the counting medium.

Let us assume that a charged particle is incident parallel to the electrodes at a distance x_0 from the anode (see figure 4.1). It produces, depending on the particle type and energy, along its track an ionization which is characteristic for the gas. The voltage U_0 in ionization chambers is chosen in such a way that there will be no gas amplification.

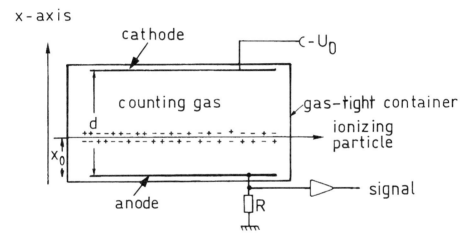

Fig. 4.1. Principle of operation of a planar ionization chamber.

The electric field strength in the chamber is constant

$$|\vec{E}| = E_x = U_0/d .\tag{4.1}$$

In the following we will assume that the produced charge is collected in the electric field completely and that there are no secondary ionization processes.

The drifting charge carriers will induce an electric charge on the capacitor plates. This flows across the working resistance R and produces a small voltage signal which can be further processed with the help of a preamplifier. The charge signal can be measured equally well with a current sensitive amplifier yielding the recorded charge by integrating over the current.

Suppose N charge carrier pairs are produced along the particle track at the distance x_0. The parallel electrodes of the ionization chamber acting as a capacitor with capacitance C are initially charged to the voltage U_0. Due to the charge carrier transport it will be discharged to a certain extent. Thereby the stored energy $\frac{1}{2}CU_0^2$ will be reduced to $\frac{1}{2}CU^2$ according to the following equations [1, 2, 63]

$$\frac{1}{2}CU^2 = \frac{1}{2}CU_0^2 - N\int_{x_0}^x qE_x dx , \qquad (4.2)$$

$$\frac{1}{2}CU^2 - \frac{1}{2}CU_0^2 = \frac{1}{2}C(U + U_0)(U - U_0) = -N \cdot q \cdot E_x \cdot (x - x_0) . \quad (4.3)$$

The voltage drop, however, will only be very small and one may approximate

$$U + U_0 = 2U_0$$

and

$$U - U_0 = \Delta U .$$

Using $E_x = U_0/d$ one can work out ΔU with the help of equation (4.3)

$$\Delta U = -\frac{N \cdot q}{C \cdot d}(x - x_0) . \qquad (4.4)$$

The signal amplitude ΔU has contributions from the fast moving electrons and the slowly drifting ions. If v^+ and v^- are the constant drift velocities of ions and electrons one obtains

$$\Delta U^+ = -\frac{Nq}{Cd}v^+\Delta t^+ ,$$
$$\Delta U^- = -\frac{N(-e)}{Cd}v^-\Delta t^- \qquad (4.5)$$

if Δt^+ or Δt^-, respectively, are the corresponding drift times. Because of $v^- \gg v^+$, the signal amplitude will rise initially linearly up to

$$\Delta U_1 = \frac{Ne}{Cd} \cdot (-x_0) \qquad (4.6)$$

(the electrons will drift to the anode which is at $x = 0$) and then will increase more slowly by the amount which originates from the movement of ions

$$\Delta U_2 = -\frac{Nq}{Cd}(d - x_0) . \qquad (4.7)$$

The total signal amplitude therefore is

$$\Delta U = \Delta U_1 + \Delta U_2 = -\frac{Ne}{Cd}x_0 - \frac{Nq}{Cd}(d - x_0) , \qquad (4.8)$$

and since $q = +e$ it follows that

$$\Delta U = -\frac{N \cdot e}{C} \, . \tag{4.9}$$

This result may also be derived from the equation describing the charge on a capacitor $\Delta Q = -N \cdot e = C \cdot \Delta U$, which means that, independently of the construction of the ionization chamber, the charge Q on the capacitor is reduced by the collected ionization ΔQ and this leads to a voltage amplitude of $\Delta U = \frac{\Delta Q}{C}$.

These considerations are only true if the charging resistor is infinitely large, or more precisely

$$RC \gg \Delta t^-, \Delta t^+ \, . \tag{4.10}$$

In practical cases RC is usually large compared to Δt^-, but smaller than Δt^+. In this case one obtains [2]

$$\Delta U = -\frac{Ne}{Cd} x_0 - \frac{Ne}{d} v^+ R (1 - e^{-\Delta t^+ / RC}) \, , \tag{4.11}$$

which reduces to equation (4.9) if $RC \gg \Delta t^+ = \frac{d - x_0}{v^+}$.

For electric field strengths of 500 V/cm and typical drift velocities of $v^- = 5$ cm/µs, collection times for electrons of $2 \, \mu$s and for ions of about $2 \, $ms are obtained for a drift path of 10 cm. If the time constant $RC \gg 2 \, $ms, the signal amplitude is independent of x_0.

For many applications this is much too long. If one restricts oneself to the measurement of the electron signal, which can be done by differentiating the signal, the total amplitude will not only be smaller, but also depend on the point of production of the ionization (see equation (4.6)). This disadvantage can be overcome by mounting a grid between anode and cathode ('Frisch grid'). If the charged particle enters the larger volume between grid and cathode the produced charged carriers will first drift through this region which is shielded from the anode. Only when the electrons penetrate through the grid will the signal on the working resistor R rise. Ions will not produce any signal on R because their effect is screened by the grid. Consequently, this type of ionization chamber with a Frisch grid measures only the electron signal which, in this configuration, is independent from the ionization production region, as long as it is between the grid and the cathode.

Figure 4.2 shows the pulse height spectrum of α-particles emitted from a mixture of radioisotopes ^{234}U and ^{238}U recorded by a Frisch grid ionization chamber [2]. ^{234}U emits α-particles with energies of 4.77 MeV (72 %) and 4.72 MeV (28 %), while ^{238}U emits mainly α-particles of energy 4.19 MeV. Although the adjacent α-energies of the ^{234}U-isotope cannot be resolved one can, however, clearly distinguish between the two different uranium isotopes. Ionization chambers can also be used in the spectroscopy of particles of higher charge because in this case the deposited energies are in general larger compared to those of singly charged minimum ionizing particles.

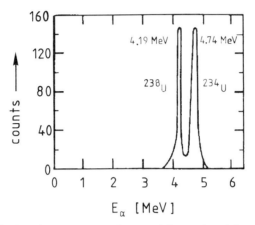

Fig. 4.2.   Pulse height spectrum of α-particles emitted from a ^{234}U$/^{238}$U isotope mixture, recorded with a Frisch grid ionization chamber [2].

Apart from planar ionization chambers, cylindrical ionization chambers are also in use. Because of the cylindrical arrangement of the electrodes the electric field in this case is no longer constant. The potential distribution in the cylindrical ionization chamber can be derived from the Laplace equation

$$\Delta V = 0 \ . \tag{4.12}$$

In cylinder coordinates, which are appropriate to this problem, the Laplace equation reads

$$\frac{1}{r}\frac{\partial}{\partial r}\left(r\frac{\partial V}{\partial r}\right) + \frac{\partial^2 V}{\partial z^2} + \frac{1}{r^2}\frac{\partial^2 V}{\partial \varphi^2} = 0 \ ; \tag{4.13}$$

z is the coordinate along the anode wire and φ the azimuthal angle. If we neglect edge effects at the end of the counting wire, equation (4.13) simplifies because of the azimuthal symmetry to

$$\frac{1}{r}\frac{\partial}{\partial r}\left(r\frac{\partial V}{\partial r}\right) = 0 \ . \tag{4.14}$$

Integrating this equation and considering the constraints $V = 0$ for $r = r_{\mathrm{a}}$ and $V = U_0$ for $r = r_{\mathrm{i}}$ (r_{a} radius of cylindrical cathode, r_{i} anode wire radius; see figure 4.3) leads to

$$V = \frac{U_0 \ln r/r_{\mathrm{a}}}{\ln r_{\mathrm{i}}/r_{\mathrm{a}}} \ . \tag{4.15}$$

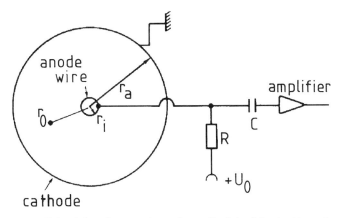

Fig. 4.3. Principle of operation of a cylindrical ionization chamber.

Using $\vec{E}(r) = -\mathrm{grad}\, V$ yields the field strength in the cylindrical ionization chamber

$$|\vec{E}(r)| = \frac{U_0}{r \ln r_\mathrm{a}/r_\mathrm{i}} \,, \tag{4.16}$$

which is a field that rises like $1/r$ to the anode wire. The field dependent drift velocity can no longer be assumed to be constant. The drift time of electrons is obtained by

$$\Delta t^- = \int_{r_0}^{r_\mathrm{i}} \frac{\mathrm{d}r}{v^-(r)} \,, \tag{4.17}$$

if the ionization has been produced locally at a distance r_0 from the anode wire (e.g., by the absorption of an X-ray photon). The drift velocity can be expressed by the mobility μ ($\vec{v}^- = \mu^- \cdot \vec{E}$), and one obtains in the approximation that the mobility does not depend on the field strength [1] ($\vec{v} \,\|\, (-\vec{E})$),

$$\begin{aligned}
\Delta t^- &= -\int_{r_0}^{r_\mathrm{i}} \frac{\mathrm{d}r}{\mu^- \cdot E} = -\int_{r_0}^{r_\mathrm{i}} \frac{\mathrm{d}r}{\mu^- \cdot U_0} r \ln r_\mathrm{a}/r_\mathrm{i} \\
&= \frac{\ln r_\mathrm{a}/r_\mathrm{i}}{2\mu^- \cdot U_0} (r_0^2 - r_\mathrm{i}^2) \,. \tag{4.18}
\end{aligned}$$

In practical cases the mobility does depend on the field strength, so that the drift velocity of electrons is not a linear function of the field strength. For this reason equation (4.18) presents only a rough approximation. The related signal amplitude can be computed in a way similar to equation (4.2) from

$$\frac{1}{2}CU^2 = \frac{1}{2}CU_0^2 - N \int_{r_0}^{r_\mathrm{i}} q \cdot \frac{U_0}{r \ln r_\mathrm{a}/r_\mathrm{i}} \mathrm{d}r \tag{4.19}$$

to

$$\Delta U^- = -\frac{Ne}{C \ln r_\mathrm{a}/r_\mathrm{i}} \ln r_0/r_\mathrm{i} \qquad (4.20)$$

with $q = -e$ for drifting electrons and C being the detector capacitance. It may clearly be seen that the signal amplitude in this case depends only logarithmically on the production point of the ionization.

The signal contribution due to the drift of the positive ions is obtained in an analogous fashion

$$\Delta U^+ = -\frac{Ne}{C}\frac{\ln r_\mathrm{a}/r_0}{\ln r_\mathrm{a}/r_\mathrm{i}} \ . \qquad (4.21)$$

The ratio of amplitudes originating from ions or electrons, respectively, is obtained as

$$\frac{\Delta U^+}{\Delta U^-} = \frac{\ln r_\mathrm{a}/r_0}{\ln r_0/r_\mathrm{i}} \ . \qquad (4.22)$$

Supposing that the ionization is produced at a distance $r_\mathrm{a}/2$ from the anode wire one gets

$$\frac{\Delta U^+}{\Delta U^-} = \frac{\ln 2}{\ln r_\mathrm{a}/(2r_\mathrm{i})} \ . \qquad (4.23)$$

Because $r_\mathrm{a} \gg r_\mathrm{i}$, we obtain

$$\Delta U^+ < \Delta U^- \ ,$$

i.e., for all practical cases (homogeneous illumination of the chamber assumed) the largest fraction of the signal in the cylindrical ionization chamber originates from the movement of electrons. For typical values of $r_\mathrm{a} = 1\,\mathrm{cm}$ and $r_\mathrm{i} = 15\,\mu\mathrm{m}$ the signal ratio is

$$\Delta U^+/\Delta U^- = 0.12 \ .$$

Ionization chambers, filled with suitable gas mixtures (e.g., 80 % Ar and 20 % CF_4) can be made very fast (pulse duration $\sim 35\,\mathrm{ns}$) and very long [128]. An ionization chamber in the form of a gas dielectric cable with a length of 3500 m is used as a beam loss monitor at SLAC [129].

For radiation protection purposes ionization chambers are frequently used in the current mode, rather than signal mode, for monitoring the personal radiation dose. These ionization dosimeters usually consist of a cylindrical air capacitor. The capacitor is charged to a voltage U_0. The charge carriers which are produced in the capacitor under the influence of radiation will drift to the electrodes and partially discharge the capacitor. The voltage reduction is a measure for the absorbed dose. The directly readable pocket dosimeters (figure 4.4) are equipped with an electrometer.

The state of discharge can be read at any time using an in-built optics [65, 116, 130].

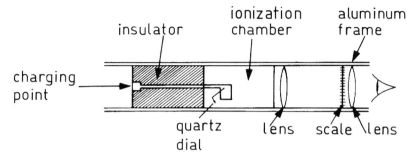

Fig. 4.4. Construction of an ionization pocket dosimeter.

4.2 Proportional counters

In ionization chambers there is no gas amplification, i.e., the number of originally produced charge carriers is not increased. The primary ionization produced by the incident particle is merely collected via the applied electric field. If, however, thin diameter anode wires are used in cylindrical chambers, or if the electric field strength is increased by using a higher anode voltage, the drifting electrons reach regions of high electric field strength close to the anode wire because of the $1/r$-dependence of the electric field. If the electrons can gain a sufficient amount of energy from the field between two collisions, they cause ionization themselves and as a consequence the number of charge carriers is increased [131]-[136]. The energy gain between two collisions is

$$
\begin{aligned}
\Delta E_{\mathrm{kin}} &= -e \int_{r_1}^{r_2} \vec{E}(r) \cdot \mathrm{d}\vec{r} \\
&= \frac{eU_0}{\ln r_\mathrm{a}/r_\mathrm{i}} \int_{r_1}^{r_2} \frac{\mathrm{d}r}{r} = \frac{eU_0}{\ln r_\mathrm{a}/r_\mathrm{i}} \cdot \ln r_2/r_1 \ .
\end{aligned} \tag{4.24}
$$

If ΔE_{kin} is larger than the ionization energy of the gas an electron avalanche can develop. The secondary electrons thus produced drift into regions of increasing field strength and multiply as in an avalanche. The signal amplitude is increased by the gas amplification factor A; it holds that (see equation (4.9))

$$
\Delta U = -\frac{eN}{C} \cdot A \ . \tag{4.25}
$$

Due to the Penning effect [104], which can occur in gas mixtures, the number of primary produced charge carriers can be increased for a fixed field strength. The Penning effect occurs if the metastable excitation level of one gas component (e.g., neon, $U^{\mathrm{exc}} = 16.53\,\mathrm{eV}$) is energetically higher compared to the ionization energy of the other gas component (e.g., argon, $U^{\mathrm{exc}} = 15.76\,\mathrm{eV}$). Therefore, excited neon atoms can ionize argon atoms in collisions according to:

$$\mathrm{Ne}^* + \mathrm{Ar} \to \mathrm{Ar}^+ + e^- + \mathrm{Ne} \ . \tag{4.26}$$

Cross sections for these reactions are of the order of some $10^{-16}\,\mathrm{cm}^2$.

The Penning effect always occurs in mixtures consisting of a noble gas and molecular vapors. This originates from the fact that the ionization potentials of molecules are smaller than the excitation levels of noble gases. However, one has to bear in mind that in complicated molecules the opposite effect can also occur, namely that by excitation of rotational and vibrational energy levels of the molecule some energy can be dissipated and lost for the production of charge carrier pairs.

The proportional range of a counter is characterized by the fact that the gas amplification factor A takes a constant value. As a consequence the measured signal is proportional to the produced ionization. Gas amplification factors of up to 10^6 are possible in the proportional mode. Typical gas amplifications are rather in the range between 10^4 up to 10^5.

The number of electron-ion pairs, which are produced per unit length by an electron during the formation of the avalanche, is called the first Townsend coefficient α. If σ_{ion} is the collision cross section, α is obtained from

$$\alpha = \sigma_{\mathrm{ion}} \cdot \frac{N_{\mathrm{A}}}{V_{\mathrm{mol}}} \ , \tag{4.27}$$

where N_{A} is the Avogadro number and V_{mol} the molar volume ($= 22.4$ l/mol for ideal gases). If N_0 primary electrons are produced, the number of particles $N(x)$ at the point x is calculated from

$$\mathrm{d}N(x) \ = \ \alpha N(x)\mathrm{d}x$$

to be

$$N(x) \ = \ N_0 e^{\alpha x} \ . \tag{4.28}$$

The first Townsend coefficient α depends on the field strength $\vec{E}$ and thereby on the position x in the gas counter. Therefore, more generally, it holds that

$$N(x) = N_0 \cdot e^{\int \alpha(x)\,\mathrm{d}x} \ , \tag{4.29}$$

where the gas amplification factor is given by

$$A = \exp\left\{\int_{r_k}^{r_i} \alpha(x)\,dx\right\}. \tag{4.30}$$

The lower integration limit is fixed by the distance r_k from the center of the gas counter, where the electric field strength exceeds the critical value E_k from which point on charge carrier multiplication starts. The upper integration limit is the anode wire radius r_i. The collision cross section for ionization σ_{ion}, or respectively, the mean free path $\lambda = 1/\alpha$ determines the first Townsend coefficient. The cross sections for ionization through collisions and photoionizations in some noble gases are shown in figures 4.5 and 4.6 [104, 137, 138].

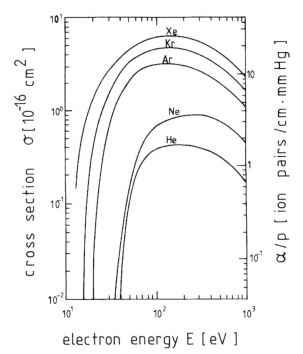

Fig. 4.5. Energy dependence of the cross section for ionization by collision [104, 139, 140].

The first Townsend coefficient for different gases is shown in figure 4.7 for noble gases, and in figure 4.8 for argon with various additions of organic vapors. The first Townsend coefficient for argon-based gas mixtures at high electric fields can be taken from the literature [142, 143].

If U_{th} is the threshold voltage for the onset of the proportional range, the gas amplification factor expressed by the detector parameters can be

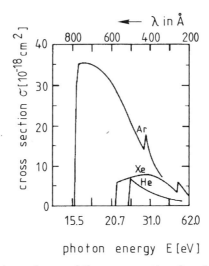

Fig. 4.6.   Energy dependence of the cross section for photoionization [104, 141].

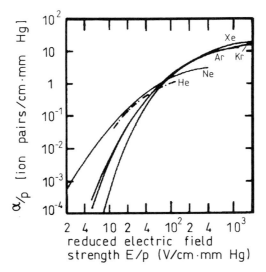

Fig. 4.7.   First Townsend coefficient for some noble gases [51, 143, 144, 145].

calculated to be [51]

$$A = \exp\left\{ 2\sqrt{\frac{kLCU_0 r_i}{2\pi\varepsilon_0}} \left[\sqrt{\frac{U_0}{U_{\text{th}}}} - 1 \right] \right\} \; ; \tag{4.31}$$

where

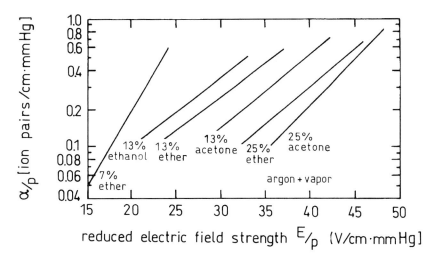

Fig. 4.8. First Townsend coefficient for argon with some organic vapor admixtures [51, 146, 147].

U_0 – applied anode wire voltage

$C = \dfrac{2\pi\varepsilon_0}{\ln r_{\mathrm{a}}/r_{\mathrm{i}}}$ – capacitance per unit length of the counter

L – number of atoms/molecules per unit volume ($\frac{N_{\mathrm{A}}}{V_{\mathrm{mol}}} = 2.69 \cdot 10^{19}/\mathrm{cm}^3$) at normal pressure and temperature; k is a gas dependent constant of the order of magnitude $10^{-17}\,\mathrm{cm}^2/\mathrm{Volt}$, which can be obtained from the relation

$$\alpha = \frac{k \cdot L \cdot E_{\mathrm{e}}}{e}\,, \qquad (4.32)$$

where E_{e} is the average electron energy (in eV) between two collisions [51].

In the case $U_0 \gg U_{\mathrm{th}}$, equation (4.31) simplifies to

$$A = \mathrm{const} \cdot e^{U_0/U_{\mathrm{ref}}}\,, \qquad (4.33)$$

where U_{ref} is a reference voltage.

Equation (4.33) shows that the gas amplification rises exponentially with the applied anode wire voltage. The detailed calculation of the gas amplification is difficult [5, 6, 131], [148]-[154]; however, it can be measured quite easily. Let N_0 be the number of primary charge carriers produced in the proportional counter which, for example, have been produced by the absorption of an X-ray photon of energy E_γ ($N_0 = E_\gamma/W$, where W is the average energy that is required for the production of one electron-ion pair). The integration of the current at the output of the proportional counter leads to the gas amplified charge

$$Q = \int i(t)\, \mathrm{d}t \, , \qquad (4.34)$$

which again is given by the equation $Q = e \cdot N_0 \cdot A$. From the current integral and the known primary ionization N_0 the gas amplification A can be easily obtained. If the current integral is evaluated only after a preamplifier its electronic amplification factor must, of course, be considered.

For very high field strengths the accelerated electrons can also liberate electrons from inner shells. This leads to excited gas atoms which are de-excited by photon emission.

The previous considerations are only true as long as photons produced in the course of the avalanche development are of no importance. These photons, however, will produce further electrons by the photoelectric effect in the gas or at the counter wall, which have an influence on the avalanche development. Apart from gas amplified primary electrons, secondary avalanches initiated by the photoelectric processes must also be taken into account. For the treatment of the gas amplification factor with inclusion of photons we will first derive the number of produced charge carriers in the different generations.

In the first generation, N_0 primary electrons are produced by the ionizing particle. These N_0 electrons are gas amplified by a factor A. If γ is the probability that one photoelectron per electron is produced in the avalanche, an additional number of $\gamma(N_0 A)$ photoelectrons is produced via photo processes. These, however, are again gas amplified so that in the second generation $(\gamma N_0 A) \cdot A = \gamma N_0 A^2$ gas amplified photoelectrons arrive at the anode wire, which again create $(\gamma N_0 A^2)\gamma$ further photoelectrons by the gas amplification process, which again are gas amplified themselves. The gas amplification A_γ under inclusion of photons, therefore, is obtained from

$$\begin{aligned} N_0 A_\gamma &= N_0 A + N_0 A^2 \gamma + N_0 A^3 \gamma^2 + \cdots \\ &= N_0 A \cdot \sum_{k=0}^{\infty} (A\gamma)^k = \frac{N_0 A}{1 - \gamma A} \end{aligned} \qquad (4.35)$$

to be

$$A_\gamma = \frac{A}{1 - \gamma A} \, . \qquad (4.36)$$

The factor γ, which determines the gas amplification with the inclusion of photons, is also called the second Townsend coefficient.

As the number of produced charges increases, they begin to have an effect on the external applied field and saturation effects occur. For $\gamma A \to 1$ the signal amplitude will be independent of the primary ionization. The proportional, or rather the saturated proportional, region is limited by gas amplification factors around $A = 10^8$.

The process of avalanche formation takes places in the immediate vicinity of the anode wire (figure 4.9, see also figure 4.27). The mean free paths of electrons are of the order of some μm so that the total avalanche formation process according to equation (4.28) requires only about $20\,\mu$m. As a consequence the effective production point of the charge (start of the avalanche process) is

$$r_0 = r_i + k \cdot \lambda \,, \tag{4.37}$$

where k is the number of mean free paths which are required for the avalanche formation.

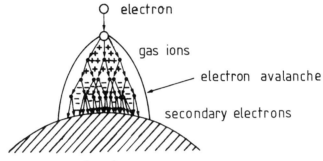

Fig. 4.9. Illustration of the avalanche formation on an anode wire in a proportional counter. By lateral diffusion a drop-shaped avalanche develops.

The ratio of signal amplitudes which are caused by the drift of positive ions or electrons, respectively, is determined to be (see equation (4.22) and [1])

$$\frac{\Delta U^+}{\Delta U^-} = \frac{-\frac{Ne}{C}\frac{\ln r_a/r_0}{\ln r_a/r_i}}{-\frac{Ne}{C}\frac{\ln r_0/r_i}{\ln r_a/r_i}} = \frac{\ln r_a/r_0}{\ln r_0/r_i} = R \,. \tag{4.38}$$

The gas amplification factor cancels in this ratio because equal numbers of electrons and ions are produced.

If $k \cdot \lambda \ll r_i$ this ratio can be approximated by

$$R = \frac{\ln r_a - \ln(r_i + k \cdot \lambda)}{\ln[(r_i + k \cdot \lambda)/r_i]} \approx \frac{\ln r_a/r_i}{k\lambda/r_i} \,. \tag{4.39}$$

With typical values of $r_a = 1\,$cm, $r_i = 30\,\mu$m and $k\lambda = 20\,\mu$m for argon at normal pressure this ratio is $R \approx 10$, which implies that in the proportional counter the signal on the anode wire is caused mainly by the ions

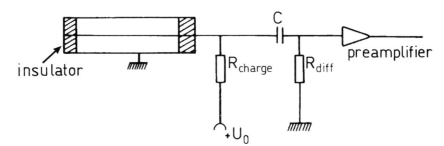

Fig. 4.10.   Readout of a proportional counter.

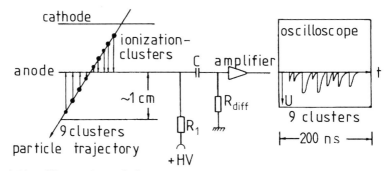

Fig. 4.11.   Illustration of the time structure of a signal in the proportional counter.

drifting slowly away from the wire and not by the electrons which quickly drift in the direction of the wire.

The rise time of the electron signal can be calculated from equation (4.18). For electron mobilities in the range between $\mu^- = 100$ to $1000\,\text{cm}^2/\text{Vs}$, an anode voltage of several hundred volts and typical detector dimensions as given above, the rise time is of the order of nanoseconds. The corresponding rise time of the signal due to the movement of the ions, however, is in the range of $10\,\text{ms}$. Differentiating the signal with an RC-combination (figure 4.10) one can restrict the measurement to the electron component alone.

If $R_{\text{diff}} \cdot C \approx 1\,\text{ns}$ is chosen, one can even resolve the time structure of the ionization in the proportional counter (figure 4.11).

Raether was the first to photograph electron avalanches (figure 4.12, [51, 155, 156]). In this case, the avalanches were made visible in a cloud chamber by droplets which had condensed on the positive ions. The size of the luminous region of an avalanche in a proportional chamber is rather small compared to different gas discharge operation modes, such as in Geiger-Müller or streamer tubes (see also figure 4.17).

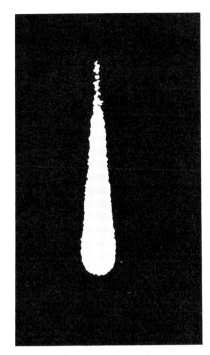

Fig. 4.12. Photographic reproduction of an electron avalanche [51, 155, 156]. The photo shows the form of the avalanche. It was made visible in a cloud chamber (see section 4.12) by droplets which have condensed on the positive ions.

Proportional counters are particularly suited to the spectroscopy of X-rays. Figure 4.13 shows the energy spectrum of 59.53 keV X-ray photons which are emitted in the α-decay $^{241}_{95}$Am $\to$ $^{237}_{93}$Np from the excited neptunium nucleus. The spectrum was measured in a xenon proportional counter. The characteristic X-ray lines of the detector material and the Xe-escape peak are also seen [157]. The escape peak is the result of the following process. The incident X-rays ionize the Xe gas in most cases in the K-shell. The resulting photoelectron only gets the X-ray energy minus the binding energy in the K-shell. If the gap in the K-shell is filled up by electrons from outer shells, X-rays characteristic of the gas may be emitted. If these characteristic X-rays are also absorbed by the photoelectric effect in the gas, a total absorption peak is observed; if the characteristic X-rays leave the counter undetected, the escape peak is formed (see also section 1.2.1).

Proportional counters can also be used for X-ray imaging. Special electrode geometries allow a two-dimensional readout with high resolution for X-ray synchrotron radiation experiments which also work at high rates

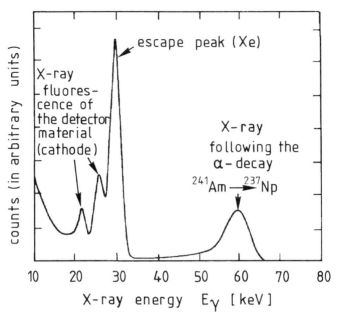

Fig. 4.13. Energy spectrum of 59.53 keV X-ray photons which are emitted in the α-decay of ^{241}Am, measured in a xenon proportional counter [157].

[158]. The electronic imaging of ionizing radiation with limited avalanches in gases has a wide field of application ranging from cosmic-ray and elementary particle physics to biology and medicine [26].

The energy resolution of proportional counters is limited by the fluctuations of the charge carrier production and their multiplication. Avalanche formation is localized to the point of ionization in the vicinity of the anode wire. It *does not* propagate laterally along the anode wire.

4.3   Geiger counters

The increase of the field strength in a proportional counter leads to a copious production of photons during the avalanche formation. As a consequence, the probability to produce further new electrons by the photoelectric effect increases. This photoelectric effect can also occur at points distant from the production of the primary avalanche. These electrons liberated by the photoelectric effect will initiate new avalanches whereby the discharge will propagate along the anode wire [159, 160] (figure 4.14).

The probability of photoelectron production γ per electron in the original avalanche becomes so large that the total number of charge carriers produced by the various secondary and tertiary avalanches increases

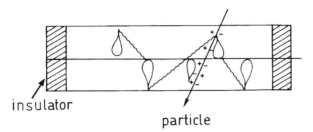

insulator

particle

Fig. 4.14. Schematic representation of the transverse avalanche propagation along the anode wire in a Geiger counter.

rapidly. As a consequence the proportionality between the signal and the primary ionization gets lost. This domain in which the liberated amount of charge does not depend on the primary ionization is called the Geiger mode. The signal only depends on the applied voltage. In this mode of operation the signal amplitude corresponds to a charge signal of 10^8 up to 10^{10} electrons per primary produced electron.

After the passage of a particle through a Geiger counter (also called Geiger-Müller counter) a large number of charge carriers is formed all along the anode wire. The electrons are quickly drained by the anode, however, the ions form a kind of flux tube which is practically stationary. The positive ions migrate with low velocities to the cathode. Upon impact with the electrode they will liberate, with a certain probability, new electrons, thereby starting the discharge anew.

Therefore, one must cause the discharge to be interrupted. This can be achieved if the charging resistor R is chosen to be so large that the momentary anode voltage $U_0 - IR$ is smaller than the threshold value for the Geiger mode (quenching by resistor).

Together with the total capacitance C the time constant RC has to be chosen in such a way that the voltage reduction persists until all positive ions have arrived at the cathode. This results in times of the order of magnitude of milliseconds, which strongly impairs the rate capability of the counter.

It is also possible to lower the applied external voltage to a level below the threshold for the Geiger mode for the ion drift time. This will, however, also cause long dead times. These can be reduced if the polarity of the electrodes is interchanged for a short time interval, thereby draining the positive ions, which are all produced in the vicinity of the anode wire, to the anode which has been made negative for a very short period.

A more generally accepted method of quenching in Geiger counters is the method of self-quenching. In self-quenching counters a quench gas is admixed to the counting gas which is in most cases a noble gas. Hydrocarbons like methane (CH_4), ethane (C_2H_6), isobutane (iC_4H_{10}), alcohols like ethylalcohol (C_2H_5OH) or methylal ($CH_2(OCH_3)_2$), or halides, like ethylbromide, are suitable as quenchers. These additions will absorb photons in the ultraviolet range (wavelength $100 - 200\,nm$) thereby reducing their range to a few wire radii ($\approx 100\,\mu m$). The transverse propagation of the discharge proceeds only along and in the vicinity of the anode wire because of the short range of the photons. The photons have no chance to liberate electrons from the cathode by the photoelectric effect because they will be absorbed before they can reach the cathode.

After a flux tube of positive ions has been formed along the anode wire, the external field is reduced by this space charge by such an amount that the avalanche development comes to an end. The positive ions drifting in the direction of the cathode will collide on their way with quench gas molecules, thereby becoming neutralized.

$$Ar^+ + CH_4 \rightarrow Ar + CH_4^+ \;. \tag{4.40}$$

The molecule ions, however, have insufficient energy to liberate electrons from the cathode upon impact. Consequently the discharge stops by itself. The charging resistor, therefore, can be chosen to be smaller, with the result that time constants of the order of $1\,\mu s$ are possible.

Contrary to the proportional mode, the discharge propagates along the whole anode wire in the Geiger mode. Therefore it is impossible to record two charged particles in one Geiger tube at the same time. This is only achievable if the lateral propagation of the discharge along the anode wire can be interrupted. This can be accomplished by stretching insulating fibers perpendicular to the anode wire or by placing small droplets of insulating material on the anode wire. In these places the electric field is so strongly modified that the avalanche propagation is stopped. This locally limited Geiger mode allows the simultaneous registration of several particles on one anode wire. However, it has the disadvantage that the regions close to the fibers are inefficient for particle detection. The inefficient zone is typically $5\,mm$ wide. The readout of simultaneous particle passages in the limited Geiger range is done via segmented cathodes.

4.4 Streamer tubes

In Geiger counters the fraction of counting gas to quenching gas is typically 90:10. The anode wires have diameters of $30\,\mu m$ and the anode voltage is around $1\,kV$. If the fraction of the quenching gas is considerably increased, the lateral propagation of the discharge along the anode wire can be completely suppressed. One again obtains, as in the proportional counter, a localized discharge with the advantage of large signals

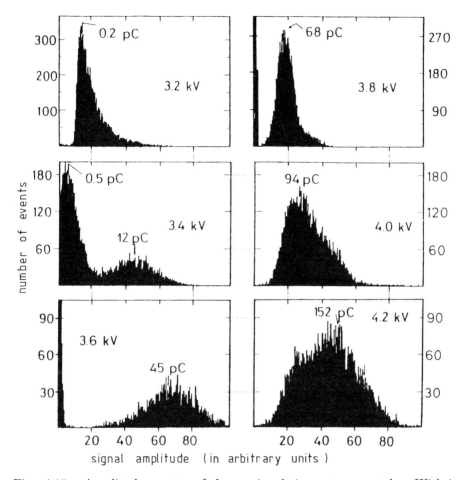

Fig. 4.15. Amplitude spectra of charge signals in a streamer tube. With increasing anode voltage the transition from the proportional to the streamer mode is clearly visible [165].

(gas amplification $\geq 10^{10}$ for sufficiently high anode voltages), which can be processed without any additional preamplifiers. These streamer tubes (Iarocci tubes, also developed by D.M. Khazins) [161, 162, 163] are operated with 'thick' anode wires between $50\,\mu$m and $100\,\mu$m. Gas mixtures with $\leq 60\,\%$ argon and $\geq 40\,\%$ isobutane can be used. Also streamer tubes operated with pure isobutane have proven to function well [164]. In this mode of operation the transition from the proportional range to streamer mode proceeds bypassing the Geiger discharges. Figure 4.15 shows at relatively low voltages of $3.2\,$kV (anode wire diameter $100\,\mu$m, gas filling argon/isobutane 60:40) the amplitude spectra of small proportional signals caused by electrons of a ^{90}Sr source [165]. At higher voltages

(3.4 kV) for the first time streamer signals with distinct higher amplitudes also occur along with the proportional signals. For even higher voltages the proportional mode completely disappears, so that from 4 kV onwards only streamer signals are observed. The charge collected in the streamer mode does not depend on the primary ionization.

The discontinuous transition from the proportional to the streamer mode can clearly be seen from figure 4.16. The streamer mode develops from the proportional mode via the large number of produced photons which are re-absorbed in the immediate vicinity of the original avalanche via the photoelectric effect and are the starting point of new secondary and tertiary avalanches which merge with the original avalanche.

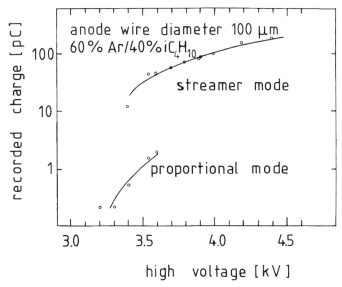

Fig. 4.16. Collected avalanche charge as a function of high voltage. Clearly visible is the discontinuous transition from the proportional to the streamer mode and the coexistence of these two discharge mechanisms in a narrow overlapping range around 3.5 kV [165].

It has been reported in the literature that at even higher voltages a further ('second') streamer mode may occur which again would represent a discontinuous transition from the first to the second streamer mode [163]. But it is also possible that this phenomenon can be explained by multiple streamer discharges [165].

The photographs in figure 4.17 [166] demonstrate the characteristic differences of discharges in the proportional counter (a), Geiger counter (b), and a self-quenching streamer tube (c). The arrows indicate in each case the position of the anode wire.

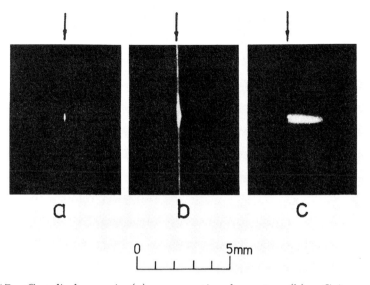

Fig. 4.17. Gas discharges in (a) a proportional counter, (b) a Geiger counter, and (c) a self-quenching streamer tube; the arrows indicate the position of the anode wire [166].

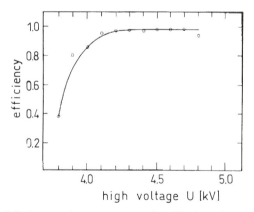

Fig. 4.18. Efficiency of a streamer tube filled with pure isobutane [165].

As has been discussed, streamer tubes have to be operated at high voltages ($\approx 5\,\mathrm{kV}$). They are, however, characterized by an extremely long efficiency plateau ($\approx 1\,\mathrm{kV}$) which enables a stable working point. Figure 4.18 shows the efficiency of a streamer tube filled with pure isobutane that was irradiated with electrons from a $^{90}\mathrm{Sr}$ source [165].

The onset of the efficiency, of course, depends on the threshold of the discriminator used. The upper end of the plateau is normally determined

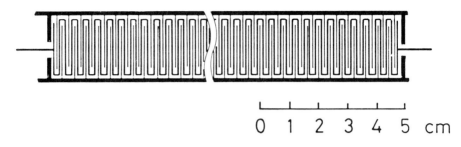

Fig. 4.19. Meandering cathode structure as delay line for the readout of the coordinate along the anode wire in a streamer tube [165].

by after-discharges and noise. It is not recommended to operate streamer tubes in this region because electronic noise and after-discharges cause additional dead times, thereby reducing the rate capability of the counter (see also figure 4.22).

If 'thick' anode wires are used, the avalanche is caused mostly by only one primary electron and the discharge is localized to the side of the anode wire which the electron approaches. The signals can be directly measured on the anode wire. Additionally or alternatively one can also record the signals induced on the cathodes. A segmentation of the cathodes allows the determination of the track position along the anode wire.

The track position along the anode wire can also be determined by a cathode constructed as a delay line [165, 167]. Figure 4.19 shows a cathode structure which allows the determination of the spatial coordinate along the anode wire. The result of such a measurement is shown in figure 4.20, where a streamer tube has been irradiated at well-defined positions by a ^{55}Fe source. In this case a spatial resolution along the wire of 2 mm was achieved [165].

Because of the simple mode of operation and the possibility of multiparticle registration on one anode wire, streamer tubes are an ideal candidate for sampling elements in calorimeters. A fixed charge signal Q_0 is recorded per particle passage. If a total charge Q is measured in a streamer tube, the number of equivalent particles passing is calculated to be $N = Q/Q_0$.

The choice of the high voltage, of the counting gas, or anode wire diameter, respectively, determines the discharge and thereby the operation mode of cylindrical counters. Figure 4.21 shows the different regions of operation in a comprehensive fashion (after [51]).

In addition to gas amplification the efficiency has been already mentioned to be a very important characteristic of gas counters. Figure 4.18 showed the efficiency for charged particles in a streamer tube as a function of the high voltage. The length of the efficiency plateau and its variation

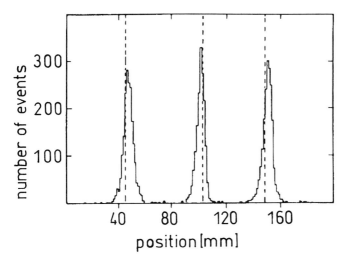

Fig. 4.20. Counting rate distributions of 5.9 keV-photons incident at different positions along the anode wire in a streamer tube [165].

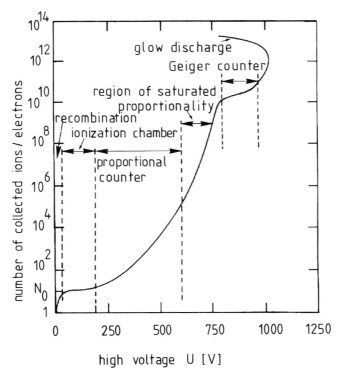

Fig. 4.21. Characterization of the modes of operation of cylindrical gas detectors (after [51]).

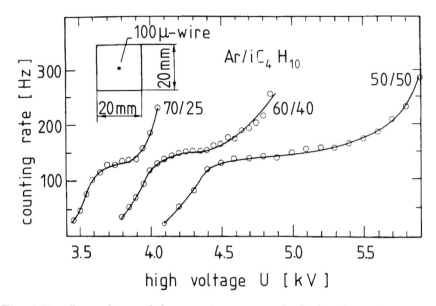

Fig. 4.22.   Dependence of the counting rate on the high voltage in a streamer tube [165].

with high voltage over the plateau region (which should be as small as possible) are criteria for the quality of the counter [165]. For extremely high voltages, paradoxically, the efficiency can reach values ≥ 1, since at these high field strengths after-discharges can occur which occasionally can produce two signals per one incident particle.

The performance of a gas counter can also be monitored via the measurement of the counting rate at constant particle flux as a function of the high voltage. The working point is chosen best closer to the end of the counting rate plateau (figure 4.22, [165]), but before the rate increases drastically due to after-discharges.

The operation mode of streamer tubes has also many applications in completely different gaseous detectors. In streamer chambers (see section 4.13) streamers develop in a very strong homogeneous pulsed electric field. The ionization trails formed by charged particles incident at right angles to the electric field are gas amplified and made visible as a sequence of streamers. For inclined incidence ($\pm\,30°$ with respect to the field), the streamers will merge into one another and form a plasma channel along the track (track spark chamber, see sections 4.13 and 4.15).

Also resistive plate chambers (see section 5.3) can be operated in the streamer mode thereby producing large signals which can easily be processed. This is also true for imaging chambers (see section 4.9) where a streamer-mode operation allows high light yields.

4.5 Particle measurements in liquids

Ionization chambers filled with liquids have the advantage compared to gas filled detectors of a density which is a factor of 1000 times higher, which means that a 1000-fold energy absorption is possible [168, 169]. The average energy for the production of an electron-ion pair in liquid argon (LAr) is 24 eV, and in liquid xenon (LXe) 16 eV. Therefore, ionization chambers filled with liquids are excellent candidates for sampling detectors in calorimeters [170]. A technical disadvantage, however, is related to the fact that noble gases only become liquid at low temperatures. Typical temperatures of operation are 85 K for LAr, 117 K for LKr, and 163 K for LXe. Liquid gases are homogeneous, and therefore have excellent counting properties. Problems may, however, arise with electronegative impurities which must be kept at an extremely low level because of the slow drift velocities in the high density liquid counting medium. To make operation possible, the mean free path λ of electrons must be very large compared to the electrode distance. This necessitates that the concentration of electronegative gases such as O_2 be reduced to the level of the order of 1 ppm ($\equiv 10^{-6}$). The drift velocity in pure liquid noble gases at field strengths around 10 kV/cm, which are typical for LAr-counters, is of the order 0.4 cm/μs. The addition of small amounts of hydrocarbons (e.g., 0.5 % CH_4) can, however, increase the drift velocity significantly. This originates from the fact that the admixture of molecular gases changes the average electron energy. The electron scattering cross section, in particular in the vicinity of the Ramsauer minimum, is strongly dependent on the electron energy. So, small changes in the energy can have a dramatic influence on the drift properties.

The ion mobility in liquids is extremely small. The induced charge due to the ion movement has a rise time so slow that it can hardly be used electronically.

In contrast to gaseous ionization chambers, liquid noble gas chambers are usually used in such a way that the charged particles to be detected cross the chamber more or less perpendicularly (see figure 4.23).

Let us assume that the total charge $-Ne$ being produced by a particle passing through the chamber is uniformly distributed along its track. In that case the charge density is $\varrho = -Ne/d$. Let us further assume that the produced charge is drained with a constant drift velocity v^-. While the charge density remains constant during the whole process of electron collection, the charge in the drift volume, of course, is permanently reduced according to

$$q(t) = \varrho \cdot (d - v^- \cdot t) \,. \tag{4.41}$$

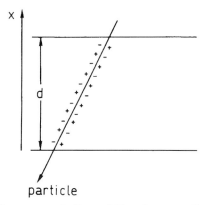

Fig. 4.23. Schematic representation of the charge collection in a liquid noble gas detector.

For $t \geq d/v^-$ all charges have been collected and the charge $q(t)$ becomes zero. Due to the electron drift, a charge is induced on the electrodes, which will reduce the stored energy on the capacitor [1]. Energy conservation requires

$$\mathrm{d}\left(\frac{1}{2}CU^2\right) = \mathrm{d}\left(\frac{1}{2}\frac{Q^2}{C}\right) = q(t) \cdot \vec{E} \cdot \mathrm{d}\vec{x} \ . \tag{4.42}$$

With the help of equation (4.41) it follows that:

$$\frac{Q\mathrm{d}Q}{C} = |\vec{E}| \cdot \left(-\frac{N \cdot e}{d}(d - v^-t)\right) \cdot v^- \mathrm{d}t \ . \tag{4.43}$$

Because

$$Q = C \cdot U = C \cdot |\vec{E}| \cdot d \ , \tag{4.44}$$

$\mathrm{d}Q$ is calculated to be

$$
\begin{aligned}
\mathrm{d}Q &= -\frac{C \cdot |\vec{E}|}{Q} \cdot N \cdot e \left(1 - \frac{v^-}{d}t\right) v^- \mathrm{d}t \\
&= -N \cdot e \left(1 - \frac{v^-}{d}t\right)\frac{v^-}{d}\mathrm{d}t \ .
\end{aligned}
\tag{4.45}
$$

Since d/v^- is the total drift time t_{D}, equation (4.45) leads to

$$\mathrm{d}Q = -Ne\left(1 - \frac{t}{t_{\mathrm{D}}}\right) \cdot \frac{\mathrm{d}t}{t_{\mathrm{D}}} \ . \tag{4.46}$$

Integrating equation (4.46) yields

$$Q(t) - Q_0 = -\frac{Ne}{t_D}\left(t - \frac{t^2}{2t_D}\right)$$

$$= -N \cdot e \left(\frac{t}{t_D} - \frac{1}{2}\left(\frac{t}{t_D}\right)^2\right). \qquad (4.47)$$

For $t = t_D$ the total charge is collected leading to

$$Q(t_D) - Q_0 = -\frac{1}{2}N \cdot e. \qquad (4.48)$$

The operation of liquid noble gas ionization chambers requires cryogenic equipment. This technical disadvantage can be overcome by the use of 'warm' liquids. The requirements for such 'warm' liquids, which are already in the liquid state at room temperature, are considerable: they must possess excellent drift properties and they must be extremely free of electronegative impurities. The molecules of the 'warm' liquid must have a high symmetry (i.e., a near spherical symmetry) to allow favourable drift properties. As 'warm' liquids petrol like substances like tetramethylsilane (TMS) or tetramethylpentane (TMP) are suited [170]-[174]. Attempts to obtain higher densities, in particular for the application of liquid ionization counters in calorimeters, have also been successful. This can be achieved, for example, if the silicon atom in the TMS-molecule is replaced by lead or tin. The flammability and toxicity problems associated with such materials can be handled in practice, if the liquids are sealed in vacuum tight containers. These 'warm' liquids show excellent radiation hardness. Due to the high fraction of hydrogen they also allow for compensation of signal amplitudes for electrons and hadrons in calorimeters (see section 7.3).

Obtaining gas amplification in liquids by increasing the working voltage has also been investigated, in an analogous fashion to cylindrical ionization chambers. This appears to have been successfully demonstrated in small prototypes; however, it has not been reproduced on a larger scale with full-size detectors [175, 176, 177].

In closing, one should remark that *solid* argon can also be used successfully as counting medium for ionization chambers [178].

4.6   Multiwire proportional chambers

A multiwire proportional chamber [26, 125, 179, 180, 181] is essentially a planar layer of proportional counters without separating walls (see figure 4.24). The shape of the electric field is somewhat modified compared to the pure cylindrical arrangement in proportional counters (see figures 4.25 and 4.26) [51], [182]-[185].

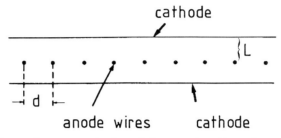

Fig. 4.24. Schematic representation of the construction of a multiwire proportional chamber.

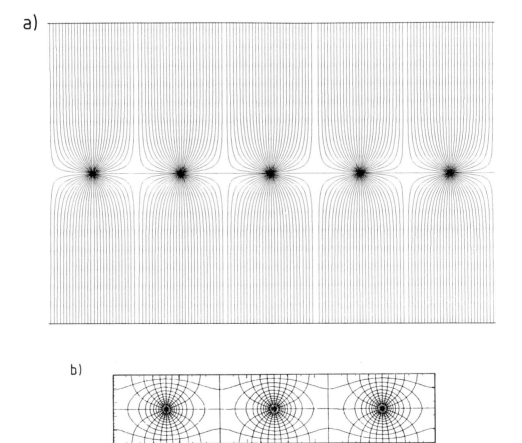

Fig. 4.25.   a) Field lines in a five-wire proportional chamber [184]. b) Field and equipotential lines in a three-wire proportional chamber [183].

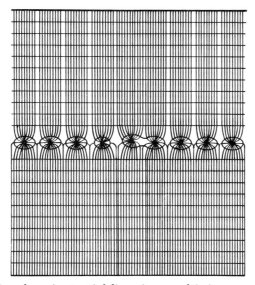

Fig. 4.26. Field and equipotential lines in a multiwire proportional chamber. The effect of a minor displacement of one anode wire on the field quality is clearly visible [51, 182].

Avalanche formation in a multiwire proportional chamber proceeds exactly in the same way as in proportional counters. Since for each anode wire the bulk charge is produced in its immediate vicinity, the signal originates predominantly from the positive ions slowly drifting in the direction of the cathode (see equation (4.39)). If the anode signal is read out with a high time resolution oscilloscope or with a fast analog-to-digital converter (flash ADC) the ionization structure of the particle track can also be resolved in the multiwire proportional chamber (see figure 4.11).

The time development of the avalanche formation in a multiwire proportional chamber can be detailed as follows (see figure 4.27 [51, 186]). A primary electron drifts towards the anode (a), the electron is accelerated in the strong electric field in the vicinity of the wire in such a way that it can gain on its path between two collisions a sufficient amount of energy so that it can ionize further gas atoms. At this moment the avalanche formation starts (b). Electrons and positive ions are created in the ionization processes essentially in the same place. The multiplication of charge carriers comes to an end when the space charge of positive ions reduces the external electric field below a critical value. After the production of charge carriers, the electron and ion clouds drift apart (c). The electron cloud drifts in the direction of the wire and broadens slightly due to lateral diffusion. Depending on the direction of incidence of the primary electron a slightly asymmetric density distribution of secondary electrons around

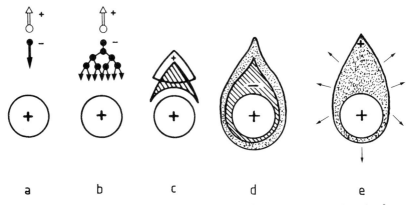

Fig. 4.27. Temporal and spatial development of an electron avalanche [51, 186].

the wire will be formed. This asymmetry is even more pronounced in streamer tubes. In this case, because of the use of thick anode wires and also because of the strong absorption of photons, the avalanche formation is completely restricted to the side of the anode wire where the electron was incident (see also figures 4.9 and 4.17) (d). In a last step the ion cloud recedes radially and slowly drifts to the cathode (e).

In most cases gold-plated tungsten wires with diameters between $10\,\mu$m and $30\,\mu$m are used as anodes. A typical anode wire distance is $2\,$mm. The distance between the anode wire and the cathode is of the order of $10\,$mm. The individual anode wires act as independent detectors. The cathodes can be made from metal foils or also as layers of stretched wires.

As counting gases, all gases and gas mixtures, which also are standard for the operation of proportional counters, namely noble gases like Ar, Xe with admixtures of CO_2, CH_4, isobutane and other hydrocarbons can be used [187, 188, 189]. Typical gas amplifications of 10^5 are achieved in multiwire proportional chambers.

In most chambers the possibility to process the analog information on the wires is not taken advantage of. Instead, only thresholds for the incoming signals are set. In this mode of operation the multiwire proportional chamber is only used as a track detector. For an anode wire distance of $d = 2\,$mm the root mean square deviation of the spatial resolution is given by (see equation (2.5))

$$\sigma(x) = \frac{d}{\sqrt{12}} = 577\,\mu\text{m}\ . \qquad (4.49)$$

Electrostatic repulsion between long anode wires, in particular, can cause problems. The wires are only stable against oscillations if the wire tension T is large compared to a value T_0, which can be calculated from the relation

$$V \leq \frac{d}{lC}\sqrt{4\pi\varepsilon_0 T_0} \, , \tag{4.50}$$

where V is the anode voltage, d the wire spacing, l the wire length and C the capacitance per unit length of the detector [34, 190]. ε_0 is the permittivity of free space ($\varepsilon_0 = 8.854 \cdot 10^{-12}$ F/m).

For a cylindrical capacitor (central conductor radius r_i, outer conductor radius r_a) the capacitance per unit length is

$$C = \frac{4\pi\varepsilon_0}{2\ln r_a/r_i} \, . \tag{4.51}$$

In multiwire proportional chambers the corresponding capacitance can be approximated by

$$C = \frac{4\pi\varepsilon_0}{2\left\{\frac{\pi L}{d} - \ln\frac{2\pi r_i}{d}\right\}} \tag{4.52}$$

if $L \gg d \gg r_i$ holds for the anode-cathode distance (see figure 4.24) [4]. Using this equation, the required wire tension for stable wires can be calculated from equation (4.50) to be

$$T_0 \geq \left(\frac{V \cdot l \cdot C}{d}\right)^2 \cdot \frac{1}{4\pi\varepsilon_0} \tag{4.53}$$

$$\geq \left(\frac{V \cdot l}{d}\right)^2 \cdot 4\pi\varepsilon_0 \left[\frac{1}{2\left\{\frac{\pi L}{d} - \ln 2\pi r_i/d\right\}}\right]^2 \, . \tag{4.54}$$

For a wire length $l = 1$ m, an anode voltage $V = 5$ kV, an anode-cathode distance of $L = 10$ mm, an anode wire spacing of $d = 2$ mm, and an anode wire diameter of $2r_i = 30\,\mu$m equations (4.53) or (4.54) yield a minimum mechanical wire tension of 0.49 N corresponding to a stretching of the wire with a mass of about 50 g.

Longer wires must be stretched with larger forces or, if they cannot withstand higher tensions, they must be supported at fixed distances. This will, however, lead to locally inefficient zones.

For a reliable operation of multiwire proportional chambers it is also important that the wires do not sag gravitationally due to their own mass [191]. A sag of the anode wire would reduce the distance from the anode to the cathode, thereby reducing the homogeneity of the electric field.

A horizontally aligned wire of length l stretched with a tension T would exhibit a sag due to the pull of gravity [192] of

$$f = \frac{\pi r_i^2}{8} \cdot \varrho \cdot g \frac{l^2}{T} = \frac{mlg}{8T} \tag{4.55}$$

(m, l, ϱ, r_i – mass, length, density and radius of the unsupported wire, g – acceleration due to gravity, and T – wire tension (in N)).

Taking our example from above, a gold-plated tungsten wire ($r_i = 15\,\mu$m; $\varrho_W = 19.3\,$g/cm^3) would develop a sag at the center of the wire of

$$f = 34\,\mu\text{m}$$

which would be acceptable if the anode-cathode distance is of the order of 10 mm.

If, instead of the classical multiwire proportional chambers with anode wires, straw chambers (see section 4.8.1) were to be used, the sag of the straws due to gravitation can be calculated for thin tubes which are fixed at both ends according to [193, 194, 195]

$$y = \frac{l^4 \cdot \varrho}{192 \cdot E \cdot R^2} \,, \tag{4.56}$$

if E is the modulus of elasticity and R the radius of the tube.

For 40 cm long aluminized mylar straws ($\varrho = 1.4\,$g/cm^3, $E = 3.4\,$GPa $\hat{=} 3.46 \cdot 10^7\,$g/cm^2) of 7 mm diameter a sag of 44 μm is obtained.

Multiwire proportional chambers provide a relatively poor spatial resolution which is of the order of $\approx 600\,\mu$m. They also give only the coordinate perpendicular to the wires and not along the wires. An improvement in the performance can be obtained by a segmentation of the cathode and a measurement of the induced signals on the cathode segments. The cathode, for example, can be constructed of parallel strips, rectangular pads ('mosaic counter'), or of a layer of wires (figure 4.28).

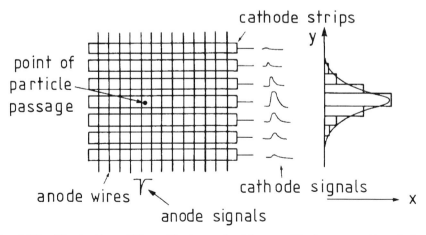

Fig. 4.28.   Illustration of the cathode readout in a multiwire proportional chamber.

In addition to the anode signals, the induced signals on the cathode strips are now also recorded. The coordinate along the wire is given by the center of gravity of the charges, which is derived from the signals induced on the cathode strips. Depending on the subdivision of the cathode, spatial resolutions along the wires of $\approx 50\,\mu\mathrm{m}$ can be achieved, using this procedure. In case of multiple tracks also the second cathode must be segmented to exclude ambiguities.

Figure 4.29 sketches the passage of two particles through a multiwire proportional chamber. If only one cathode were segmented the information from the anode wires and cathode strips would allow the reconstruction of four possible track coordinates, two of which, however, would be 'ghost coordinates'. They can be excluded with the help of signals from a second segmented cathode plane. A larger number of simultaneous particle tracks can be successfully reconstructed if cathode pads instead of cathode strips are used. Naturally this results also in an increased number of electronic channels.

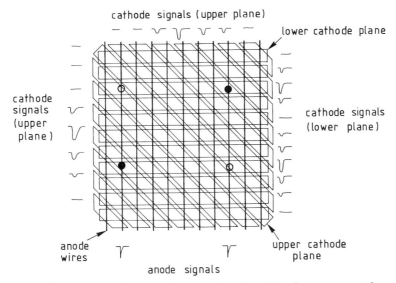

Fig. 4.29. Illustration of the resolution of ambiguities for two-particle registering in a multiwire proportional chamber.

The construction of multiwire proportional chambers would be simplified and their stability and flexibility would be greatly enhanced if it were possible to realize anodes in the form of strips or dots on dielectric surfaces instead of stretching anode wires in the counter volume. If anode strips are used on dielectrics the field quality in the proportional chamber will be influenced because of positive ion attachment on the surface of the dielectric. Nevertheless, detectors with anode structures on insulating or semiconducting surfaces have been successfully operated[196]-[207][†]. So far chambers of this construction have been built with relatively small dimensions.

These microstrip gas detectors are miniaturized multiwire proportional chambers, in which the dimensions are reduced by about a factor of ten in comparison to conventional chambers (figure 4.30). This has been made possible because the electrode structures can be reduced with the help of electron lithography. The wires are replaced by strips which are evaporated onto a thin substrate. Cathode strips arranged between the anode strips allow for an improved field quality. The segmentation of the otherwise planar cathodes in the form of strips or pixels [202, 208] also permits two-dimensional readout. Instead of mounting the electrode structures on ceramic substrates, they can also be arranged on thin plastic foils. In this way, even light, flexible detectors can be constructed which exhibit a high spatial resolution. Possible disadvantages lie in the electrostatic charging-up of the insulating plastic structures which can lead to time-dependent amplification properties because of the modified electric fields [209]-[214].

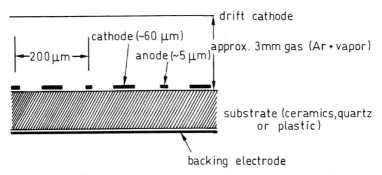

Fig. 4.30.   Schematic arrangement of a microstrip gas detector.

The obvious advantages of these microstrip detectors — apart from their excellent spatial resolution — are the low dead time (the positive ions being produced in the avalanche will drift the very short distance to the cathode strips in the vicinity of the anodes), the reduced radiation damage (because of the smaller sensitive area per readout element), and the high rate capability. Therefore, microstrip gas chambers are excellent candidates for track detectors at high rate accelerators like the

---

[†] In passing it should be mentioned that the attachment of positive ions on dielectric surfaces occasionally also can be taken advantage of to improve the field quality for certain chamber types or even achieve the desired field quality (see 'electrodeless drift chambers', section 4.7).

LHC (Large Hadron Collider) and at 'particle factories' (B, τ, ϕ-factories) [215, 216].

Microstrip proportional chambers can also be operated in the drift mode (see section 4.7).

The microgap chamber is a position sensitive proportional gas counter built using microelectronics technology [217]. In this detector the separation between the electrodes collecting the avalanche charges (the anode-cathode gap) is only a few microns. Therefore, the collection time for the positive ions is very short (≈ 10 ns) allowing extremely high rate capabilities ($> 10^7$ counts/mm s).

Wire chambers and in particular multiwire proportional chambers have been and still are extensively used in elementary particle physics [215, 216, 218, 219, 220] but also find interesting applications in other fields like, for example, medicine [221].

4.7 Planar drift chambers

A certain time Δt is required for the drift of the primary produced electrons to reach the anode wire. This time Δt between the moment of the particle passage through the chamber and the arrival time of the charge cloud at the anode wire depends on the point of passage of the particle through the chamber (figure 4.31). If v^- is the constant drift velocity of the electrons, the following linear relation holds

$$x = v^- \cdot \Delta t \tag{4.57}$$

or, if the drift velocity varies along the drift path

$$x = \int v^-(t) \mathrm{d}t \ . \tag{4.58}$$

In order to produce a suitable drift field, potential wires are introduced between neighbouring anode wires.

The measurement of the drift time allows the number of anode wires in a drift chamber to be reduced considerably in comparison to a multiwire proportional chamber, or by using small anode wire spacings to improve significantly the spatial resolution. Normally both advantages can be achieved at the same time [222]. Taking a drift velocity of $v^- = 5$ cm/μs and a time resolution of the electronics of $\sigma_t = 1$ ns, spatial resolutions of $\sigma_x = v^- \sigma_t = 50\,\mu$m can be achieved. However, the spatial resolution not only has contributions from the time resolution of the electronics, but also from the diffusion of the drifting electrons and the fluctuations of the statistics of primary ionization processes. The latter are most important in the vicinity of the anode wire (figure 4.32 [51, 223]).

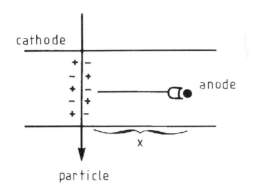

Fig. 4.31. Working principle of a drift chamber.

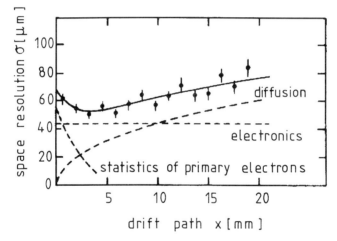

Fig. 4.32. Spatial resolution in a drift chamber as a function of the drift path [51, 223].

For a particle trajectory perpendicular to the chamber, electron-ion pairs are produced statistically along the particle track. The electron-ion pair closest to the anode wire is not necessarily produced on the connecting line between anode and potential wire. Spatial fluctuations of charge carrier production result in large drift path differences for particle trajectories close to the anode wire while they have only a minor effect for distant particle tracks (see figure 4.33).

Naturally, the time measurement cannot discriminate between particles having passed the anode wire on the right or on the left-hand side. A double layer of drift cells where the layers are staggered by half a cell width can resolve this left-right ambiguity (figure 4.34).

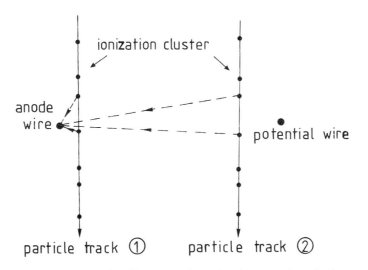

Fig. 4.33. Illustration of different drift paths for 'near' and 'distant' particle tracks to explain the dependence of the spatial resolution on the primary ionization statistics.

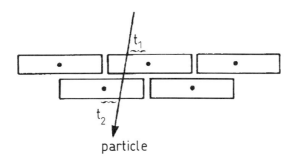

Fig. 4.34. Resolution of the left-right ambiguity in a drift chamber.

In multiwire proportional chambers there are regions of low field strength between the anode wires (see figures 4.25 and 4.26). As has been already indicated the field quality can be considerably improved by introducing potential wires at negative potential between two neighbouring anode wires (figure 4.35).

Drift chambers can be made very large [225, 226, 227]. For larger drift volumes the potential between the anode wire position and the negative potential on the chamber ends is divided linearly by using cathode strips connected to a chain of resistors (see figure 4.36).

The maximum achievable spatial resolution for large-area drift chambers is limited primarily by mechanical tolerances. For large chambers

a)

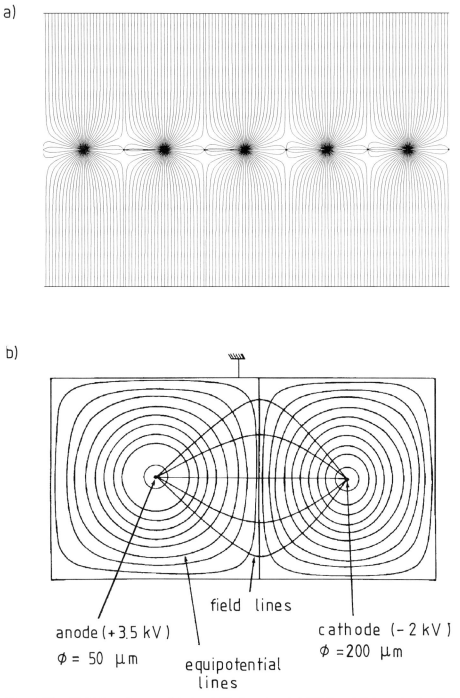

b)

field lines

anode (+3.5 kV)

Φ = 50 μm

equipotential
lines

cathode (−2 kV)

Φ =200 μm

Fig. 4.35.   Field lines in a multiwire drift chamber (a) [184] and (b) equipotential
and field lines in a multiwire drift chamber cell [51, 224].

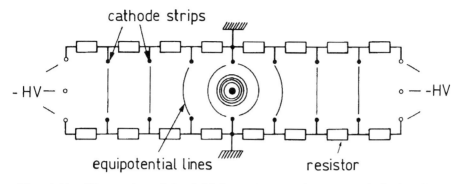

Fig. 4.36. Illustration of the field formation in a large-area drift chamber.

typical values of $200\,\mu$m are obtained. In small chambers ($10 \times 10\,\text{cm}^2$) spatial resolutions of $20\,\mu$m have been achieved. In the latter case the time resolution of the electronics and the diffusion of electrons on their way to the anode are the main limiting factors. The determination of the coordinate along the wires can again be performed with the help of cathode pads.

The relation between the drift time t and the drift distance in a large-area ($80 \times 80\,\text{cm}^2$) drift chamber with only one anode wire is shown in figure 4.37 [227]. The chamber was operated with a gas mixture of $93\,\%$ argon and $7\,\%$ isobutane.

Field formation in large-area drift chambers can also be achieved by the attachment of positive ions on insulating chamber surfaces. In these chambers an insulating foil is mounted on the large-area cathode facing the drift space (figure 4.38). In the time shortly after the positive high voltage on the anode wire has been switched on, the field quality is insufficient to expect a reasonable electron drift with good spatial resolution over the whole chamber volume (figure 4.39a). The positive ions which have been produced by the penetrating particle now start to drift along the field lines to the electrodes. The electrons will be drained by the anode wire, but the positive ions will get stuck on the inner side of the insulator on the cathode thereby forcing the field lines out of this region. After a certain while ('charging-up time') no field lines will end on the covers of the chamber and an ideal drift field configuration will have been formed (figure 4.39b [228, 229]). If the chamber walls are not completely insulating, i.e., their volume or surface resistance is finite, some field lines will still end on the chamber covers (figure 4.39c). Although, in this case, no ideal field quality is achieved, an overcharging of the cathodes is avoided since the chamber walls have a certain conductivity or transparency to allow for a removal of surplus surface charges.

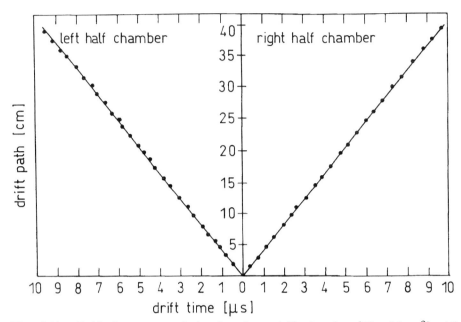

Fig. 4.37. Drift time space relation in a large drift chamber ($80 \times 80\,\text{cm}^2$) with only one anode wire [227].

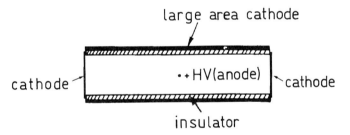

Fig. 4.38. Principle of construction of an electrodeless drift chamber.

Initial difficulties with long charging times ($\approx 1\,\text{hr}$) and problems of overcharging of the insulators at high rates can be overcome by a suitable choice of dielectrics on the cathodes [230]. Based on this principle chambers of very different geometry (rectangular chambers, cylindrical chambers, drift tubes, ...) even with long drift paths ($> 1\,\text{m}$) have been constructed [231]-[234].

The principle of electron drift in drift chambers can be used in many different ways. The introduction of a grid into a drift chamber enables the separation of the drift volume proper from the gas amplification region.

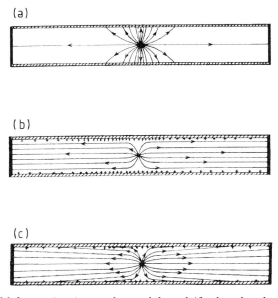

Fig. 4.39. Field formation in an electrodeless drift chamber by ion attachment [228, 229].

The choice of suitable gases and voltages allows very low drift velocities in the drift volume so that the ionization structure of a track of a charged particle can be electronically resolved without large expense (principle of a time expansion chamber) [235, 236]. The use of very small anode wire distances also allows high counting rates per unit area because the rate per wire in this case stays within reasonable limits.

The induction drift chamber [237, 238, 239] also allows high spatial resolutions by using anode and potential wires with small relative distances. The formation of an electron avalanche on the anode will induce charge signals on neighbouring pick-up electrodes which allow at the same time the determination of the angle of incidence of a particle and the resolution of the right-left ambiguity. Because of the small anode spacing the induction drift chamber is also an excellent candidate for high rate experiments, for example, for the investigation of electron-proton interactions in a storage ring at high repetition frequencies (e.g., in HERA, the hadron-electron storage ring at the German electron synchrotron DESY). Particle rates up to 10^6 per mm^2 and s can be processed.

The finite drift time can also be taken advantage of to decide whether or not an event in a detector is of interest. This, for example can be realized in the multistep avalanche chamber. Figure 4.40 shows the principle of operation [240]. The detector consists of two multiwire proportional chambers (MWPC 1 and 2), whose gas amplifications are arranged to be relatively small ($\approx 10^3$). All particles penetrating the detector will produce in both proportional chambers relatively small signals. Electrons from the avalanche in MWPC 1 can be transferred with a certain probability into the drift region situated between the two chambers. Depending on the width of the drift space these electrons require several hundred nanoseconds to arrive at the second multiwire proportional chamber. The end of the drift space is formed by a wire grid which is only opened by a voltage signal if some external logic signals an interesting event. In this case the drifting electrons are again multiplied by a gas amplification factor 10^3 so that a gas amplification of $10^6 \cdot \varepsilon$ in MWPC 2 is obtained, where ε is the mean transfer probability of an electron produced in chamber 1 into the drift space. If ε is sufficiently large (e.g., > 0.1), the signal in chamber 2 will be large enough to trigger the conventional readout electronics of this chamber. These 'gas delays', however, are nowadays mainly realized by purely electronic delay circuits.

Experiments at electron–positron storage rings and at future proton–proton colliders require large-area chambers for muon detection. There are many candidates for muon chambers, such as layers of streamer tubes. For the accurate reconstruction of decay products of the searched-for Higgs particles, for example, excellent spatial resolutions over very large areas are compulsory. These conditions can be met with modular drift chambers [241, 242].

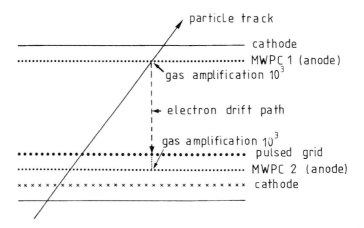

Fig. 4.40. Principle of operation of a multistep avalanche chamber [240].

4.8 Cylindrical wire chambers

For storage ring experiments cylindrical detectors have been developed which fulfill the requirement of a maximum solid-angle coverage, i.e., hermeticity. In the very first experiments cylindrical multigap spark chambers [104] (see section 4.15) and multiwire proportional chambers were used, however, recently drift chambers have been almost exclusively adopted for the measurement of particle trajectories and the determination of the specific ionization of charged particles.

There are several types of such detectors: cylindrical drift chambers whose wire layers form cylindrical surfaces, jet chambers, where the drift spaces are segmented in azimuthal direction, and time projection chambers, which are in the sensitive volume free of any material (apart from the counting gas), and where the information on particle trajectories is drifted to circular endplate detectors.

Cylindrical drift chambers operated in a magnetic field allow the determination of the momenta of charged particles. The transverse momentum p (in GeV/c) of charged particles is calculated from the axial magnetic field (in Tesla) and the bending radius of the track ϱ (in meters) to be (see section 4.11)

$$p = 0.3 \, B \cdot \varrho \, . \tag{4.59}$$

4.8.1 Cylindrical proportional and drift chambers

Figure 4.41 shows the principle of construction of a cylindrical drift chamber. All wires are stretched in an axial direction (in the z-direction, the direction of the magnetic field). Layers of anode wires are interspersed with layers of potential wires. For cylindrical drift chambers a potential wire is stretched between two anode wires. Two neighbouring layers of anode wires are separated by a cylindrical layer of potential wires. In the most simple configuration the individual drift cells are trapezoidal where the boundaries are formed by eight potential wires. Figure 4.41 shows a projection in the $r\varphi$ plane, where r is the distance from the center of the chamber and φ the azimuthal angle. Apart from this trapezoidal drift cell other drift cell geometries are also in use [1].

In so-called open trapezoidal cells every second potential wire on the potential wire planes is left out (figure 4.42).

The field quality can be improved by using closed cells (figure 4.43) at the expense of a larger number of wires. The compromise between the above mentioned drift cell configurations is a hexagonal structure of the cells (figure 4.44). In all these configurations the potential wires are of larger diameter ($\phi \approx 100 \, \mu$m) compared to the anode wires ($\phi \approx 30 \, \mu$m).

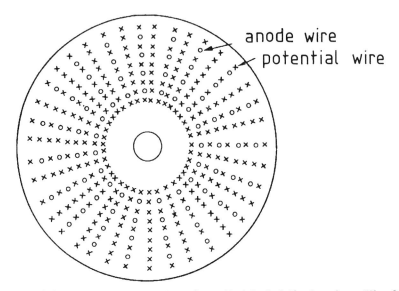

Fig. 4.41. Schematic representation of a cylindrical drift chamber. The figure shows a view of the chamber perpendicular to the wires.

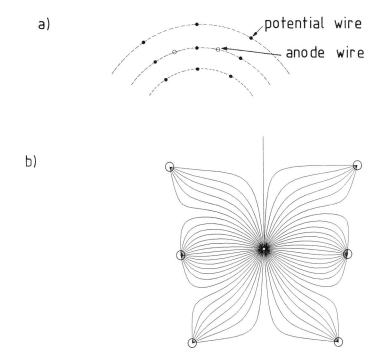

Fig. 4.42. a) Illustration of an open drift cell geometry. b) Field lines in an open drift cell [184].

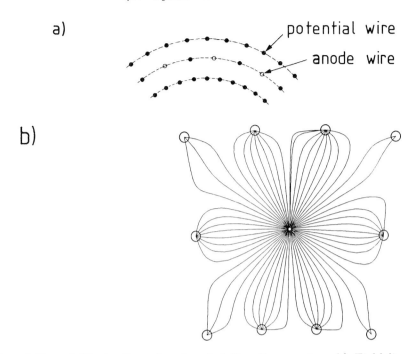

Fig. 4.43.   a) Illustration of a closed drift cell geometry. b) Field lines in a closed drift cell [184].

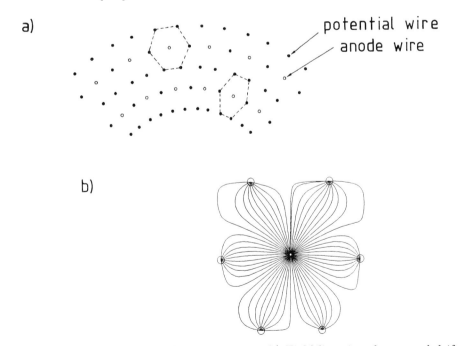

Fig. 4.44.   a) Hexagonal drift cell geometry. b) Field lines in a hexagonal drift cell [184].

All wires are stretched between two endplates which must take the whole wire tension. For large cylindrical wire chambers with several thousand anode and potential wires this tension can amount to several tons.

The configurations described so far do not allow a determination of the coordinate along the wire. Since it is impossible to segment the cathode wires in these configurations, other methods to determine the coordinate along the wire have been adopted. One method of determining the z-coordinate consists of the measurement of the signal currents I_1 and I_2 at the ends of each anode wire [243, 244]. The ratio $(I_1 - I_2)/(I_1 + I_2)$ determines the position of the avalanche and thereby the point of particle intersection (charge division method). Equally well, the propagation times of signals on the anode wires can be measured at both ends. The charge division technique allows accuracies of the order of 1 % of the wire length. This precision can also be obtained with fast electronics applied to the propagation time technique.

Another method for measuring the position of the avalanche along the sense wire uses spiral wire delay lines, of diameter smaller than 2 mm, stretched parallel to the sense wire [245]. This technique, which is mechanically somewhat complicated for large detector systems, allows accuracies of the order of 0.1 % along the wires. If the delay line is placed between two closely spaced wires, it also solves the left–right ambiguity. More sophisticated delay line readouts allow even higher spatial resolutions [246, 247].

However, there is also a fourth possibility by which one can determine the z-coordinate along the wire. In this case, some anode wires are stretched not exactly parallel to the cylinder axis, but are tilted by a small angle with the respect to this axis ('stereo wires'). The spatial resolution $\sigma_{r,\varphi}$ measured perpendicular to the anode wires is then translated into a resolution σ_z along the wire according to

$$\sigma_z = \frac{\sigma_{r,\varphi}}{\sin\gamma} \, , \tag{4.60}$$

if γ is the 'stereo angle' (see figure 4.45). For typical $r\varphi$ resolutions of 200 μm z resolutions of the order of $\sigma_z = 3$ mm are obtained, if the stereo angle is $\gamma \approx 4^0$. In this case, the z-resolution does not depend on the wire length. The magnitude of the stereo angle is limited by the maximum allowed transverse cell size. Cylindrical drift chambers with stereo wires are also known as hyperbolic chambers, because the tilted stereo wires appear to sag hyperbolically with respect to the axial anode wires.

In all these types of chambers, where the drift field is perpendicular to the magnetic field, special attention must be paid to the Lorentz angle.

Figure 4.46 shows the drift trajectories of electrons in an open rectangular drift cell with and without an axial magnetic field [248, 249].

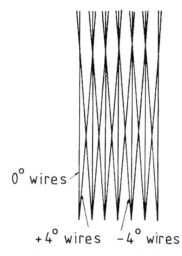

0° wires

+4° wires   −4° wires

Fig. 4.45. Illustration of the determination of the coordinate along the anode wire by use of stereo wires.

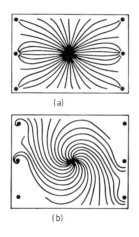

(a)

(b)

Fig. 4.46. Drift trajectories of electrons in an open rectangular drift cell a) without and b) with magnetic field [248, 249].

Figure 4.47 shows the $r\varphi$ projections of reconstructed particle tracks from an electron-positron interaction (PLUTO) in a cylindrical multiwire proportional chamber [250]. Figure 4.47a shows a clear two-jet structure which originated from the process $e^+e^- \rightarrow q\bar{q}$ (production of a quark-antiquark pair). Part (b) of this figure exhibits a particularly interesting event of an electron-positron annihilation from the aesthetic point of view. The track reconstruction in this case was performed using only the fired anode wires and the cathode strips without making use of drift time information (see section 4.6). The spatial resolutions obtained in this

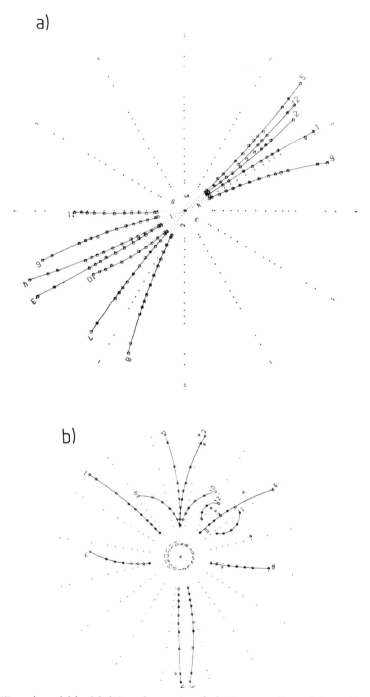

Fig. 4.47.   a) and b). Multitrack events of electron-positron interactions measured in the PLUTO central detector [250].

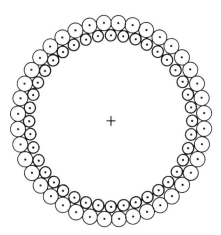

Fig. 4.48. Cylindrical configuration of thin-wall straw-tube chambers [251, 195] (©1989 IEEE).

way, of course, cannot compete with those that can be reached in drift chambers.

Cylindrical multiwire proportional chambers can also be constructed from layers of so-called 'straw chambers' (figure 4.48) [195, 251, 252, 253]. Such straw-tube chambers are frequently used as vertex detectors in storage ring experiments [254, 255]. Due to the construction of these chambers the risk of broken wires is minimized. In conventional cylindrical chambers a single broken wire can disable large regions of a detector [256]. In contrast, in straw-tube chambers only the straw with the broken wire is affected.

These straw chambers are made from thin aluminized mylar foils. The straw tubes have diameters of between 5 and 10 mm and are frequently operated at overpressure. These detectors allow for spatial resolutions of $30 \, \mu$m.

Because of their small size straw-tube chambers are candidates for high-rate experiments [257]. Due to the short electron drift distance they can also be operated in high magnetic fields without significant deterioration of the spatial resolution [258].

Very compact configurations with high spatial resolution can also be obtained with multiwire drift modules (figure 4.49) [255, 259, 260].

In the example shown, 70 drift cells are arranged in a hexagonal structure of 30 mm diameter only. Figure 4.50 shows the structure of electric field and equipotential lines for an individual drift cell [259]. Figure 4.51 shows a single particle track through such a multiwire drift module [259].

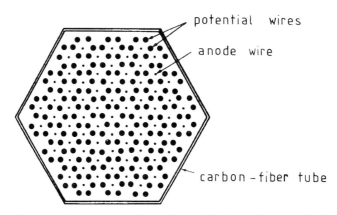

Fig. 4.49. Schematic representation of a multiwire drift module. In this hexagonal structure each anode wire is surrounded by six potential wires. Seventy drift cells are incorporated in one container of 30 mm diameter only which is made from carbon fiber material [259].

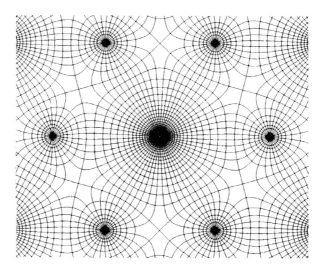

Fig. 4.50. Calculated electric field and equipotential lines in one individual hexagonal drift cell of the multiwire drift module [259].

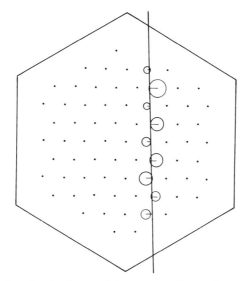

Fig. 4.51. Example of a single particle passage through a multiwire drift module. The circles indicate the measured drift times of the fired anode wires. The particle track is a tangent to all drift circles [259].

4.8.2 Jet drift chambers

Cylindrical drift chambers typically have 10 to 15 layers of anode wires. If one wants to identify particles by measuring the specific energy loss with the anode wires, e.g., to discriminate charged pions against kaons, this provides an insufficient number of dE/dx measurements.

Particle discrimination or particle identification requires an accurate measurement of the momentum

$$p = mv = \gamma m_0 \beta c . \tag{4.61}$$

The energy loss determination, if it is sufficiently accurate, measures according to the Bethe-Bloch relation ($-\frac{dE}{dx} \propto \frac{1}{\beta^2}$, if $\beta\gamma \ll 4$ and $\frac{dE}{dx} \propto \ln(a\gamma)$ if $\beta\gamma \gg 4$; a is a parameter which depends on the particle species and the target material) the velocity β and thereby allows with the help of equation (4.61), for a known momentum, a determination of the particle mass. In jet drift chambers an accurate measurement of the energy loss by ionization is performed by determining the specific ionization on as large a number of anode wires as possible. The central detector of the JADE experiment [261, 262] at PETRA determines the energy loss of charged particles on 48 wires, which are stretched parallel to the magnetic field. The cylindrical volume of the drift chamber is subdivided into 24 radial segments. Figure 4.52 sketches in principle the

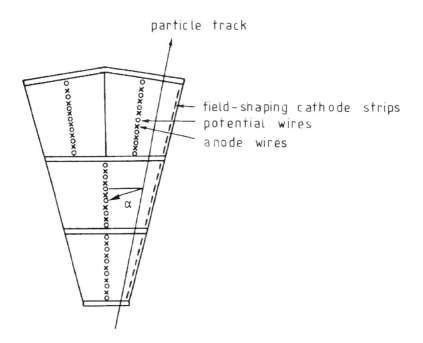

Fig. 4.52. Segment of a jet drift chamber (after [1, 261, 262, 263]). The field-forming cathode strips are only shown on one side of the segment (for reasons of simplicity and not to overload the figure the two inner rings 1 and 2 show only five and the outer ring 3 only six anode wires).

arrangement of one of these sectors, which is itself again subdivided into smaller drift regions of 16 anode wires each.

The field formation is made by potential strips at the boundaries between two sectors. The electric field is perpendicular to the counting wire planes and also perpendicular to the direction of the magnetic field. For this reason the electron drift follows the Lorentz angle which is determined from the electric and magnetic field strengths and the drift velocity. For the solenoidal $\vec{B}$-field (see figure 8.10) of 0.45 Tesla in JADE a Lorentz angle of $\alpha = 18.5°$ is obtained. To reach a maximum accuracy for an individual energy loss measurement the chamber is operated under a pressure of four atmospheres. This overpressure also suppresses the influence of primary ion statistics on the spatial resolution. However, it is important not to increase the pressure to too high a value since the logarithmic rise of the energy loss, which is the basis for particle separation, may be reduced by the onset of the density effect.

The determination of the coordinate along the wire is done by using the charge division method.

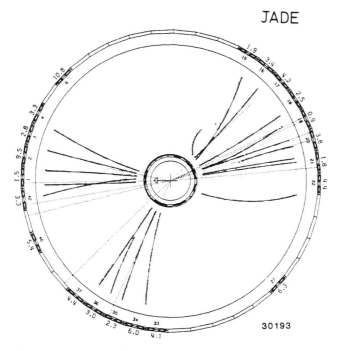

Fig. 4.53. The $r\varphi$ projection of interaction products from an electron-positron collision in the JADE central detector [261, 262, 267]. The bent tracks correspond to charged particles and the dotted tracks to neutral particles which are not affected by the magnetic field (and are not registered in the chamber).

The $r\varphi$ projection of trajectories of particles from an electron-positron interaction in the JADE drift chamber is shown in figure 4.53 [261, 262]. The 48 coordinates along each track originating from the interaction vertex can clearly be recognized. The left-right ambiguity in this chamber is solved by staggering the anode wires (see also figure 4.54). An even larger jet drift chamber is mounted in the OPAL-detector at the large electron-positron storage ring LEP at CERN [264].

The structure of the new MARK II jet chamber (figure 4.54) is very similar to that of the JADE chamber [265, 266]. The ionization produced by particle tracks in this detector is collected on the anode wires. Potential wires between the anodes and layers of field-forming wires produce the drift field. The field quality at the ends of the drift cell is improved by additional potential wires. The drift trajectories in this jet chamber in the presence of the magnetic field are shown in figure 4.55 [265, 266].

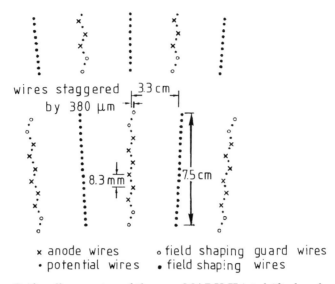

wires staggered
by 380 μm

3.3 cm

8.3 mm

7.5 cm

× anode wires   ○ field shaping guard wires
• potential wires   • field shaping wires

Fig. 4.54.   Drift cell geometry of the new MARK II jet drift chamber [265, 266].

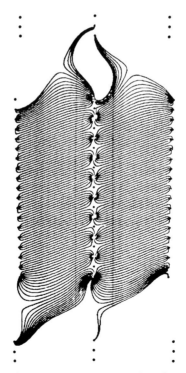

Fig. 4.55.   Calculated drift trajectories in a jet chamber drift cell in the presence of a magnetic field [265, 266].

4.8.3 Time projection chambers (TPC)

The *crème de la crème* of track recording in cylindrical detectors (also suited for other geometries) at the moment is realized with the time projection chamber [268]. Apart from the counting gas this detector contains no other constructional elements and thereby represents the optimum as far as minimizing multiple scattering and photon conversions are concerned [269]. A side view of the construction principle of a time projection chamber is shown in figure 4.56.

The chamber is divided into two halves by means of a central electrode. A typical counting gas is a mixture of argon and methane (90:10).

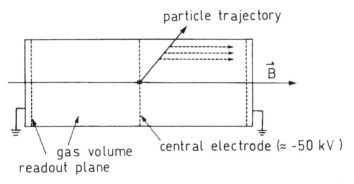

Fig. 4.56. Working principle of a time projection chamber (TPC) [268].

The primary ionization produced by charged particles drifts in the electric field — which is typically parallel to the magnetic field — in the direction of the endplates of the chamber, which in most cases consist of multiwire proportional detectors. The magnetic field suppresses the diffusion perpendicular to the field. This is achieved by the action of the magnetic forces on the drifting electrons which, as a consequence, spiral around the direction of the magnetic field. For typical values of electric and magnetic field strengths, Larmor radii below $1\,\mu$m are obtained. The arrival time of primary electrons at the endplates supplies the z-coordinate along the cylinder axis. The layout of one endplate is sketched in figure 4.57.

The gas amplification of the primary ionization takes place at the anode wires, which are stretched in azimuthal direction. The radial coordinate r can in principle be obtained from the fired wire (for short wires). To obtain three-dimensional coordinates the cathodes of endcap multiwire proportional chamber segments are usually structured as pads. Therefore the radial coordinate is also provided by reading the position of the fired pad. In addition, the pads supply the coordinate along the anode wire

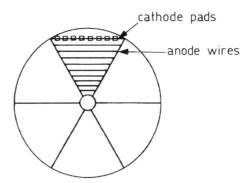

Fig. 4.57.   Principle of operation of pad readout in an endcap multiwire proportional chamber. The anode wires and some cathode pads are shown for one sector.

resulting in a determination of the azimuthal angle φ. Therefore, the time projection chamber allows the determination of the coordinates r, φ and z, i.e., a three-dimensional space point, for each cluster of primary electrons produced in the ionization process.

The analog signals on the anode wires provide information on the specific energy loss and can consequently be used for particle identification. Typical values of the magnetic field are around 1.5 Tesla, and around 20 kV/m for the electric field. Since in this construction electric and magnetic field are parallel the Lorentz angle is zero and the electrons drift parallel to $\vec{E}$ and $\vec{B}$ (there is no '$\vec{E} \times \vec{B}$-effect').

A problem, however, is caused by the large number of positive ions which are produced in the gas amplification process at the endplates and which have to drift a long way to the central electrode. The strong space charge of the drifting positive ions causes the field quality to deteriorate. This can be overcome by introducing an additional grid ('gate') between the drift volume and the endcap multiwire proportional chamber (see figure 4.58).

The gate is normally closed. It is only opened for a short period of time if an external trigger signals an interesting event. In the closed state the gate prevents ions from drifting back into the drift volume. Thereby the quality of the electric field in the sensitive detector volume remains unchanged [1]. This means that the gate serves a dual purpose. On the one hand electrons from the drift volume can be prevented from entering the gas amplification region of the endcap multiwire proportional chamber. On the other hand — for gas amplified interesting events — the positive ions are prevented from drifting back into the detector volume proper. Figure 4.59 shows the operation principle of the gate in the ALEPH-TPC [270].

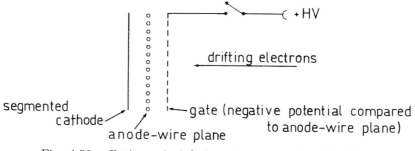

Fig. 4.58. Gating principle in a time projection chamber.

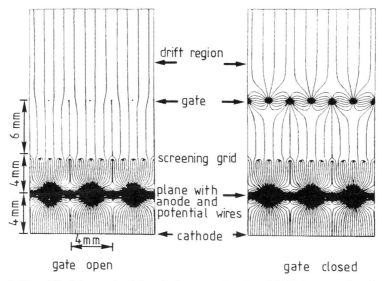

Fig. 4.59. Working principle of the gate in the ALEPH-TPC [270]. For an open gate the ionization electrons are not prevented from entering the gas amplification region. The closed gate, however, confines the positive ions to the gas amplification region. A closed gate also stops electrons in the drift volume from entering the gas amplification region. For an event of interest the gate is first opened to allow the primary electrons to enter the gas amplification region and then it is closed to prevent the positive ions produced in the avalanche process from drifting back into the detector volume.

Time projection chambers can be made very large (diameter $\geq 3\,\mathrm{m}$, length $\geq 5\,\mathrm{m}$). They contain a large number of analog readout channels (number of anode wires ≈ 5000 and cathode pads $\approx 50\,000$). Several hundred samples can be obtained per track, which ensure an excellent determination of the radius of curvature and allow an accurate measurement of the energy loss, which is essential for particle identification [270, 271, 272]. The drawback of the time projection chamber is the fact that high particle rates cannot be handled, because the drift time of the electrons in the detector volume amounts to $40\,\mu$s (for a drift path of $2\,$m) and the readout of the analog information also requires several μs.

In large time projection chambers typical spatial resolutions of $\sigma_z = 1\,$mm and $\sigma_{r,\varphi} = 160\,\mu$m are obtained. In particular, the resolution of the z-coordinate requires an accurate knowledge of the drift velocity. This, however, can be calibrated and monitored by UV-laser-generated ionization tracks.

Figure 4.60 shows the $r\varphi$ projection of an electron-positron annihilation in the ALEPH time projection chamber [270, 271].

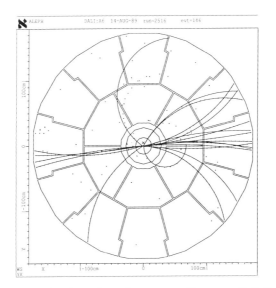

Fig. 4.60. The $r\varphi$ projection of an electron-positron annihilation in the ALEPH time projection chamber [270, 271]. The endplate detector is structured into two rings which are made from six (inner ring) and twelve (outer ring) multiwire proportional chamber segments.

Time projection chambers can also be operated with liquid noble gases. Such liquid argon time projection chambers represent an electronic replacement for bubble chambers with the possibility of three-dimensional event reconstruction. In addition, they can serve simultaneously as a calorimetric detector (see chapter 7), are permanently sensitive and can intrinsically supply a trigger signal by means of the scintillation light produced in the liquid noble gas (see section 5.2) [273]-[278]. The electronic resolution of the bubble chamber-like pictures is of the order of $100\,\mu$m.

The operation of large liquid argon time projection chambers, however, requires ultrapure argon (contaminants $< 0.1\,\mathrm{ppb}$ ($1\,\mathrm{ppb} \equiv 10^{-9}$)) and high performance low noise preamplifiers since no gas amplification occurs in the counting medium. Multi-kiloton liquid argon TPCs appear to be good candidates to study rare phenomena in underground experiments ranging from the search for nucleon decay to solar neutrino observations [279, 280].

Self-triggering time projection chambers have also been operated successfully with liquid xenon [281, 282].

4.9 Imaging chamber

The working principle of the imaging chamber is very similar to the time projection chamber. Its geometrical form does not have to be cylindrically symmetric. Because of its similarity to the TPC the imaging chamber is described under the heading of cylindrical chambers. After all, time projection chambers need not necessarily be cylindrical in shape. The geometry of multiwire drift chambers has to be chosen in such a way as to fit the application in question best.

Like the time projection chamber the imaging chamber consists of a large detector volume filled with gas that is operated in a homogeneous electric field. Tracks are registered in the sensitive volume of the chamber and the ionization drifts — like in the TPC — to the endplates of the chamber. The drift information is evaluated in this endcap, which can be a planar multiwire proportional chamber or some other planar chamber type with parallel electrodes. However, in contrast to the TPC, the track information is not read out electronically. With a large gas amplification, leading to strong photon multiplication in the avalanche formation, an optical picture of the track can be obtained. The photon emission from the electron avalanches is photographed in projection in a similar fashion as it is done in streamer chambers (see section 4.13). Figure 4.61 shows the principle of operation of such an imaging chamber.

Photons are produced copiously during avalanche formation, but predominantly at short wavelengths where the absorption length is rather short. The chamber gas is hardly transparent for these short wavelength photons. Therefore, it is important for the successful operation of an imaging chamber to find a gas mixture which allows a sufficiently large number of photons in the optical wavelength range to be produced during avalanche formation. A possible gas mixture is argon-methane-TEA, which exhibits a maximum of emission around $300\,\mathrm{nm}$ (TEA $= (\mathrm{C_2H_5})_3\mathrm{N}$) [283, 265].

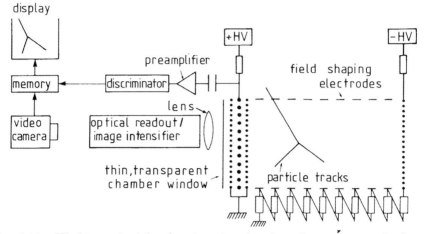

Fig. 4.61. Working principle of an imaging chamber. Ionization tracks formed in the sensitive volume of the detector drift to the endcaps. An optical readout system supplies projected tracks [265, 283].

Large light yields are obtained if the endcap detector is operated in the streamer mode. However, in this case the proportionality between the energy loss of the detected particle and the recorded light or charge is lost. If gas amplification factors which preserve this proportionality are used, the optical readout system must be supplied with a larger amplification factor (image intensifier). In such a case the energy loss of the charged particles can also be measured by reading out the charge signal on the chamber wires via analog-to-digital converters (ADCs). If in addition to the charge profile of tracks spatial coordinates of tracks are of interest, too, the signals must also be read out via time-to-digital converters (TDCs). Like in the TPC — in this case — an external trigger signal, e.g., derived from a coincidence of scintillation counters, is also required.

4.10   Ageing effects in wire chambers

Avalanche formation in multiwire proportional or drift chambers can be considered as a microplasma discharge. In the plasma of an electron avalanche, chamber gases, vapor additions and possible contaminants are partially decomposed, with the consequence that aggressive radicals may be formed (molecule fragments). These free radicals can then form long chains of molecules, i.e., polymerization can set in. These polymers may be attached to the electrodes of the wire chamber, thereby reducing the gas amplification for a fixed applied voltage: the chamber ages. After a certain amount of charge deposited on the anodes, the chamber properties deteriorate so much that the detector can no longer be used for accurate measurements (e.g., energy loss measurements for particle identification).

Which processes now are of importance for the premature ageing of drift chambers and which steps can be taken to increase the lifetime of the chamber?

Ageing processes are very complex. Different experimental results concerning the question of ageing are extremely difficult to compare, since ageing phenomena depend on a large number of parameters and each experiment usually has different sets of parameters. Nevertheless, some clear conclusions can be drawn though a detailed understanding of ageing processes has yet to come [284]-[292]. The main parameters which are related to wire chamber ageing are characterized below [284, 286].

A multiwire proportional or drift chamber is typically filled with a mixture of a noble gas and one or several vapor additions. Contaminants, which are present in the chamber gas or enter it by outgassing of detector components, cannot be completely avoided. The electron avalanche, which forms in such a gas environment in the immediate vicinity of the anode wire, produces a large number of molecules. The energy required for the break-up of covalent molecule bonds is typically a factor of three lower than the ionization potential. If electrons or photons from the avalanche break up a gas molecule bond, radicals that normally have quite a large dipole moment are formed. Because of the large electric field strength in the vicinity of the wire these radicals are attracted by the anode and may form in the course of time a poorly or non-conducting anode coating, which can cause the wires to be noisy. Conducting anode deposits increase the anode diameter, thereby reducing the gas amplification. Because of the relatively large chemical activity of radicals, different compounds can be produced on the anode in this way. The rate of polymerization is expected to be proportional to the density of radicals which in itself is proportional to the electron density in the avalanche. Polymerization effects, therefore, will increase with increasing charge deposition on the anode. However, not only the anode is affected. In the course of polymer formation (e.g., positive) polymers may be formed which migrate slowly to the cathode. This is confirmed by patterns of 'wire shadows' which can be formed by deposits on planar cathodes [284, 286].

Typical deposits consist of carbon, thin oxide layers or silicon compounds. Thin metal oxide layers are extremely photosensitive. If such layers are formed on cathodes even low energy photons can free electrons from the cathodes via the photoelectric effect. These photoelectrons are gas amplified thus increasing the charge deposition on the anode, thereby accelerating the ageing process. Deposits on the electrodes can even be caused during the construction of the chamber, e.g., by finger prints. Also the gases which are used, even at high purity, can be contaminated in the course of the manufacturing process by very small oil droplets or silicon dust (SiO_2). Such contaminants at the level of several ppm can cause significant ageing effects.

Once a coating on the electrodes has been formed by deposition, high electric fields between the deposit layer and the electrode can be produced by secondary electron emission from the electrode coating ('Malter effect' [293]). As a consequence of this, these strong electric fields may cause field electron emission from the electrodes, thereby reducing the lifetime of the chamber.

Which are now the most sensitive parameters that cause ageing or accelerate ageing, and which precautions have to be considered for chamber construction? In addition, it is an interesting question whether there are means to clean up ('rejuvenate') aged wires.

Generally, it can be assumed that pure gases free of any contaminants will delay ageing effects. The gases should be as resistant as possible to polymerization. It only makes sense, however, to use ultrapure gases if it can be guaranteed that contamination by outgassing of chamber materials or gas pipes into the detector volume can be prevented.

Apart from undesired contaminants there are a number of additions which make a positive influence to ageing phenomena. Consequently it is important to avoid detrimental contaminants which may lead to polymerization and to add favourable lifetime extending additions to the chamber gas.

Favourable admixtures have proven to be atomic oxygen and organic compounds which contain oxygen (like –COOH, –CO–, –OCO–, –OH, –O–). Atomic oxygen reacts with hydrocarbons to give the end products CO, CO_2, H_2O, and H_2, i.e., stable and volatile molecules which can easily be removed from the chamber volume by a steady gas flow. Oxygen-containing compounds have a very small tendency to polymer formation. However, oxygen together with contaminants of silicon can produce various silicates. Favourable additions are H_2O, alcohols (methanol: CH_3OH; ethanol: C_2H_5OH; isopropanol: $(CH_3)_2CHOH$), ether (dimethylether: $(CH_3)_2O$) and methylal ($CH_2(OCH_3)_2$). These compounds have large cross sections for the absorption of ultraviolet light thereby suppressing the lateral avalanche propagation. If these oxygen-containing molecules are broken up by electron collisions the electronegativity of oxygen frequently repairs the break-up. Water has the additional advantage that it improves the conductivity of already existing deposits thus increasing the lifetime of the chamber.

Additions of hydrogen also appear to have a positive effect since radicals like CH_2 produced by molecule decomposition are transformed easily back into their original form (CH_4).

Unfavourable contaminants which normally accelerate the ageing of chambers are carbon, carbon-containing polymers, silicon compounds, halides and sulphur-containing compounds. Quite frequently gas bottles are already contaminated by halide compounds. Carbon-containing polymers are found in oils which may be present already in traces in commercial gas. They can also be introduced into the chamber volume by gas flow systems which contain oil. Many rubber sealings, tubes ('silicon tubes'), greases ('silicon grease' for sealing purposes) contain detrimental silicon compounds. PVC-tubes contain chlorine and thereby also cause the gas quality to deteriorate.

The unfavourable effects of these contaminants can be understood in the following fashion [284, 286]: halide compounds ($C - Cl$, $C - Br$) are more weakly bound compared to carbon compounds. Therefore, halogenated hydrocarbons like CF_2Cl_2, CH_3Cl, C_2H_3Cl, ... are more easily decomposed into radicals compared to, e.g., methane (CH_4). Even small amounts of hydrocarbon compounds which contain chlorine, bromine or fluoride can significantly accelerate the rate of polymerization of hydrocarbons like CH_4, C_2H_2, C_2H_6, etc.

Silicon — as the most frequently occurring element on earth — is contained in many materials which are used for chamber construction (like G–10 (glass fiber reinforced epoxy resin), various oils and molecular sieves). Silicon is frequently contained in gas bottles in form of silane (SiH_4) or tetrafluorsilane (SiF_4). Silicon can, together with hydrocarbons, form silicon carbide; this, together with oxygen silicates, which have — because of their high mass — a low volatility and almost are impossible to remove from the chamber volume will be preferentially deposited on the electrodes.

Apart from avoiding unfavourable contaminants in the chamber gas, and by carefully selecting components for chamber construction and the gas system, some constructional features can also be recommended to suppress ageing affects.

Larger cathode surfaces normally have smaller electric fields at their surface compared to layers of cathode wires. Therefore, continuous cathodes have a reduced tendency for deposition compared to cathode wires. The effect of deposits on thin anode wires is quite obviously enhanced compared to thick anode wires. Also careful selection of the electrode material can be of major influence on the lifetime of the chamber. Gold-plated tungsten wires are quite resistant against contaminants while wires of high resistance material (Ni/Cr/Al/Cu-alloys) tend to react with contaminants or their derivatives and this may lead to drastic ageing effects.

Certain contaminants and deposits can be dissolved at least partially by additions of, e.g., water vapor or acetone. Macroscopic deposits on wires can be 'burnt off' by deliberately causing sparks. On the other hand sparking may also lead to the formation of carbon fibers ('whiskers') which significantly reduce the lifetime of chambers, and can even induce wire breaking [256].

Wire chambers which are filled with a mixture of a noble gas and a hydrocarbon (e.g., Ar/CH_4) show significant ageing effects for charge depositions greater than 0.05 Coulomb per cm anode wire. If hydrocarbons are replaced by CO_2, the lifetime of the chamber may be increased by a factor of ten. Quite obviously low gas amplification is favourable for avoiding ageing effects.

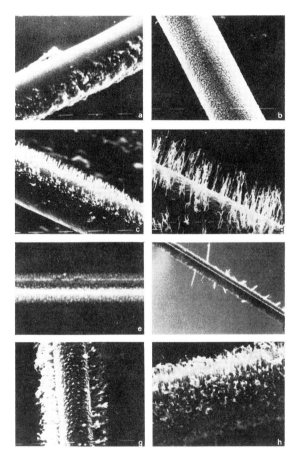

Fig. 4.62. Examples of depositions on anode wires [286].

Figure 4.62 shows some examples of deposits on anode wires [286]. On the one hand, one can see more or less continuous anode coatings which may alter the surface resistance of the anode. On the other hand, also hair-like polymerization structures are visible which will decisively

Fig. 4.63 a. Silicon depositions on an anode wire ($\phi = 30\,\mu$m) [294].

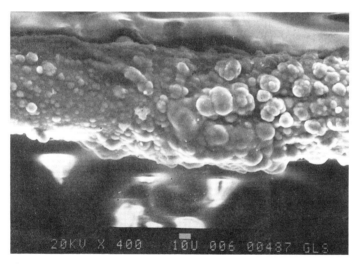

Fig. 4.63 b. Coatings consisting of silicon, chlorine and copper on a cathode wire ($\phi = 100\,\mu$m) [294].

deteriorate the field quality in the vicinity of the anode wire and also may lead to sparking.

Figure 4.63a shows deposits on a $30\,\mu$m thick anode wire in 880-fold magnification and figure 4.63b shows granular structures on a $100\,\mu$m thick cathode wire in a drift chamber in 400-fold magnification. It was shown by mass spectroscopic methods that the depositions on the anode mainly contain silicon while on the cathode chlorine and copper were detected in addition [294].

4.11   Bubble chambers

Bubble chambers [295]-[300], like cloud chambers, belong to the class of visual detectors and, therefore, require optical recording of events. This method of observation includes the tedious analysis of bubble chamber pictures which certainly limit the possible statistics of experiments. However, the bubble chamber allows the recording and reconstruction of events of high complexity with high spatial resolution. Therefore, it is perfectly suited to study rare events (e.g., neutrino interactions); still, the bubble chamber has now been superseded by detectors with a purely electronic readout.

In a bubble chamber the liquid gas (H_2, D_2, Ne, C_3H_8, Freon, ...) is held in a pressure container close to the boiling point. Before the expected event the chamber volume is expanded by retracting a piston. The expansion of the chamber leads to a reduction in pressure thereby exceeding the boiling temperature of the bubble chamber liquid. If in this superheated liquid state a charged particle enters the chamber, bubble formation sets in along the particle track.

The positive ions produced by the incident particles act as nuclei for bubble formation. The lifetime of these nuclei is only 10^{-11} s to 10^{-10} s. This is too short to trigger the expansion of the chamber by the incoming particles. For this reason the superheated state has to be reached *before* the arrival time of the particles. Bubble chambers, however, can be used at accelerators where the arrival time of particles in the detector is known and, therefore, the chamber can be expanded in time ('synchronization').

In the superheated state the bubbles grow until the growth is stopped by a termination of the expansion. At this moment the bubbles are illuminated by light flashes and photographed. Figure 4.64 shows the principle of operation of a bubble chamber [2]. The inner walls of the container have to be extremely smooth so that the liquid 'boils' only in those places where bubble formation should occur, namely along the particle trajectory, and not on the chamber walls.

Depending on the size of the chamber repetition times down to 100 ms can be obtained with bubble chambers.

The bubble chamber pressure before expansion is several atmospheres. To transform the gases into the liquid state they generally must be strongly cooled. Because of the large amount of stored gases, the experiments with hydrogen bubble chambers can be potentially dangerous

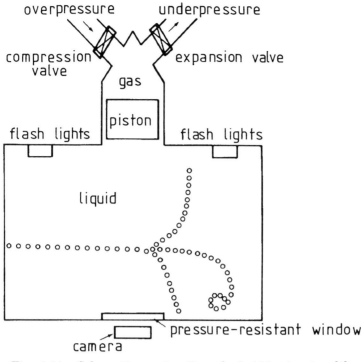

Fig. 4.64. Schematic construction of a bubble chamber [2].

because of the possible formation of explosive oxyhydrogen gas, if the chamber gas leaks from the bubble chamber. Also operation with organic liquids, which must be heated for the operation, represents a risk because of their flammability. Bubble chambers are usually operated in a high magnetic field (several Tesla). Momenta of charged particles are obtained by measuring their radius of curvature ϱ in the magnetic field of strength $\vec{B}$ according to

$$q|\vec{v} \times \vec{B}| = \frac{mv^2}{\varrho} , \qquad (4.62)$$

$$|\vec{p}| = q \cdot \varrho \cdot |\vec{B}| , \qquad (4.63)$$

if $\vec{B} \perp \vec{p}$. Furthermore, the bubble density along the track is proportional to the energy loss dE/dx by ionization. For $p/m_0 c = \beta\gamma \ll 4$ the energy loss can be approximated by (see equation (1.12))

$$\frac{dE}{dx} \propto \frac{1}{\beta^2} . \qquad (4.64)$$

If the momentum of the particle is known and if the velocity is determined from an energy loss measurement, the particle can be identified:

$$m_0 = \frac{p}{\gamma \beta c} = \frac{\sqrt{1 - \beta^2}}{\beta c} \, p \; . \qquad (4.65)$$

Figure 4.65 shows the spiraling track of an electron in a small bubble chamber in a transverse magnetic field [301]. The progressive energy loss of the electron in the bubble chamber liquid leads to an increased bending of the electron track thereby causing it to spiral inwards. The increase of the energy loss towards the end of the track is also visible.

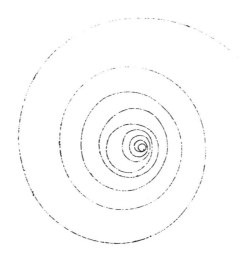

Fig. 4.65.   Inward spiraling track of an electron in a bubble chamber [301].

The choice of certain bubble chamber fillings is dictated by the physics aim of the experiment. Bubble chambers are interaction target and detector at the same time (active target).

For the investigation of photoproduction on protons naturally the best choice is a pure hydrogen filling. Results on photoproduction off neutrons can be obtained from ^{2}D-fillings, because no pure neutron liquid exists (maybe with the exception of neutron stars). The photonuclear cross section on neutrons can be determined according to

$$\sigma(\gamma, n) = \sigma(\gamma, d) - \sigma(\gamma, p) \; . \qquad (4.66)$$

If, e.g., the production of neutral pions is to be investigated, a bubble chamber filling with small radiation length X_0 is required, because the π^0 decays in two photons which have to be detected via the formation of electromagnetic showers. In this case, xenon or Freon can be chosen as chamber gas.

Table 4.1 lists some important gas fillings for bubble chambers along with their characteristic parameters [1, 2].

Table 4.1. *Characteristic properties of bubble chamber liquids [1, 2]*

bubble-chamber filling	T [°K]	vapor pressure [bar]	density g/cm^3	X_0 [cm]	absorption length λ_a [cm]
^{4}He	3.2	0.4	0.14	1027	437
^{1}H	26	4	0.06	1000	887
^{2}D	30	4.5	0.14	900	403
^{20}Ne	36	7.7	1.02	27	89
C_3H_8	333	21	0.43	110	176
CF_3Br (Freon)	303	18	1.5	11	73

If one wants to study nuclear interactions with bubble chambers, the absorption length λ_a should be as small as possible. In this case heavy liquids like Freon are indicated.

Bubble chambers are an excellent device if the main purpose of the experiment is to analyze complex and rare events. For example, the Ω^- — after first hints from experiments in cosmic rays — could be unambiguously discovered in a bubble chamber experiment.

Figure 4.66 [302, 303] shows the production and the decay of an Ω^- in a K^--beam according to the following reaction:

$$
\begin{aligned}
K^- + p \;\rightarrow\; &\Omega^- + K^+ + K^0 \\
&\hookrightarrow \Xi^0 + \pi^- \\
&\quad\hookrightarrow \pi^0 + \Lambda^0 \\
&\qquad\qquad \hookrightarrow \pi^- + p \\
&\qquad \hookrightarrow \gamma + \gamma \\
&\qquad\qquad \hookrightarrow e^+ e^- \\
&\qquad \hookrightarrow e^+ e^-
\end{aligned}
\tag{4.67}
$$

To ease the recognition of the production and decay of the Ω^- a schematic representation of the reaction chain is added to the bubble chamber photograph.

In recent times the application of bubble chambers has, however, been superseded by other detectors like electronic devices. The reasons for this originate from some serious intrinsic drawbacks of the bubble chamber, as follows.

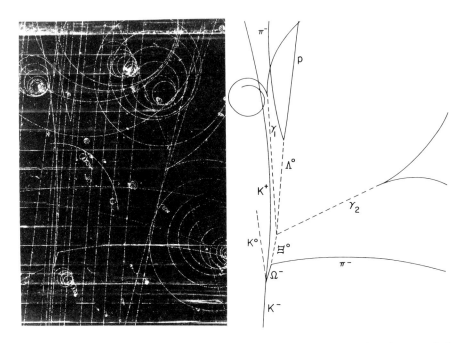

Fig. 4.66. Ω^--production in a bubble chamber. In the right-hand part of the figure the production and decay of the Ω^- is represented schematically [302, 303].

- Bubble chambers cannot be triggered.

- They cannot be used in storage ring experiments because it is difficult to achieve a 4π-geometry with this type of detector. Also the 'thick' entrance windows required for the pressure container prevents good momentum resolution because of multiple scattering.

- For high energies the bubble chamber is not sufficiently massive to stop the produced particles. This precludes an electron and hadron calorimetry — not to mention the difficult and tedious analysis of these cascades — because shower particles will escape from the detector volume.

- The identification of muons with momenta above several GeV/c in the bubble chamber is impossible because they look almost exactly like pions as far as the specific energy loss is concerned. Only by use of additional detectors (external muon counters) can a π/μ-separation be achieved.

– The lever arm of the magnetic field is generally insufficient for an accurate momentum determination of high momentum particles.

– Experiments with high statistics are not really practicable because of the time consuming analysis of bubble chamber pictures.

However, bubble chambers are still used in experiments with external targets (fixed target experiments). Because of their high intrinsic spatial resolution of several μm bubble chambers can serve as vertex detectors in these experiments.

Figure 4.67 shows the production and decay of charmed mesons in an extremely small bubble chamber (BIBC = Berne Infinitesimal Bubble Chamber) with dimensions of 6.5 cm diameter and 3.5 cm depth. This chamber is filled with a heavy liquid [304].

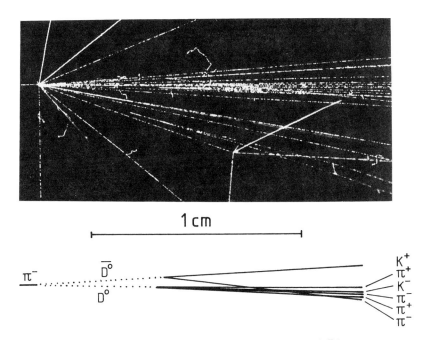

Fig. 4.67. Production and decay of charmed mesons ($D^0 \bar{D}^0$-production) in a high resolution bubble chamber [304].

The secondary vertices which indicate the D^0 and $\bar{D}^0$-decays are not clearly visible in the photographic record. Guided by the schematic representation of the decays in the lower part of the figure it is, however, possible also to identify these vertices in the photograph. Such chambers are well suited for the measurement of short lived particles [304]. Using heavy liquids as filling (e.g., Freon) the variation of the production and interaction behaviour of charmed mesons with the target mass can also be studied. Experiments on this topic, usually done with hydrogen bubble chambers, can be supplemented in this way.

To be able to measure short lifetimes in bubble chambers the size of the bubbles must be limited. This means that the event under investigation must be photographed relatively soon after the onset of bubble formation when the bubble size is relatively small, thereby guaranteeing a good spatial resolution, and, as a consequence, also good time resolution. In any case the bubble size must be small compared to the decay length of the particle. Figure 4.68 shows 'new' and 'old' tracks in the infinitesimal BIBC where the new tracks are characterized by bubble diameters of typically $30\,\mu$m, while old tracks of previous events have grown to considerable size during the expansion cycle [305].

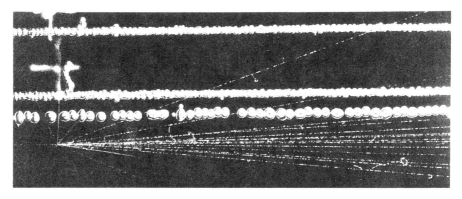

Fig. 4.68. 'Old' and 'new' tracks in a high resolution bubble chamber [305].

By use of the technique of holographic recording a three-dimensional event reconstruction can be achieved [306]. With these high resolution bubble chambers, e.g., the lifetimes of short-lived particles can be determined precisely. For a spatial resolution of $\sigma_x = 6\,\mu$m time measurement errors of σ_τ

$$\sigma_\tau = \frac{\sigma_x}{c} = 2 \cdot 10^{-14}\,\text{s} \qquad (4.68)$$

are reachable.

Bubble chambers have contributed significantly to the field of high energy hadron collisions and neutrino interactions [307].

4.12 Cloud chambers

The cloud chamber ('Wilson chamber') is one of the oldest detectors for track and ionization measurement [308]-[311]. In 1932 Anderson discovered the positron in cosmic rays by operating a cloud chamber in a strong magnetic field (2.5 Tesla). Five years later Anderson, together with Neddermeyer, discovered the muon again in a cosmic ray experiment with cloud chambers.

A cloud chamber is a container filled with a gas-vapor mixture (e.g., air-water vapor, argon-alcohol) at the vapor saturation pressure. If a charged particle traverses the cloud chamber it produces an ionization trail. The lifetime of positive ions produced in the ionization process in the chamber gas is relatively long ($\approx$ ms). Therefore after the passage of the particle a trigger signal, for example, can be derived from a coincidence of scintillation counters, which initiates a fast expansion of the chamber. By means of adiabatic expansion the temperature of the gas mixture is lowered and the vapor gets supersaturated. It condenses on nuclei, which are represented by the positive ions yielding droplets marking the particle trajectory. The track consisting of droplets is illuminated and photographed. A complete expansion cycle in a cloud chamber is shown in figure 4.69 [2].

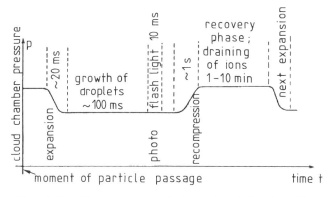

Fig. 4.69. Expansions cycle in a cloud chamber [2].

The characteristic times, which determine the length of a cycle, are the lifetime of nuclei produced by the ionization ($\approx$ 10 ms), the time required for the droplets to grow to a size where they can be photographed ($\approx$ 100 ms), and the time which has to pass after the recording of an event until the chamber is recycled to be ready for the next event. The last time can be very long since the sensitive volume of the chamber must be cleared of the slowly moving positive ions. In addition, the cloud chamber must be transformed into the initial state by recompression of the gas-vapor mixture.

In total, cycle times from one up to ten minutes can occur, limiting the application of this chamber type to rare events in the field of cosmic rays.

Figure 4.70 shows two electron cascades initiated by cosmic ray muons in a multiplate cloud chamber [312, 313].

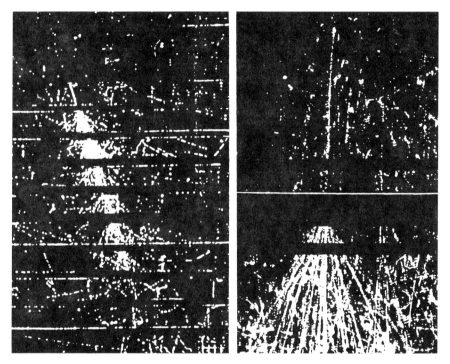

Fig. 4.70. Two electromagnetic cascades initiated by cosmic ray muons in a multiplate cloud chamber [312, 313].

A multiplate cloud chamber is essentially a sampling calorimeter with photographic readout (see chapter 7). The introduction of lead plates into a cloud chamber, which in this case was used in an extensive air shower experiment (see section 11.8), serves the purpose of obtaining an electron-hadron-muon separation by means of the different interaction behaviours of these elementary particles.

In contrast to the expansion cloud chamber, a diffusion cloud chamber is permanently sensitive. Figure 4.71 shows schematically the construction of a diffusion cloud chamber [2, 314, 315, 316]. The chamber is, like the expansion cloud chamber, filled with a gas-vapor mixture. A constant temperature gradient provides a region where the vapor is in a permanently supersaturated state. Charged particles entering this region produce a trail automatically without any additional trigger requirement. Zone widths (i.e., regions in which trails can form) with supersaturated vapor of 5 to 10 cm can be obtained. A clearing field removes the positive ions from the chamber.

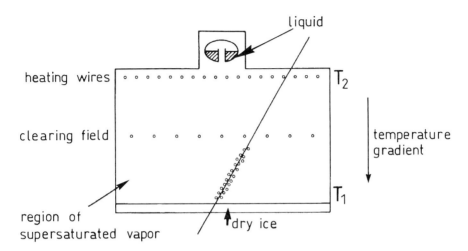

Fig. 4.71. Schematic representation of the construction of a diffusion cloud chamber [2].

The advantage of permanent sensitivity is obtained at the expense of small sensitive volumes. Since the chamber cannot be triggered, all events, even background events without interest, are recorded.

Because of the long repetition time for triggered cloud chambers and the disadvantage of photographic recording this detector type is rarely used nowadays.

4.13   Streamer chambers

In contrast to streamer *tubes*, which are a particular mode of operation for special cylindrical counters, streamer *chambers* are large volume detectors in which events are normally recorded photographically [4, 317]. In streamer chambers the volume between two planar electrodes is filled with a counting gas. After passage of a charged particle, a high voltage pulse of high amplitude, short rise time, and limited duration, is applied to the electrodes. Figure 4.72 sketches the principle of operation of such a detector.

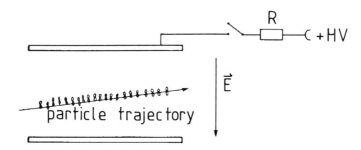

Fig. 4.72. Principle of construction of a streamer chamber.

In the most frequent mode of operation, particles are incident approximately perpendicular to the electric field into the chamber. Each individual ionization electron will start an avalanche in the homogeneous, very strong electric field in the direction of the electrodes. Since the electric field is time dependent (amplitude of the high voltage pulse, $\approx 500 \, \text{kV}$; rise and decay time, $\approx 1 \, \text{ns}$; pulse duration, several ns), the avalanche formation is interrupted after the decay time of the high voltage pulse. The high amplitude of the voltage pulse leads to large gas amplifications ($\approx 10^8$) like in streamer tubes; however, the streamers can only extend over a very small region of space. Naturally, in the course of the avalanche development large numbers of gas atoms are excited and subsequently de-excite leading to light emission. Luminous streamers are formed. Normally, these streamers are not photographed in the side view as sketched in figure 4.72, but through one electrode which can be made from a transparent wire mesh. In this projection the longish streamers appear as luminous dots characterizing the track of the charged particle (figure 4.73).

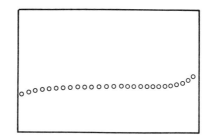

Fig. 4.73. Principle of streamer chamber photography.

The art of the operation of streamer chambers lies in the production of a high voltage signal with the required properties. The rise time must be extremely short (ns), otherwise the leading edge of the pulse would displace the ionization electrons from the original track with a field strength lower than the critical field required for the onset of gas amplification. A slow leading edge of the pulse would act as a clearing field resulting in a displacement of the particle track. Streamer development proceeds in very large electric fields ($\approx 30\,\mathrm{kV/cm}$). It must, however, be interrupted after a short time, so that the streamers will not grow too large or may even reach the electrodes. Streamers that are too large imply a poor spatial resolution. A suitable high voltage pulse can be obtained using a Marx generator connected by a suitable circuit (transmission line, Blumlein circuit, spark gaps) to the streamer chamber providing short signals of high amplitude [4, 318]. In Marx generators (spark discharge generators) a bank of n capacitors connected in parallel is charged up to a voltage U_0 via a chain of resistors. By triggering spark gaps the capacitors are connected in series so that a voltage of nU_0 is produced across the chain of capacitors. The purpose of the Blumlein line is to transmit the high voltage signal from the spark discharge generator to the streamer chamber in such a way that no losses, e.g., in form of reflections, occur.

For fast repetition rates the large number of electrons produced in the course of streamer formation poses a problem. It would take too long a time to remove these electrons from the chamber volume by means of a clearing field. Therefore, electronegative components are added to the counting gas to which the electrons are attached. Electronegative quenchers like SF_6 or SO_2 have proven to be good. These quenchers allow cycle times of several $100\,\mathrm{ms}$. The positive ions produced during streamer formation do not present a problem because they can never start new streamer discharges due to their low mobility.

Streamer chambers provide pictures of excellent quality. Also targets can be mounted in the chamber to obtain the interaction vertex in the sensitive volume of the detector. Figure 4.74 shows the interaction of a ^{32}S-ion with a stationary gold target at a beam energy of $200\,\mathrm{GeV}$ per nucleon or $6400\,\mathrm{GeV}$ per sulphur nucleus, respectively [319, 321]. To quote the energy per nucleus only makes sense if accelerator experiments are to be compared with investigations from cosmic rays in which the energy is determined by calorimetric methods, and the identity of the nucleus is frequently — at least at very high energies — unknown.

Figure 4.75 shows a stereo photograph of a proton-antiproton interaction in the $7.5\,\mathrm{m}$ long streamer chamber of the UA5-experiment at a center of mass energy of $\sqrt{s} = 900\,\mathrm{GeV}$ [320].

The spatial resolution in a streamer chamber is fundamentally limited by the diffusion of primary produced charge carriers. By means of a trick [323] the diffusion of electrons during the time delay between particle passage and high voltage pulse can to a large extent be suppressed. This is done by mixing a certain amount of oxygen with the streamer chamber gas. Since oxygen is strongly electronegative, oxygen molecules,

Fig. 4.74. Streamer chamber photograph of a heavy ion interaction (^{32}S) with a stationary gold target at a beam energy of 6400 GeV [319, 321].

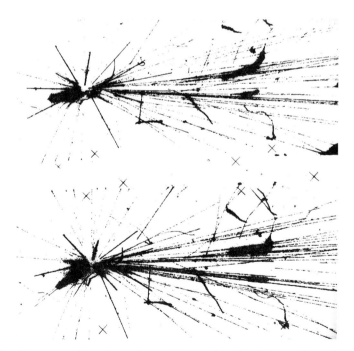

Fig. 4.75. Stereographic view of a proton-antiproton interaction at a center of mass energy of 900 GeV (UA5-experiment) [320, 322].

depending on the concentration, will attach the ionization electrons under formation of O_4^-. Attachment times of the order of 20 ns can easily be obtained. According to equation (1.108) the track width caused by diffusion is proportional to the root of the attachment time. The ionization track is now stored in form of almost immobile O_4^--ions. Within a trigger delay of about 3 μs they are displaced by less than 1 μm by the electric field from the original point of production. To make the latent track visible again, the electrons are detached from the O_4^--molecules by UV laser light via the photoelectric effect. The time delay between the laser pulse and the high voltage signal must be short compared to the attachment time (< 20 ns) so that the electrons can grow streamers. Figure 4.76 shows this diffusion suppression on the example of an electron track in a streamer chamber. The upper track is not diffusion suppressed. Within the trigger delay (1.2 μs) the ionization electrons have migrated considerably from the original point of production. The lower track has been diffusion suppressed by an addition of oxygen with a partial pressure of 275 mbar. In streamer chambers spatial resolutions of about 30 μm are obtained.

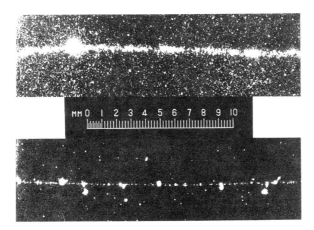

Fig. 4.76. Example of the diffusion suppression for an electron track in a streamer chamber. The upper track is not diffusion suppressed. It was recorded in a neon-helium mixture at 33 atmospheres. The lower track was diffusion suppressed by means of the addition of oxygen (≈ 275 mbar) [323].

In a different mode of operation of the streamer chamber, particles are incident within $\pm 30°$ with respect to the electric field into the detector. Exactly as mentioned before very short streamers will develop which now, however, merge into one another and form a plasma channel along the particle track. (This variant of the streamer chamber is also called track spark chamber [104, 324].) Since the high voltage pulse is very short, no spark between the electrodes develops. Consequently only a very low current is drawn from the electrodes [2, 4, 104].

Streamer chambers are well suited for the recording of complex events; they have, however, the disadvantage of a time consuming analysis.

4.14   Neon flash tube chamber

The neon flash tube chamber also is a discharge chamber [300], [325]-[328]. Neon or neon/helium-filled glass tubes, glass spheres ('Conversi tubes') or polypropylene extruded plastic tubes with rectangular cross section are placed between two metal electrodes (see figure 4.77).

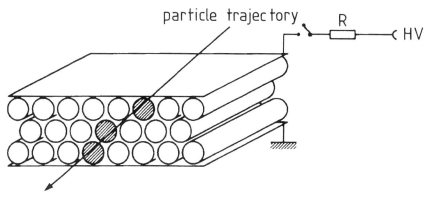

Fig. 4.77.   Working principle of a neon flash tube chamber.

After a charged particle has passed through the neon flash tube stack, a high voltage pulse is applied to the electrodes that initiates a gas discharge in those tubes, which have been passed by the particle. This gas discharge propagates along the total length of the tube and leads to a glow discharge in the whole tube. Typical tube lengths are around 2 m with diameters between 5 and 10 mm. The glow discharge can be intensified by after pulsing with high voltage so that the flash tubes can be photographed end on. But purely electronic recording with the help of pick-up electrodes at the faces of the neon tubes can also be applied ('Ayre-Thompson technique', see figure 4.78 [329, 330]). These pick-up electrodes supply large signals which can be directly processed without additional preamplifiers.

Depending on the tube diameter, spatial resolutions of several millimeters can be obtained. The memory time of this detector lies in the range around 20 μs; the dead time, however, is rather long at $30 - 1000$ ms. For

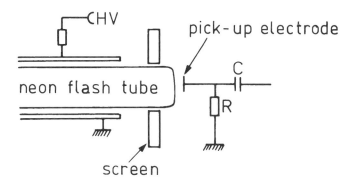

Fig. 4.78. Electronic readout of neon flash tubes [329].

reasons of geometry, caused by the tube walls, the efficiency of one layer of neon flash tubes is limited to $\approx 80\%$. To obtain three dimensional coordinates crossed layers of neon flash tubes are required.

Because of the relatively long dead time of this detector it is mainly used in cosmic ray experiments, the search for nucleon decay, or in neutrino experiments. Figure 4.79 shows a shower of parallel cosmic ray muons in a neon flash tube chamber [331, 332].

Fig. 4.79. Shower of parallel muons in a flash tube chamber [331, 332].

Fig. 4.80. Photographic record of an extensive air shower in a large matrix of Conversi tubes [333]. The extensive air shower is incident perpendicular to the flash tube layer. Each fired Conversi tube indicates the passage of a particle.

A variant of the neon flash tube is the spherical Conversi tube [326, 327]. These are spherical neon tubes of approximately 1 cm diameter. Figure 4.80 shows the photographic record of an extensive air shower (see section 11.8) which has traversed a horizontal layer of Conversi tubes [333]. The neon tubes are embedded in a matrix arranged between two electrodes, one of which is made as a transparent grid. After the passage of an extensive air shower through this arrangement a high voltage signal is applied to the chamber and causes those tubes which have been hit by shower particles to flash.

4.15 Spark chambers

Before multiwire proportional and drift chambers were invented, the spark chamber was the most commonly used track detector which could be triggered ([334]-[338], [300] and references therein).

In a spark chamber a number of parallel plates is mounted in a gas filled volume. Typically, a mixture of helium and neon is used as counting gas. Alternatingly, the plates are either grounded or connected to a high voltage supply (figure 4.81). The high voltage pulse is normally triggered, to every second electrode, by a coincidence between two scintillation counters placed above and below the spark chamber. The gas amplification is chosen in such a way that a spark discharge occurs at the point of the passage of the particle. This is obtained for gas amplifications between 10^8 and 10^9. For lower gas amplifications sparks will not develop, while for larger gas amplifications sparking at unwanted positions (e.g., at spacers which separate the plates) can occur. The discharge channel follows the electric field. Up to an angle of $30°$ the conducting plasma channel can, however, follow the particle trajectory [104] as in the track spark chamber (see section 4.13).

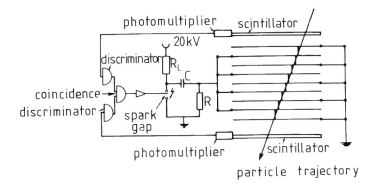

Fig. 4.81. Principle of operation of a multiplate spark chamber.

Between two discharges the produced ions are removed from the detector volume by means of a clearing field. If the time delay between the passage of the particle and the high voltage signal is less than the memory time of about 1 ms the efficiency of the spark chamber is close to 100 %. A clearing field, of course, removes also the primary ionization from the detector volume. For this reason the time delay between the passage of the particle and the application of the high voltage signal has to be chosen as short as possible to reach full efficiency. Also the rise time of the high voltage pulse must be short because otherwise the leading edge acts as a clearing field before the critical field strength for spark formation is reached.

Figure 4.82 shows the track of a cosmic ray muon in a multiplate spark chamber [2, 339]. Figure 4.83 shows a stopping particle in a multiplate spark chamber. Here one can clearly see the increase of ionization towards the end of the range of the particle indicated by the increasing spark brightness (here seen via the increased spark width) [340].

Fig. 4.82.   Track of a cosmic ray muon in a multiplate spark chamber [339].

Fig. 4.83.   Stopping particle in a multiplate spark chamber [340].

If several particles penetrate the chamber simultaneously, the probability that all particles will form a spark trail decreases drastically with increasing number of particles. This is caused by the fact that the first spark discharges the charging capacitor to a large extent so that less voltage or energy, respectively, is available for the formation of further sparks. This problem can be solved by limiting the current drawn by a spark. In current-limited spark chambers glass plates are mounted in front of the metallic electrodes which prevent a high current spark discharge. In such glass spark chambers a high multitrack efficiency can be obtained [341, 342]. Figure 4.84 shows a photographic record of an electron cascade in such a current-limited multiplate spark chamber [343, 344, 345]. In this case it is clearly visible that the sparks follow the tracks of the shower particles up to a certain angle ($\lesssim 30°$). One part of such a cascade (figure 4.85) with four particle tracks clearly demonstrates the high multitrack efficiency. The plasma discharges ('sparks') in a glass spark chamber differ from streamer discharges in such a way that they — in contrast to the usually very short streamers — connect both electrodes like in a track spark chamber (see section 4.13). The apparent 'feet' of the discharges (see figure 4.85) are caused by reflection and scattering of the light on the glass plates.

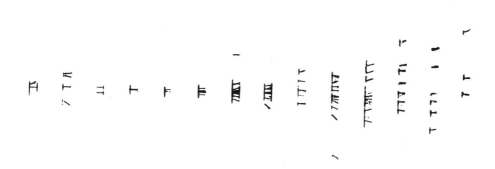

Fig. 4.84. Photograph of an electron cascade in a multiplate glass spark chamber [343, 344, 345].

Apart from the photographic recording in spark chambers, which has to be made stereoscopically to allow three dimensional event reconstruction, a purely electronic readout is also possible.

If the electrodes are made from layers of wires, the track coordinate can be obtained like in the multiwire proportional chamber by identifying the discharged wire. This method would require a large number of wires to obtain a high spatial resolution. On the other hand the track reco儿-

Fig. 4.85.   Four-track event in a stack of three glass spark chambers [345].

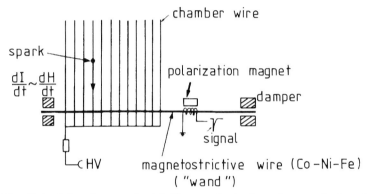

Fig. 4.86.   Principle of magnetostrictive readout of a wire spark chamber [104, 346].

struction can be simplified with the help of a magnetostrictive readout. The principle of operation of this method is shown in figure 4.86.

The spark discharge represents a time dependent current dI/dt. The current signal propagates along the chamber wire and reaches a magnetostrictive delay line stretched perpendicular to the chamber wires. This magnetostrictive delay line is positioned directly on the chamber wires without having ohmic contact to them. The current signal, along with its associated time dependent magnetic field $d\vec{H}/dt$, produces in the magnetostrictive delay line a magnetostriction, i.e., a local variation of the length, which propagates in time and space with its characteristic velocity of sound. In a pick-up coil the mechanical signal of magnetostriction is converted back into a time dependent magnetic field signal $d\vec{H}/dt$ leading to a voltage pulse of

$$U = -\frac{d\phi}{dt} = -\mu_0 \cdot \frac{d}{dt} \int \vec{H} \cdot d\vec{A} \qquad (4.69)$$

(ϕ is a magnetic flux through the area $\vec{A}$ and μ_0 the permeability of free space). The measurement of the propagation time of the sound wave on the magnetostrictive delay line can be used to identify the number and hence the spatial coordinate of the discharged wire. Typical sound velocities of $\approx 5\,\mathrm{km/s}$ lead to spatial resolutions of the order of $200\,\mu\mathrm{m}$ [104]. The sound velocity of the signal depends on the modulus of elasticity E and the density ϱ of the material used for the magnetostrictive delay line according to

$$v = \sqrt{E/\varrho}\,. \qquad (4.70)$$

For typical alloys (e.g., Fe − Ni or Cu − Fe) the modulus of elasticity E takes values around $2 \cdot 10^5\,\mathrm{N/mm}^2$.

A clearing field which is necessary to remove the positive ions from the detector volume causes dead times of several milliseconds.

The magnetostrictive delay lines are mostly made from cobalt-nickel-iron alloys. Because of their very sensitive and critical positioning with respect to the chamber wires, the magnetostrictive delay lines are also called 'wands'.

A somewhat older method of identifying discharged wires in wire spark chambers uses ferrite cores to localize the discharged wire. In this method each chamber wire runs through a small ferrite core (figure 4.87) [104]. The ferrite core is in a well-defined state. A discharging spark chamber wire causes the ferrite core to flip. The state of ferrite cores is recorded by a readout wire. After reading out the event the flipped ferrite cores are reset into the initial state by a reset wire.

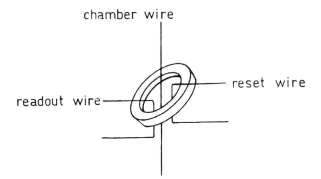

Fig. 4.87. Principle of ferrite core readout of a wire spark chamber [104].

4.16   Nuclear emulsions

Tracks of charged particles in nuclear emulsions can be recorded by the photographic method [347]-[351]. Nuclear emulsions consist of fine-grained silver halide crystals (AgBr and AgCl), which are embedded in a gelatine substrate. A charged particle produces a latent image in the emulsion (figure 4.88). Due to the free charge carriers liberated in the ionization process some halide molecules are reduced to metallic silver in the emulsion.

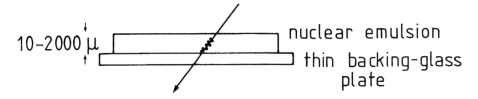

10-2000 μ   nuclear emulsion
thin backing-glass plate

Fig. 4.88.   Measurement principle of a nuclear emulsion.

In the subsequent development process the silver halide crystals are chemically reduced. This affects preferentially those microcrystals (nuclei) which are already disturbed and partly reduced. These are transformed into elemental silver. The process of fixation dissolves the remaining silver halide and removes it. Thereby the charge image, which has been transformed into elemental silver particles, remains stable.

The evaluation of the emulsion is usually done under a microscope by eye, but it can also be performed by using a Charged Coupled Device (CCD)-camera and a semi-automatic pattern recognition device. Fully automated emulsion analysis systems have also been developed [352].

The sensitivity of the nuclear emulsion must be high enough so that the energy loss of minimum ionizing particles is sufficient to produce individual silver halide microcrystals along the track of a particle. Commercially available photoemulsions do not have this property. Furthermore the silver grains which form the track and also the silver halide microcrystals must be sufficiently small to enable a high spatial resolution [2, 63]. The requirements of high sensitivity and low grain size are in conflict and, therefore, demand a compromise. In most nuclear emulsions the silver grains have a size of 0.1 to $0.2\,\mu$m and so are much smaller than in commercial films ($1-10\,\mu$m). The mass fraction of the silver halide (mostly AgBr) in the emulsion amounts to approximately 80 %.

Because of the high density of the emulsion ($\varrho = 3.8\,\mathrm{g/cm^3}$) and the related short radiation length ($X_0 = 2.9\,\mathrm{cm}$), stacks of nuclear emulsions are perfectly suited to detect electromagnetic cascades. On the other hand, hadron cascades hardly develop in such stacks because of the much larger absorption length ($\lambda_a = 35$ cm).

The efficiency of emulsions for single or multiple particle passages is close to 100 %. Emulsions are permanently sensitive but they cannot be triggered. They have been and still are in use in many cosmic ray experiments [350]. They are, however, also suited for accelerator experiments as vertex detectors with high spatial resolution ($\sigma_x \approx 2\,\mu$m) for the investigation of decays of short-lived particles.

Figure 4.89 [27] shows an early record of the interaction of a neutral cosmic ray particle which produces a 'star' consisting of secondary particles in an emulsion. The specific ionization of a particle can be determined from the darkness of the track. For low energy particles the observed multiple scattering also allows the momentum to be determined and thereby an identification of the particle (see equation (1.47)).

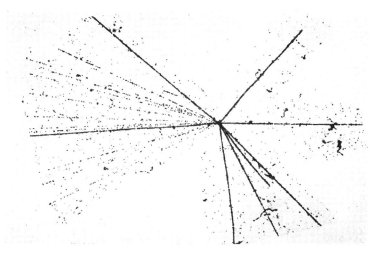

Fig. 4.89. A neutral cosmic ray particle produces a 'star' with eight strongly and eleven weakly ionizing particles [27].

Figure 4.90 shows α-particles emanating from a grain of radium salt which was placed on a nuclear emulsion. The central region of darkness has a diameter of about 100 μm ([351], [353], reprinted in [301]).

Figure 4.91 exhibits the interaction of a 6.4 TeV (= 6400 GeV) sulphur ion with a nucleus in a photographic emulsion. Apart from weakly ionizing particles, the heavily ionizing tracks of fission fragments can also be recognized [354]. The large multiplicity of projectile and target fragments is also clearly visible in the interaction of a 228.5 GeV-uranium nucleus in a nuclear emulsion (figure 4.92) [355].

Fig. 4.90.   Tracks of α-particles in a nuclear emulsion, which are emanating from a grain of radium salt ([351], [353], reprinted in [301]).

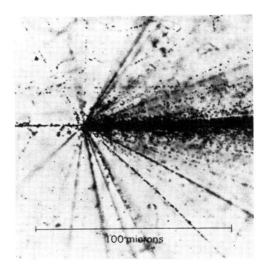

Fig. 4.91.   Interaction of a 6.4 TeV-sulphur ion with a nucleus in a photographic emulsion [354].

Figure 4.93 shows the interaction of a cosmic ray carbon nucleus with a proton in a nuclear emulsion ([351], reprinted in [301]). The carbon nucleus fragments into three α-particles which escape into the forward direction. The target proton receives a transverse momentum and is scattered to the left.

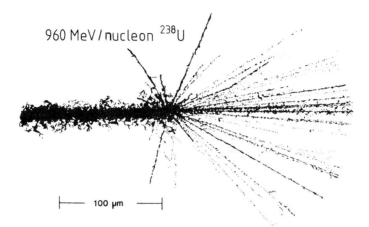

960 MeV/nucleon ^{238}U

├─── 100 μm ───┤

Fig. 4.92. Interaction of an uranium nucleus of energy 228.5 GeV in a nuclear emulsion [355].

Finally, figure 4.94 shows tracks of nuclei of different charge in an emulsion ([351], reprinted in [301]). The ionization density increases with the square of the nuclear charge.

Occasionally, nuclear emulsions are used in accelerator experiments as vertex detectors. Other detectors in the experiment (e.g., drift chambers) provide tracking with moderate accuracy. A back extrapolation of these tracks into the emulsion allows the identification of the interaction vertices and they may be correlated event by event with tracks measured in the rest of the experiment. The knowledge of this correlation simplifies the evaluation of the emulsion because only the vertex region determined from the other detectors must be analyzed. Figure 4.95 shows the layout of such an experiment.

In these hybrid experiments, however, one must consider that events which are electronically recorded in the drift chambers must be associated correctly with tracks in the nuclear emulsions which are integrated over a certain period of time.

Among others, the emulsion technique has contributed significantly in past decades to the fields of cosmic ray physics, high energy heavy ion collisions, hypernuclear physics, neutrino oscillations and to the study of charm and bottom particles [356].

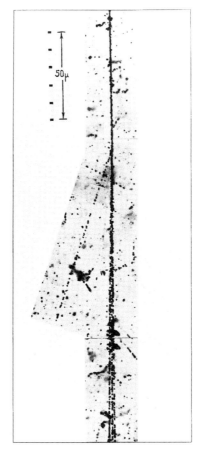

Fig. 4.93. A cosmic ray carbon nucleus collides with a proton in a nuclear emulsion. The carbon nucleus disintegrates into three α-particles. The recoil proton is scattered to the left ([351], reprinted in [301]).

4.17 Silver halide crystals

The disadvantage of nuclear emulsions is that the sensitive volume of the detector is usually very small. The production of large area AgCl-crystals has recently been made possible. This has allowed the construction of another passive detector similar to emulsions. Charged particles produce along their track in a AgCl-crystal Ag^+-ions and electrons. The mobility of Ag^+-ions in the lattice is very limited. They usually occupy positions between regular lattice atoms thereby forming a lattice defect. Free electrons from the conduction band reduce the Ag^+-ions to metallic silver. These Ag-atoms attach further Ag^+-ions: the formation of silver clusters starts. To stabilize these silver clusters the crystal must be illuminated

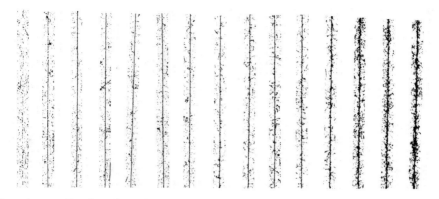

Fig. 4.94. Tracks of nuclei in nuclear emulsions. The ionization density of tracks increases with a square of the nuclear charge ([351], reprinted in [301]). The 'fuzziness' of the tracks is caused by the emission of δ-rays.

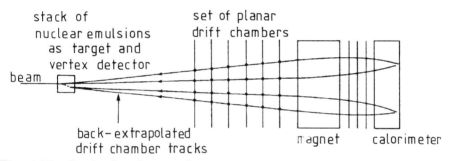

Fig. 4.95. Stack of nuclear emulsions as vertex detector in an accelerator experiment. Primary and possibly secondary vertices are approximated by an extrapolation of tracks from the drift chambers. The precise determination of these vertices is done by a careful analysis of the emulsion data.

during or shortly after the passage of the particle providing the free electrons required for the reduction of the Ag^+-ions (storage or conservation of particle tracks). This is frequently done by using light with a wavelength of around 600 nm [357]. If this illumination is not done during data taking the tracks will fade away. In principle, this illumination can also be triggered by an external signal which would allow the separation of interesting events from background. To this extent, the event recording in an AgCl-crystal — in contrast to nuclear emulsions or plastic detectors (see section 4.21) — can be triggered.

Even in the unirradiated state there are a certain number of Ag-ions occupying places between the regular lattice positions. A small admixture of cadmium chloride serves to reduce this unwanted silver concentration. This reduces the formation of background silver nuclei on lattice defects, thereby decreasing the 'noise' in AgCl-crystals.

To allow the Ag-clusters to grow to microscopically visible size, the AgCl-crystal is irradiated by short wavelength light during the development process. This provides further free electrons in the conduction band which in turn will help to reduce the Ag^+-ions as they attach themselves to the already existing clusters.

This process of track amplification ('decoration') produces a stable track which can then be evaluated under a microscope.

Silver chloride detectors show — just like plastic detectors — a certain threshold effect. The energy loss of relativistic protons is too small to produce tracks which can be developed in the crystal. The AgCl-detector, however, is well suited to measure tracks of heavy nuclei ($Z \geq 3$).

The tedious evaluation of nuclear tracks under a microscope can be replaced by automatic pattern recognition methods similar to those which are in use for nuclear emulsions and plastic detectors [358, 359, 360]. The spatial resolution which can be achieved in AgCl-crystals is comparable to that of nuclear emulsions.

4.18   X-ray films

Emulsion chambers, i.e., stacks of nuclear emulsions when used in cosmic ray experiments, are frequently equipped with additional large area X-ray films [361, 362, 363]. These industrial X-ray films allow the detection of high energy electromagnetic cascades (see chapter 7), and the determination of the energy of electrons or photons initiating these cascades, by photometric methods. This is done by constructing a stack of X-ray films alternating with thin lead sheets. The longitudinal and lateral development of electromagnetic cascades can be inferred from the structure of the darkness in the X-ray films.

X-ray films used in cosmic ray experiments have low grain size and are mainly used for the detection of photons and electrons in the TeV-range. Hadronic cascades are harder to detect in stacks with X-ray films. They can, however, be recorded via the π^0-fraction in the hadron shower $(\pi^0 \to \gamma\gamma)$. This is related to the fact that photons and electrons initiate narrowly collimated cascades producing dark spots on the X-ray film, whereas hadronic cascades, because of the relatively large transverse momenta of secondary particles, spread out over a larger area on the film thus not exceeding the threshold required for a blackening of the film.

Saturation effects in the region of the maximum of shower development (central blackening) cause the relation between the deposited energy E and the photometrically measured blackening D not to be linear [364]. For typical X-ray films which are used in the TeV-range one gets

$$D \propto E^{0.85} \; . \tag{4.71}$$

The radial distribution of the blackening allows the determination of the point of particle passage with relatively high precision.

4.19   Thermoluminescence detectors

Thermoluminescence detectors are used in the field of radiation protection [365, 366, 367] and also in cosmic ray experiments.

Particle detection in thermoluminescence detectors is based on the fact that in certain crystals ionizing radiation causes electrons to be transferred from the valence band to the conduction band where they may occupy stable energy states [65]. In the field of radiation protection media, which retain the dose information, such as manganese- or titanium-activated calcium fluoride (CaF_2) or lithium fluoride (LiF) crystals are used. The stored energy caused by irradiating the crystal is proportional to the absorbed dose. Heating the thermoluminescence dosimeter to a temperature between 200 and 400 °C can liberate this energy by emission of photons. The number of produced photons is proportional to the absorbed energy dose.

In cosmic ray experiments thermoluminescence films (similar to X-ray films) are used for the measurement of high energy electromagnetic cascades. A thermoluminescence detector is made by coating a glass or metal surface with a layer of thermoluminescent powder. The smaller the grain size of microcrystals on the film the better the spatial resolution that can be reached. The ionizing particles in the electron cascade produce stable thermoluminescence centers. The determination of where the energy is deposited on the film can be achieved by scanning the film with an infrared laser. During the process of scanning the intensity of emitted photons must be measured with a photomultiplier. If the spatial resolution is not limited by the radial extension of the laser spot, resolutions of the order of a few micrometers can be obtained [368].

Apart from doped calcium fluoride or lithium fluoride crystals and storage phosphors commonly used in the field of radiation protection, cosmic ray experiments utilize mainly $BaSO_4$, Mg_2SiO_4 and $CaSO_4$ as thermoluminescent agents. While thermoluminescence dosimeters measure the integrated absorbed energy dose, in cosmic ray experiments the measurement of individual events is necessary.

In such experiments thermoluminescence films are stacked similar to

X-ray films or emulsions alternatingly with lead absorber sheets. The hadrons, photons or electrons to be detected initiate hadronic or electromagnetic cascades in the thermoluminescence calorimeter. Neutral pions produced in hadronic cascades (see section 7.3) decay relatively quickly (in $\approx 10^{-16}$ s) into two photons thereby initiating electromagnetic subcascades (see section 7.2).

In contrast to hadronic cascades with a relatively large lateral width, the energy in electromagnetic cascades is deposited in a relatively small region thereby enabling a recording of these showers. That is why electromagnetic cascades are directly measured in such a detector type while hadronic cascades are detected via their π^0 content. Thermoluminescence detectors exhibit an energy threshold for the detection of particles. This threshold is approximately $1 \, \text{TeV}$ per event in europium-doped $BaSO_4$-films [368].

4.20   Radiophotoluminescence detectors

Silver-activated phosphate glass, after having been exposed to ionizing radiation, emits fluorescence radiation in a certain frequency range if irradiated by ultraviolet light. The intensity of the fluorescence radiation is a measure for the energy deposition by the ionizing radiation. The Ag^+-ions produced by ionizing particles in the glass represent stable photoluminescence centers. Reading out the energy deposition with ultraviolet light does not erase the information of the energy loss in the detector [65]. Yokota glass is mostly used in these phosphate glass detectors. It consists of $45\,\%$ $AlPO_3$, $45\,\%$ $LiPO_3$, $7.3\,\%$ $AgPO_3$ and $2.7\,\%$ B_2O_3 and has a typical density of $2.6 \, \text{g/cm}^3$ for a silver mass fraction of $3.7\,\%$. Such phosphate glass detectors are mainly used in the field of radiation protection for dosimetric measurements.

By scanning a two-dimensional radiophotoluminescence sheet with a UV-laser it is possible to determine the spatial dependence of the energy deposit by measuring the position-dependent fluorescence light yield. If individual events are recorded, a threshold energy of the order of $1 \, \text{TeV}$ is required just as in thermoluminescence detectors. The spatial resolution that can be obtained is limited also in this case by the resolution of the scanning system.

4.21   Plastic detectors

Particles of high electric charge destroy the local structure in a solid along their tracks. This local destruction can be intensified by etching and thereby made visible. Solids such as inorganic crystals, glasses, plastics,

minerals or even metals can be used for this purpose [369, 370, 371].
The damaged parts of the material react with the etching agent more
intensively than the undamaged material and characteristic etch cones
will be formed. Figure 4.96 shows the time development of etch cones in
a plastic foil.

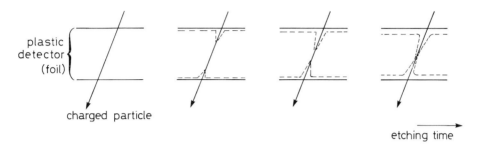

Fig. 4.96.   Time dependent development of etch cones in a plastic detector.

If the etching process is not interrupted, the etch cones starting from
the surface of the plastic will merge and form a hole at the point of the
particle track. The etching procedure will also remove some part of the
surface material.

Figure 4.97 [372] shows a view into an etch cone which was caused
by an iron nucleus in a 1 mm thick polycarbonate foil (Lexan). For inclined incidence the etch cones exhibit an elliptical form, and a part of
the lower lying track will be visible as a diffuse edge. Tracks of fission
fragments are shown in figure 4.98 [372], also in Lexan. The Lexan foils
had been etched in NaOH at 70 °C. The width of an individual track is
approximately $3\,\mu$m.

The determination of the energy of heavy ions is frequently done in
stacks containing a large number of foils (see figure 4.99). The radiation
damage of the material — just as the energy loss of charged particles —
is proportional to the square of their charge, and depends also on the
velocity of the particles.

Plastic detectors show a threshold effect: the minimum radiation damage caused by protons and α-particles is frequently insufficient to produce
etchable tracks. The detection and measurement of heavy ions, e.g., in
primary cosmic rays $(Z \geq 3)$, will consequently not be disturbed by a
high background of protons and α-particles. The size of the etch cones
(for a fixed etching time) is a measure of the energy loss of the particles.
It allows, therefore, if the velocity of the particles is known, a determination of the charge of the nuclei. A stack of plastic detectors, flown in
a balloon in a residual atmosphere of several grams per square centime-

Fig. 4.97. Etch cone produced by an iron nucleus in a plastic detector [372]. The track width amounts to 193 μm.

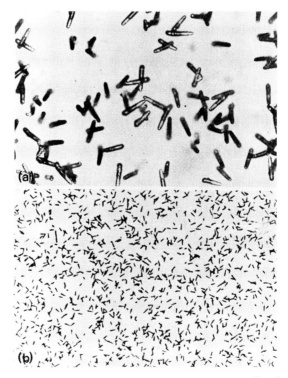

Fig. 4.98. Tracks of fission fragments in a plastic detector [372]. a) 180-fold, b) 40-fold magnification.

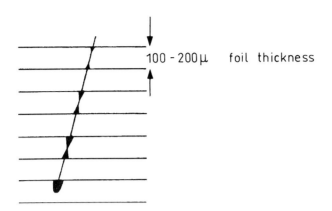

Fig. 4.99. Schematic representation of a stack consisting of plastic detectors.

ter, thus permits a determination of the elemental abundance in primary cosmic rays.

Plastic detectors are also utilized in the search for magnetic monopoles which, according to theory, should cause strong ionization. Such experiments can also be performed on proton storage rings because the high background of singly charged particles does not impair the search for monopoles due to the threshold behaviour of the plastic material.

In a similar way to plastic detectors, minerals also conserve a local radiation damage over a long period of time. This leads to the possibility of dating uranium-containing minerals by counting the number of spontaneous fission events. If the minerals are time calibrated in this way the number of tracks initiated by cosmic radiation in these minerals indicates that the intensity of cosmic rays has not varied significantly ($\leq 10\,\%$) over the past 10^6 years.

The evaluation of plastic detectors under the microscope is very tiresome. The information on particle tracks in a plastic sheet can, however, also be digitized by means of a CCD-camera looking through a microscope onto the foil. The digitized event is subsequently processed with a program for automatic pattern reconstruction [373]-[376].

A nuclear detector with super-high spatial resolution is provided, for example, by a small chip of MoS_2. High energy nuclei penetrating the MoS_2 sample produce craters on its surface due to local radiation damage. Analyzing these craters by scanning tunneling microscopy allows spatial resolutions of the order of $10\,\text{Å}$ and two-track resolutions of $30-50\,\text{Å}$ [377].

4.22 Comparison of detectors for track and ionization measurement

Depending on the application, different detectors for track and ionization measurement should be used. In the realm of high energy physics fast detectors like multiwire proportional and drift chambers are employed. Currently there is a trend to shorter and shorter repetition times. Meanwhile one aims to process events at a rate of 10^9 per second at future proposed proton-proton storage rings. Whether or not gas detectors can be considered for these short cycle times — 1 ns corresponds to a drift path of $50\,\mu$m — will be seen in the future. The microstrip gas chamber (see section 4.6) is certainly a possible candidate for these applications.

For experiments at electron-positron storage rings with low event rates, cylindrical drift chambers — in particular time projection chambers — are ideally suited.

Detectors using optical recording like cloud chambers, bubble chambers and spark chambers are very rarely used in real experiments nowadays. Even if the events are automatically evaluated, the repetition times of these detectors are too long. They provide, however, a spectacular impression of particles and their interactions in demonstration experiments.

Spark chambers and, in particular, streamer chambers are a source of unwanted noise signals which can have a detrimental influence on other detectors in an experiment, since they use high voltage pulses of very high amplitude and very short duration. Passive detectors like nuclear emulsion and plastic detectors, which require at least partial manual analysis, are only used for special applications. However, because of their solidity and robustness they have large advantages in balloon and space-borne experiments.

Still, it would be premature to consider certain detectors as completely superseded, because there are always improvements which may open a new field of application for older detectors. The holographic readout of bubble chambers is an example.

In table 4.2 the detectors for track and ionization measurement are listed along with some of their characteristic features [1, 104]. Typical values of the spatial resolution, dead time, sensitive time, and readout time are also given. For individual detectors entirely different values for the characteristic properties may occur. The 'sensitive times' for multiwire proportional, microstrip and drift chambers refer to an electronic gate. The listed readout times are mostly due to the electronics and do not depend on the working principle of the detector.

Table 4.2. *Comparison of characteristic properties of some detectors for track and ionization measurement [1, 104]*

track detector	track resolution [μm]	dead time [ms]	sensitive time [ns]	readout time [μs]	comments
multiwire proportional chamber	200	$< 10^{-5}$	50	10	high time resolution
micro-strip gas chamber	30	$< 10^{-5}$	20	5	high-rate capability radiation hard
drift chamber	100	$< 10^{-5}$	500	10	high spatial resolution, economic operation
bubble chamber	20	100	10^6	10^4	cannot be triggered, analysis of complex events
streamer chamber	30	10	10^3	10^4	can be triggered, multiple track resolution
flash chamber	1000	10	10^3	10^3	low cost
spark chamber	200	5	10^3	10^4	simple construction
cloud chamber	300	10^5	10^7	10^6	simple construction, shows many details
nuclear emulsion	3	0	∞	10^9	high spatial resolution
plastic detector	5	0	∞	10^9	low cost, high spatial resolution
scintillating fiber systems (see chapter 5)	35	$< 10^{-5}$	20	1	high rate capability
silicon microstrip detectors (see chapter 7)	10	$< 10^{-5}$	50	1	high spatial resolution

5

Time measurement

The main detector to be described in this chapter is the scintillation counter with its readout via light sensitive systems. The application of scintillation counters is manifold, e.g., as a trigger counter to trigger complex detector systems or as a time measurement device. High resolution timing can also be performed with planar spark counters or resistive plate chambers. First, the main readout instrument for scintillators will be described; this is the photomultiplier or secondary electron multiplier.

5.1   Photomultiplier

The most commonly used instrument for the measurement of fast light signals is the photomultiplier. Light in the visible or ultraviolet range — e.g., from a scintillation counter — liberates electrons from an alkali-metal photocathode via the photoelectric effect. For most photomultipliers a negative high voltage is applied to the photocathode. The photoelectrons are focussed by an electric field onto the first dynode, which is part of the multiplication system. The anode is normally at ground potential. The voltage between the photocathode and anode is subdivided by a chain of resistors. This voltage divider supplies further dynodes between the photocathode and anode so that the applied negative high voltage is subdivided linearly (see figure 5.1).

An important parameter of a photomultiplier is its quantum efficiency, i.e., the mean number of photoelectrons produced per incident photon. For bialkali cathodes (Cs–K with Sb) the quantum efficiency reaches values around 25 % for a wavelength of about 400 nm. Figure 5.2 shows the quantum efficiency for bialkali cathodes as a function of the wavelength [378]. The quantum efficiency decreases for short wavelengths because the transparency of the photomultiplier window decreases with increasing frequency, i.e., shorter wavelength. The range of efficiency can only be

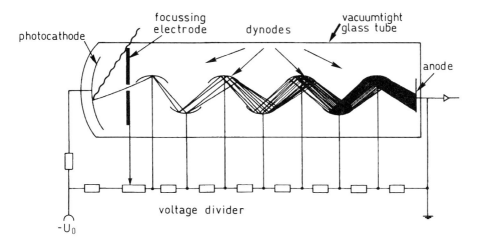

Fig. 5.1. Working principle of a photomultiplier. The electrode system is mounted in an evacuated glass tube. The photomultiplier is usually shielded by a mu-metal cylinder made from high permeability material to shield against stray magnetic fields (e.g., the magnetic field of the earth).

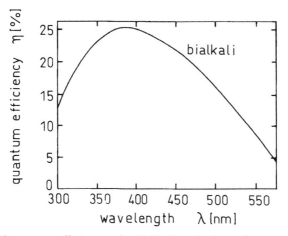

Fig. 5.2. Quantum efficiency of a bialkali cathode as function of the wavelength [378].

extended to higher frequencies by using UV-transparent quartz windows. The dynodes must have a high secondary electron emission coefficient (BeO or Mg–O–Cs). For electron energies from around 100 up to 200 eV, which correspond to typical acceleration voltages between two dynodes, approximately three to five secondary electrons are emitted [1].

For an n-step photomultiplier (i.e., with $n-1$ dynodes) with a secondary emission coefficient p, the current amplification is given by

$$A = p^{n-1} \,. \tag{5.1}$$

For typical values of $p = 4$ and $n = 14$ one obtains $A = 4^{13} = 7 \cdot 10^7$.

The charge arriving at the anode

$$Q = eA = 1.1 \cdot 10^{-11}\,\mathrm{C} \tag{5.2}$$

is collected within approximately 5 ns leading to an anode current of

$$i = \frac{\mathrm{d}Q}{\mathrm{d}t} = 2.2\,\mathrm{mA} \,. \tag{5.3}$$

If the photomultiplier is terminated with a $50\,\Omega$ resistor a voltage signal of

$$\Delta U = R \cdot \frac{\mathrm{d}Q}{\mathrm{d}t} = 110\,\mathrm{mV} \tag{5.4}$$

is obtained. The rise time of the photomultiplier signal is typically 2 ns. This time has to be distinguished from the time required for the electrons to traverse the photomultiplier. This transit time depends on the type of phototube and is typically ≈ 40 ns.

The time jitter in the arrival time of electrons at the anode poses a problem for reaching a high time resolution. This time jitter originates, e.g., from the variation in the velocity of the photoelectrons. On the other hand — depending on the type of phototube — the path length from the production point of the photoelectrons to the first dynode can be subject to large fluctuations.

The time jitter caused by different velocities of photoelectrons can easily be calculated [1]. Let us assume that two electrons start from the photocathode in the direction of the first dynode. One of these electrons starts from rest (kinetic energy zero), the other with a velocity v in the direction of the first dynode (kinetic energy E_k). Both electrons are accelerated by the focussing electric field E. The electron initially at rest traverses the path s in a time which can be calculated from

$$s = \frac{1}{2}\ddot{x} \cdot t_1^2 = \frac{1}{2}\frac{eE}{m} \cdot t_1^2 \,. \tag{5.5}$$

The electron with the initial kinetic energy E_k reaches the first dynode somewhat more quickly,

$$s = \frac{1}{2}\frac{eE}{m}t_2^2 + v \cdot t_2 \,. \tag{5.6}$$

Since the velocity is given by

$$v = \sqrt{2E_k/m} \tag{5.7}$$

it follows that

$$\frac{1}{2}\frac{eE}{m}t_1^2 = \frac{1}{2}\frac{eE}{m}t_2^2 + \sqrt{\frac{2E_k}{m}}\,t_2\,, \tag{5.8}$$

$$t_1^2 - t_2^2 = (t_1 + t_2)(t_1 - t_2) = \frac{\sqrt{2E_k/m}}{\frac{1}{2}eE/m}\cdot t_2\,. \tag{5.9}$$

Approximating $t_1 + t_2 \approx 2t$ and $t_1 - t_2 = \delta t$, one obtains for the arrival time difference

$$\delta t = \frac{\sqrt{2mE_k}}{eE}\,. \tag{5.10}$$

For $E_k = 1\,\mathrm{eV}$ and $E = 200\,\mathrm{V/cm}$ a time jitter of $\delta t = 0.17\,\mathrm{ns}$ is obtained.

The arrival time difference based on path length variations depends strongly on the size and shape of the photocathode. For an XP2041 phototube with planar photocathode and a cathode diameter of 100 mm this time difference amounts to 1 ns [379]. For large photomultipliers the achievable time resolution is limited essentially by path length differences. The photomultipliers with a 20 inch cathode diameter used in the KamiokaNDE nucleon decay and neutrino experiment [380, 381] show path length differences of up to 5 ns. For this phototube the distance between photocathode and first dynode is so large that the earth's magnetic field has to be well shielded so that the photoelectrons can reach the first dynode. Figure 5.3 shows a photograph of an eight-inch photomultiplier [382].

The typical spectral sensitivity of a bialkali photocathode with boron silicate glass window is shown in figure 5.4 [379]. For wavelengths around 450 nm maximum values of about 85 mA/W are reached. In this case the spectral sensitivity also is reduced for photons of short wavelength because of the decreasing transparency of the phototube window for UV-radiation. Using cathode windows from quartz one can detect photons down to 200 nm.

The path length fluctuations can be significantly reduced in microchannel photomultipliers (channel plates). The principle of such channel plates is shown in figure 5.5 [383]. A voltage of about 1000 V is applied to a thin glass tube (diameter 10 to 50 μm, length 5 to 10 mm) which is coated on the inside with a resistive layer. Incident photons produce photoelectrons on a photocathode or on the inner wall of the microchannel. These are, like in the normal phototube, multiplied at the — in this case — continuous dynode. Channel plates contain a large number (10^4 to 10^7) of such channels which are implemented as holes in a lead glass plate. Figure 5.6

Fig. 5.3. Photograph of an eight-inch photomultiplier (type R 4558) [382].

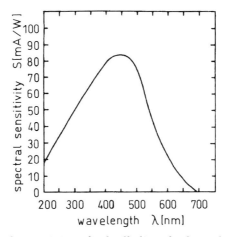

Fig. 5.4. Spectral sensitivity of a bialkali cathode with boron silicate window [379].

shows a microphotographic record of such channels with a diameter of 12.5 μm [384]. Because of the short mean path lengths of electrons in the longitudinal electric field, path length fluctuations are drastically reduced compared to a normal photomultiplier. Transit time differences below 100 ps for multiplication factors between 10^5 and 10^6 are obtained.

While normal photomultipliers practically cannot be operated in magnetic fields (or if so only heavily shielded), the effect of magnetic fields on

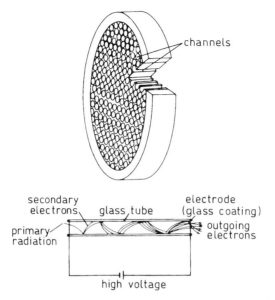

Fig. 5.5.   Working principle of a channel plate [383].

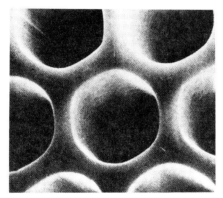

Fig. 5.6.   Microphotograph of microchannels [384].

channel plates is comparatively small. This is related to the fact that in channel plates the distance between cathode and anode is much shorter. There are, however, recent developments of conventional photomultipliers with transparent wire-mesh dynodes, which can withstand moderate magnetic fields.

A problem with channel plates is the flux of positive ions produced by electron collisions with the residual gas in the channel plate that migrate in the direction of the photocathode. The lifetime of channel plates would be extremely short if the positive ions were not prevented from reaching the photocathode. By use of extremely thin aluminium windows of $\approx 7\,\text{nm}$ thickness (transparent for electrons) mounted between photo-cathode and channel plate, the positive ions are absorbed. In this way, the photocathode is shielded against the ion bombardment.

Figure 5.7 shows the working principle of a three-step microchannel-plate photomultiplier.

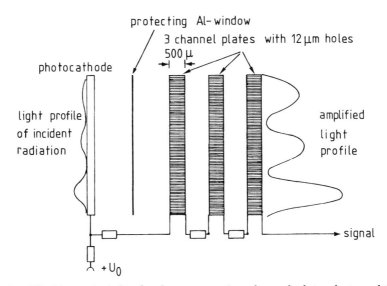

Fig. 5.7. Working principle of a three-step microchannel-plate photomultiplier.

In high altitude or satellite experiments also 'open' channel plates without photocathode are used. Mostly single individual channels are employed which are bent in order to increase the amplification and to prevent a purely accidental straight passage of radiation through the channel. The diameter of these single channel photomultipliers lies in the millimeter range. These single channel multipliers can also be used for incident low energy charged particles. Amplifications around 10^8 can be achieved.

Figure 5.8 [379] shows the layout of a single channel electron multiplier.

Image intensifiers often take advantage of channel plates. A special camera equipped with channel plates is capable of producing precise pictures of high brilliance even in a moonless clear night using only starlight.

In case larger amounts of light are available, low gain photosensitive detectors like photodiodes or phototransistors can be employed.

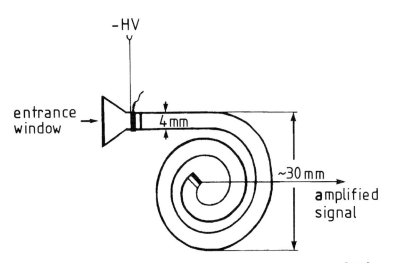

Fig. 5.8. Layout of a single channel electron multiplier [379].

5.2 Scintillation counters

The scintillator is one of the oldest particle detectors for nuclear radiation. Before this, charged particles had been detected by light flashes emitted when the particles impinged on a zinc-sulphate screen. This light was registered with the naked eye. It has been reported that the sensitivity of the human eye can be significantly increased by a cup of strong coffee possibly with a small dose of strychnine.

After a longer period of accommodation in complete darkness, the human eye is capable of recognizing approximately 15 photons as a light flash, if they are emitted within one tenth of a second and if their wavelength is matched to the maximum sensitivity of the eye.

The time span of a tenth of a second corresponds roughly to the time constant of the visual perception [63]. Chadwick [385] refers occasionally to a paper by Henri and Bancels [386], where it is mentioned that an energy deposit of approximately 3 eV corresponding to a single photon in the green spectral range should be recognizable by the eye [387].

The measurement principle of scintillation counters has remained essentially unchanged. The function of a scintillator is twofold. First, it should convert the excitation of, e.g., the crystal lattice caused by the energy loss of a particle into visible light and, second, it should transfer this light either directly or via a light guide to an optical receiver (photomultiplier, channel plate, phototransistor, photodiode, etc.) [388, 389, 390].

Scintillator materials can be inorganic crystals, organic compounds and gases. The scintillation mechanism in these scintillator materials is fundamentally different.

Inorganic scintillators are mostly crystals (NaI(Tl), CsI(Tl), LiI(Eu), etc.) [391] doped with impurities (colour centers, activator centers), while organic scintillators are polymerized solids, liquids or sometimes also crystals.

The scintillation mechanism in inorganic substances can be understood by considering the energy bands in crystals. Halide crystals, which are most commonly used, are insulators. The valence band is completely occupied, but the conduction band is normally empty (figure 5.9, [2]). The energy difference between both bands amounts to about 5 to 10 eV. Impurities which act as activator centers are deliberately introduced into the crystal lattice. These impurities are energetically localized between the valence and conduction band thereby creating additional energy levels.

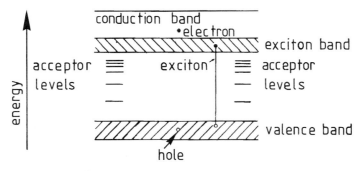

Fig. 5.9.   Energy bands in a crystal.

Electrons are transferred from the valence band to the conduction band by the energy deposited by an incident charged particle. In the conduction band they can move freely through the crystal lattice. In this excitation process a hole remains in the valence band. Due to this generated electron-hole pair the crystal gains a certain electrical conductivity. If the electron recombines again with a hole the energy liberated in this process can be emitted as a photon.

It can also happen, however, that the electron that has absorbed part of the energy loss of the incident particle does not reach the conduction band. In this case the electron is bound electrostatically to the hole. Such electron-hole bound states (excitons) migrate just like free electrons and holes through the crystal until they hit an activator center to which their binding energy may be transferred. The excitation energy of the activator center is handed over to the crystal lattice in form of lattice vibrations (phonons) or is emitted as light. A certain fraction of the energy deposited in the crystal is thereby emitted as luminescence radiation. This radiation can be converted to a voltage signal by a photosensitive detector. The decay time of the scintillator depends on the lifetimes of the excited levels.

Table 5.1 shows the characteristic parameters of some inorganic scintillators [1, 3, 392]. Inorganic scintillators have decay times in the microsecond range and are therefore relatively slow. Only cerium fluoride (CeF_3) and cerium-doped gadolinium silicate (GSO(Ce) = Gd_2SiO_5) have shorter decay times. These two inorganic crystals and barium fluoride (which also has a fast decay component in the subnanosecond range) are in addition particularly radiation resistant [392, 393, 394].

Table 5.1. *Characteristic parameters of some inorganic scintillators [1, 3, 392]*

scintillator	density $\varrho[g/cm^3]$	decay time[μs]	photons per MeV	radiation length $X_0[cm]$	$-\frac{dE}{dx}\big|_{min}$ $[\frac{MeV}{cm}]$
NaI(Tl)	3.67	0.23	$4 \cdot 10^4$	2.59	4.8
LiI(Eu)	4.06	1.3	$1.4 \cdot 10^4$	2.2	5.1
CsI(Tl)	4.51	1.0	$5.5 \cdot 10^4$	1.86	5.6
$Bi_4Ge_3O_{12}$	7.13	0.35	$2.8 \cdot 10^3$	1.12	9.2
BaF_2	4.9	0.62	$6.5 \cdot 10^3$	2.1	6
CeF_3	6.16	0.03	$\approx 5 \cdot 10^3$	1.7	7.7
GSO	6.71	≈ 0.05	$\approx 10^4$	1.38	8.3

Organic scintillators have much shorter decay times. Typically these lie in the nanosecond range. The scintillation mechanism in this case is not an effect of the lattice. Organic scintillators are normally three-component mixtures. A primary fluorescence agent is excited by the energy loss of particles. In the decay of these excited states ultraviolet light is emitted. The absorption length of this ultraviolet light, however, is rather short: the fluorescent agent is opaque for its own light. The extraction of the light is performed by adding a second fluorescent agent to the scintillator which absorbs the primary fluorescent light and re-emits light of lower frequency isotropically ('wavelength shifter'). The emission spectrum of the second component is normally matched to the spectral sensitivity of the light receiver [1].

The two active components in an organic scintillator are either dissolved in an organic liquid or are mixed with an organic material to form a polymerizing structure. In this way liquid or plastic scintillators can be produced in almost any geometry. In most cases scintillator sheets of 1 mm up to 30 mm thickness are made. Table 5.2 [1] lists some primary fluorescent agents and wavelength shifter materials. Figure 5.10 shows the emission spectra of a primary fluorescent agent and of a wavelength shifter along with the typical spectral sensitivity of the photocathode of a commonly used photomultiplier [1, 395].

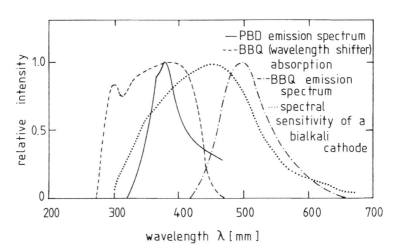

Fig. 5.10. Emission spectra of a primary fluorescent substance and wavelength shifter in comparison to the spectral sensitivity of the photocathode of a commonly used photomultiplier [1, 395].

Gas scintillation counters use light which is produced when charged particles excite atoms in interactions and these atoms decay subsequently into the ground state by light emission [388],[396]-[399]. The lifetime of the excited levels lies in the nanosecond range. Because of the low density, the light yield in gas scintillators is relatively low. However, gas scintillators can also be operated with liquid noble gases [400].

A gas scintillation counter can also be combined with a drift chamber. In this case the electrons produced in the ionization process will drift to the anode wire where, in the process of gas amplification, the number of photons is increased during the formation of an avalanche. This secondary scintillation light is, of course, much more intense than the primary scintillation light. The primary light can be used for self-triggering of such a gas scintillation drift chamber (occasionally also called an electro-luminescence drift chamber). Furthermore, the amount of primary and also secondary light is proportional to the energy loss of the incident particle if the chamber is operated in the proportional mode.

Figure 5.11 shows the light and charge signals measured in a gas scin-

Table 5.2. *Primary organic fluorescence substances and wavelength shifters. The light yield normalized to that of NaI refers to equal amounts of energy loss in all materials* [1]

fluorescence substance	$\lambda_{\max}$[nm] emission	decay time [ns]	$\dfrac{\text{yield}}{\text{NaI}}$
naphtalene	348	96	0.12
anthracene	440	30	0.5
p-therphenyl	440	5	0.25
PBD	360	1.2	
wavelength shifter			
POPOP	420	1.6	
bis-MSB	420	1.2	

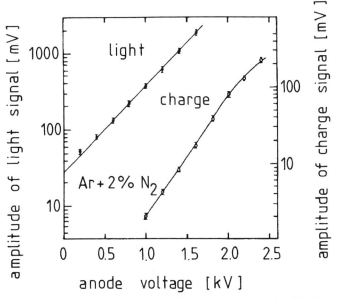

Fig. 5.11. Light and charge signal amplitudes in a gas scintillation drift chamber. The chamber is filled with argon $+2\%$ nitrogen (N_2) and was irradiated with α-particles from a ^{241}Am-source [401]. The absolute amplitudes of light and charge signals are not directly comparable because they have been obtained using different electronic readout systems.

tillation drift chamber as a function of the applied anode voltage [401].
The light intensity measured by photomultipliers provides already at very
low anode voltages reasonable signals. For an anode voltage of zero volts
only the direct primary light is detected.

Large area scintillation counters are frequently used to trigger other
detectors which may provide more detailed information. One of the most
important applications of scintillation counters is as the sampling element
in calorimeters. In these devices (see chapter 7) it is important that the
scintillation counters have a very high uniformity, which means that the
light yield should not depend on the particle's point of passage. This
is difficult to obtain, however, because the finite attenuation length (λ)
of the light in scintillators, wavelength shifters and light guides is of the
order of $\lambda \propto 1\,\mathrm{m}$. The absorption mainly takes place at short wavelengths
of the emission spectrum.

The light absorption in a 3 mm thick BBQ-wavelength shifter rod is
shown in figure 5.12 [402].

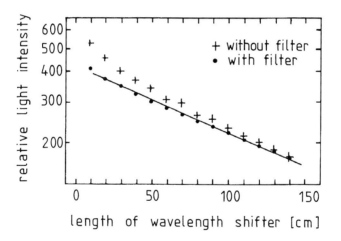

Fig. 5.12.   Light absorption in a 3 mm thick BBQ-wavelength shifter rod [402].

The uniformity of light collection can be significantly improved by dis-
pensing with the detection of the short wavelength component of the
emission spectrum by use of filters in front of the photomultipliers. For
this purpose yellow filters are commonly used.

Normally large area scintillators are read out with several photomultipliers. The relative pulse heights of these photomultipliers can be used to
determine the particle's point of passage and thereby enabling a correction
of the measured light yield for absorption effects.

Scintillators used as sampling detectors in calorimeters are usually in
the form of scintillator plates. The scintillation light emerges from the

edges of these plates and has to be guided to a photomultiplier and also matched to the usually circular geometry of the photosensing device. This matching is performed with light guides. In the most simple case (figure 5.13) the light is transferred via a triangular light guide ('fish-tail') to the photocathode of a photomultiplier. A complete light transfer, i.e., light transfer without any losses, using fish-tail light guides is impossible. Only by using complicated light guides can the end face of a scintillator plate be imaged onto the photocathode without appreciable loss of light ('adiabatic light guides'). Figure 5.14 shows the working principle of an adiabatic light guide ($dQ = 0$, i.e., no loss of light). The individual parts of the light-guide system can be only moderately bent because otherwise the light, which is normally contained in the light guide by internal reflection, will be lost at the bends.

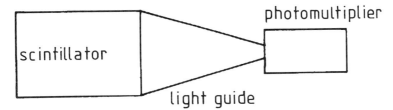

Fig. 5.13.   Light readout with a 'fish-tail' light guide.

Fig. 5.14.   Photograph of an adiabatic light guide [403].

The scintillator end face cannot be focussed without light loss onto a photocathode with smaller area because of Liouville's theorem, which states: 'The volume of an arbitrary phase space may change its form in the course of its temporal and spatial development, its size, however, remains constant'.

The time resolution of large scintillators is not so much limited by the time resolution of the photomultiplier but rather by the time difference of light paths in the scintillator itself (1 m is equivalent to 5 ns). For long scintillators and special readout electronics time resolutions of the order of 200 ps can be obtained.

If, however, as is the case in calorimeters, a sufficient amount of light is available, the light emerging from the end face of a scintillator plate can be absorbed in an external wavelength shifter rod. This wavelength shifter re-emits the absorbed light isotropically at a larger wavelength and guides it to a photosensitive device (figure 5.15).

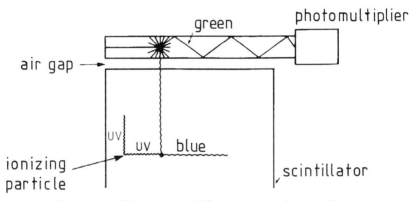

Fig. 5.15.   Wavelength shifter readout of a scintillator.

It is very important that a small air gap remain between the scintillator face and the wavelength shifter rod. Otherwise, the frequency shifted, isotropically re-emitted light would not be contained in the wavelength shifter rod by internal reflection. This method of light transfer normally entails an appreciable loss of light; typical conversion values are around 1 to 5 %.

The absorption of electromagnetic cascades in calorimeters frequently yields a very large amount of light so that even a two-step wavelength shifting method can be employed. This, quite obviously, will further reduce the light yield. The compactness of calorimeter modules, however, can be significantly improved (see figure 5.16). For these purposes scintillators and wavelength shifters have been developed, where the emission and absorption spectra are particularly well matched.

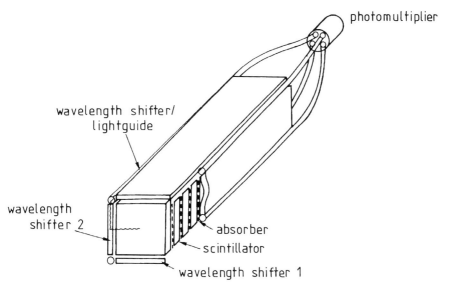

photomultiplier

wavelength shifter/
lightguide

wavelength
shifter 2

absorber

scintillator

wavelength shifter 1

Fig. 5.16. Two-step wavelength shifter readout of a calorimeter.

Typical organic scintillators use a primary fluorescent material such as naphtalene (15 %) mixed with an internal wavelength shifter from butyl PBD (1 %) dissolved in a polymerizing substance like PMMA (polymethyl-methacralate = 'Plexiglas'). It is common practice to employ BBQ-absorbers/emitters (benzimidazo-benzisochinoline-7-one) also dissolved in PMMA in external wavelength shifter rods. Multiple wavelength shifting of the primary scintillation light, however, transfers it to the long wavelength range of the spectrum. The spectral sensitivity of the phototube has to be matched correspondingly.

The separate readout, for example, of the four wavelength shifter rods in figure 5.16 allows a rough spatial resolution by comparing the relative pulse heights from the individual wavelength shifter rods. The origin of light emission in the scintillator can be determined in this way with spatial resolutions of the order of several centimeters.

A normal sampling calorimeter of absorber plates and scintillator sheets can also be read out by wavelength shifter rods or fibers running through the scintillator plates perpendicularly [404, 405, 406].

The scintillation counters used in calorimeters do not have to have the form of scintillator sheets alternating with absorber layers. They can also be embedded as scintillating fibers, for example, in a lead matrix [407, 408, 409]. In this case the readout is greatly simplified because the scintillating fibers can be bent rather strongly without loss of internal reflection. Scintillating fibers can either be read out directly or via light

fibers by photomultipliers ('spaghetti calorimeter'). A separate readout
of the individual scintillating fibers also provides an excellent spatial resolution which can even exceed the spatial resolution of drift chambers
[407],[410]-[413]. Analogously, thin capillaries ('macaronis') filled with
liquid scintillator can be used for tracking charged particles [414, 415]. In
this respect, these scintillating fiber calorimeters or more generally, light-
fiber systems, can also be considered as tracking detectors. In addition,
they represent, because of the short decay time of the light, a genuine alternative to gas discharge detectors, which are rather slow because of the
intrinsically slow electron drift. Figure 5.17 shows the track of a charged
particle in a stack of scintillating fibers. The fiber diameter in this case
amounts to 1 mm [416].

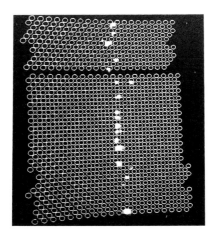

Fig. 5.17. Particle track in a stack of scintillating fibers; fiber diameter $\phi =$
1 mm [416].

Scintillating fibers, however, can also be produced with much smaller
diameters. Figure 5.18 shows a microphotograph of a bundle consisting of scintillating fibers with 60 μm diameter. Only the central fiber is
illuminated. A very small fraction of the light is scattered into the neighbouring fibers [417, 418]. The fibers are separated by a very thin cladding
(3.4 μm). The optical readout system for such light fiber systems, however, has to be made in such a way that the granular structure of the light
fibers is resolved with sufficient accuracy, e.g., with optical pixel systems
[419].

Arrangements of such fiber bundles are excellent candidates for tracking detectors in experiments with high particle rates requiring high time
and spatial resolution. Figure 5.19 shows different patterns of bundles
of scintillating fibers from different companies [420]. Figure 5.20 shows

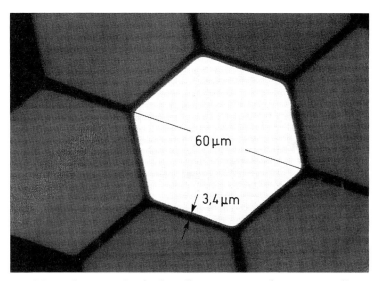

Fig. 5.18. Microphotograph of a bundle consisting of seven scintillating fibers. The fibers have a diameter of 60 μm. Only the central fiber is illuminated. The fibers are optically separated by a very thin cladding (3.4 μm) of lower refractive index to trap the light by total reflections. Only a small amount of light is scattered into the neighbouring fibers [417, 418].

the spatial resolution for charged particles obtained in a stack consisting of 8000 scintillating fibers (30 μm diameter). A single-track resolution of 35 μm and a two-track resolution of 83 μm is achieved [421].

The transparency of scintillators can deteriorate in a high radiation environment [422]. There exist, however, scintillator materials with a substantial radiation hardness [423, 424].

5.3 Planar spark counters

Planar spark counters consist of two planar electrodes to which a constant voltage exceeding the static break-down voltage at normal pressure is applied. The chambers are normally operated with slight overpressure. The planar spark counter consequently is essentially a spark chamber which is not triggered. Just as in a spark chamber the ionization of a charged particle, which has passed through the chamber, causes an avalanche, which develops into a conducting plasma channel connecting the electrodes. The rapidly increasing anode current can be used to generate a voltage signal of very short rise time via a resistor. This voltage pulse can serve as a very precise timing signal for the arrival time of a charged particle in the spark counter.

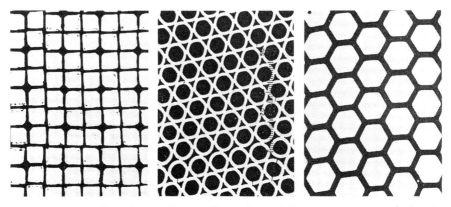

Fig. 5.19.   Bundles of scintillating fibers from different companies (left: $\phi =$ 20 μm; Schott (Mainz); center: $\phi = 20$ μm; US Schott (Mainz); right: $\phi = 30$ μm plastic fibers; Kyowa Gas (Japan)) [420].

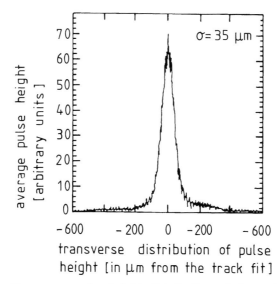

Fig. 5.20.   Transverse pulse height distribution of charged particles in a stack of 8000 scintillating fibers with 30 μm diameter [421].

Figure 5.21 shows the working principle of a planar spark counter [1, 425].   If metallic electrodes are used, the total capacitance of the chamber will be discharged in one spark.   This may lead to damage of the metallic surface and also causes a low multitrack efficiency.   If, however, the electrodes are made from a material with high specific bulk resistivity [426], only a small part of the electrode area will be discharged via the sparks. These do not cause surface damage because of the reduced current in the spark. A high multitrack efficiency is also guaranteed in this way, just as in the glass spark chamber.

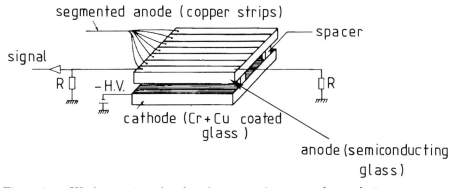

Fig. 5.21. Working principle of a planar spark counter [1, 425]. In many cases the anode is either coated with a semiconducting material or with a material of high specific bulk resistivity.

In addition to determining the arrival time of charged particles the chamber also allows for a rough spatial resolution if the anode is segmented. Noble gases with quenchers which suppress secondary spark formation are commonly used as gas filling.

Planar spark counters allow excellent time resolutions ($\sigma_t \leq 30\,\mathrm{ps}$) if properly constructed. This, however, requires narrow electrode gaps of the order of $200\,\mu$m. The production of large-area spark counters, therefore, requires very precise machining to guarantee parallel electrodes with high surface quality [427, 428].

By coating the surfaces facing the gas volume with boron, planar spark counters can also be used for neutron detection (see section 6.1) [429].

Planar spark counters can also be operated at lower gas amplifications, which are then called resistive plate chambers (RPC), if, for example, instead of semiconducting electrode materials, graphite-covered glass plates are used. These chambers are most commonly operated in the streamer or in the avalanche mode [427]-[432]. Instead of graphite-covered glass plates other materials with suitable surface resistivity can also be used. These resistive plate chambers also supply very fast signals and can — just as scintillation counters — be used for triggering with high time resolution. If the electrodes of the resistive plate chambers are segmented, they may also provide an excellent position resolution.

Planar spark counters and resistive plate chambers generally do not permit high counting rates. If the gas amplification is further reduced to values around 10^5, neither sparks nor streamers can develop. This mode of

operation characterizes a parallel-plate avalanche chamber (PPAC) [433]-
[437]. These parallel-plate avalanche chambers, with typical electrode distances of the order of 1 mm, also exhibit a high time resolution ($\approx 500\,\text{ps}$)
and, if they are operated in the proportional mode, have as well an excellent energy resolution [438]. An additional advantage of parallel-plate
avalanche chambers, compared to spark counters and resistive plate chambers, is that they can be operated at high counting rates because of the
low gas amplification involved. This is also true for glass spark counters
[439] and glass spark chambers, where the current drawn per discharge
is reduced (see section 4.15). Likewise, certain operating conditions of
resistive plate chambers also allow high counting rates [440].

All these chamber types have in common that they provide excellent
timing resolution due to the small electrode gaps.

6

Particle identification

One of the standard tasks of particle detectors is, apart from measuring characteristic values like momentum and energy, to determine the identity of particles. This implies the determination of the mass and charge of a particle. In general, this is achieved by combining information from several detectors.

For example, the radius of curvature ϱ of a charged particle in a magnetic field supplies information on the momentum p and the charge z via the relation

$$\varrho \propto \frac{p}{z} = \frac{\gamma m_0 \beta c}{z} \ . \tag{6.1}$$

The velocity $\beta = v/c$ can be obtained by time-of-flight measurements using

$$\tau \propto \frac{1}{\beta} \ . \tag{6.2}$$

The determination of the energy loss by ionization and excitation can approximately be described by

$$-\frac{\mathrm{d}E}{\mathrm{d}x} \propto \frac{z^2}{\beta^2} \ln(a\gamma\beta) \ , \tag{6.3}$$

where a is a material dependent constant. An energy measurement yields

$$E_{\text{kin}} = (\gamma - 1)m_0 c^2 \ , \tag{6.4}$$

since normally only the kinetic energy and not the total energy is measured.

Equations (6.1) to (6.4) contain three unknown quantities, namely m_0, β and z; the Lorentz factor γ is related to the velocity β according to $\gamma = \frac{1}{\sqrt{1-\beta^2}}$. Three of the above mentioned four measurements are in principle sufficient to positively identify a particle. In the field of elementary particle physics one mostly deals with singly charged particles

($z = 1$). In this case, two different measurements are sufficient to determine the particle's identity. For particles of high energy, however, the
determination of the velocity does not provide sufficient information, since
for all relativistic particles, independent of their mass, β is very close to
1 and therefore cannot discriminate between particles of different mass.

Nearly all detectors are based either on the ionization of charged particles or the production of light by charged particles. Therefore, particles
that do not ionize or produce light in scintillators first require a conversion
into charged particles. For the detection of photons the conversion processes have been described in chapter 1, namely the photoelectric effect,
Compton scattering and electron pair production. Other neutral particles
like neutrons or neutrinos have to be considered separately.

6.1   Neutron counters

Depending on the energy of neutrons, different detection techniques must
be employed. Common to all methods is that charged particles have to
be produced in neutron interactions, which then are seen by the detector
via 'normal' interaction processes like, e.g., ionization or the production
of light in scintillators [65, 441, 442].

For low-energy neutrons ($E_n^{kin} < 20\,\mathrm{MeV}$) the following conversion reactions can be used:

$$n + {}^6\mathrm{Li} \rightarrow \alpha + {}^3\mathrm{H} \tag{6.5}$$

$$n + {}^{10}\mathrm{B} \rightarrow \alpha + {}^7\mathrm{Li} \tag{6.6}$$

$$n + {}^3\mathrm{He} \rightarrow p + {}^3\mathrm{H} \tag{6.7}$$

$$n + p \rightarrow n + p \tag{6.8}$$

The cross sections for these reactions depend strongly on the neutron
energy. They are plotted in figure 6.1 [442].

For energies between $20\,\mathrm{MeV} \leq E_n \leq 1\,\mathrm{GeV}$ the production of recoil
protons via the elastic (n, p)-scattering can be used for neutron detection
(equation (6.8)). Neutrons of high energy ($E_n > 1\,\mathrm{GeV}$) produce hadron
cascades in inelastic interactions which are easy to identify.

To be able to distinguish neutrons from other particles, a neutron
counter basically always consists of an anti-coincidence counter, which
vetos charged particles, and the actual neutron detector.

Thermal neutrons ($E_n \approx \frac{1}{40}\,\mathrm{eV}$) are easily detected with ionization
chambers or proportional counters, filled with boron trifluoride gas (BF_3).
To be able to detect higher energy neutrons as well in these counters, the
neutrons first have to be moderated, since otherwise the neutron interaction cross section would be too small (see figure 6.1). The moderation of
non-thermal neutrons can best be done with substances containing many

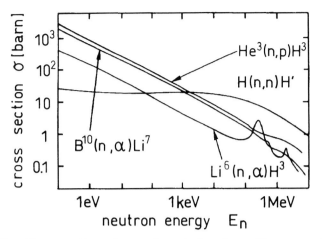

Fig. 6.1. Cross sections for neutron induced reactions as a function of the neutron energy $(1 \, \text{barn} = 10^{-24} \, \text{cm}^2)$ [442].

protons, because neutrons can transfer a large amount of energy to collision partners of the same mass. In collisions with heavy nuclei essentially only elastic scattering with small energy transfers occur. Paraffin or water are preferred moderators. Neutron counters for non-thermal neutrons are therefore covered with these substances. With BF_3-counters, neutron detection efficiencies of the order of 1 % can be achieved.

Thermal neutrons can also be detected via a fission reaction (n, f) (f = fission). Figure 6.2 shows two special proportional counters which are covered on the inside with either a thin boron or uranium coating to induce the neutrons to undergo either (n, α) or (n, f)-reactions [65]. To moderate fast neutrons these counters are mounted inside a paraffin barrel.

Thermal or quasi-thermal neutrons can also be detected with solid state detectors. For this purpose, a lithium-fluoride (^{6}LiF) coating is evaporated onto the surface of a semiconductor counter in which, according to equation (6.5), α-particles and tritons are produced. These can easily be detected by the solid state detector.

Equally well europium-doped lithium-iodide scintillation counters LiI(Eu) can be used for neutron detection since α-particles and tritons produced according to equation (6.5) can be measured via their scintillation light. Neutrons with energies in the MeV range can be detected in multiwire proportional chambers filled with a gas mixture of ^{3}He and Kr at high pressure by means of the reaction (6.7). The elastic recoil reaction (6.8) can be used in multiwire proportional chambers containing hydrogen-rich components (e.g., CH_4 + Ar). The size of a neutron

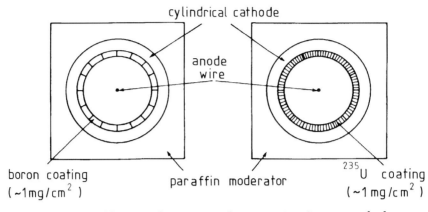

Fig. 6.2. Neutron detection with proportional counters [65].

counter should be large compared to the maximum range of the recoil protons; 10 cm in typical gases [37] (see figure 6.3). In solids the range of protons is reduced approximately in reverse proportion to the density (see figure 6.4).

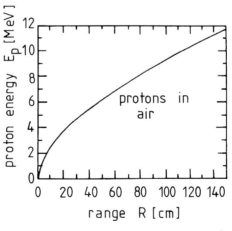

Fig. 6.3. Range of protons in air [37].

Neutrons in the energy range $1 - 100$ MeV can also be detected in organic scintillation counters via the production of recoil protons according to equation (6.8). However, the cross section for the (n, p)-reaction decreases rapidly with increasing neutron energy so that the neutron detection efficiency is reduced.

If σ is the cross section for the (n, p)-process, the cross section per unit mass ϕ in $(g/cm^2)^{-1}$ (sometimes also incorrectly called 'interaction

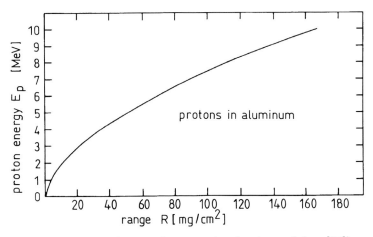

Fig. 6.4. Range of protons in aluminum (after [37]).

probability') for a (n, p)-reaction is given by

$$\phi[\text{cm}^2/\text{g}] = \sigma[\text{cm}^2] \cdot N_A \ [\text{g}^{-1}] \ , \tag{6.9}$$

where N_A is Avogadro's number.

For neutrons of 10 MeV with a cross section of about 1 barn one gets $\phi = 0.60$ per g/cm^2 (compare figure 6.1, curve H(n, n)H′). For an organic scintillator of 1 cm thickness (density $\varrho = 1.2$ g/cm^3 assumed) with a 30 % fraction of free protons, neutron detection efficiencies in this case of about 20 % are obtained. Generally one can say that the efficiency for neutron detection increases with the product of density and thickness $\varrho \cdot \text{d}x$.

In some applications — e.g., in the field of radiation protection — the measurement of the neutron energy is of great importance because the relative biological efficiency of neutrons is strongly energy dependent. The measurement of the neutron energy is frequently carried out with threshold detectors. Such a detector consists of a carrier foil covered with an isotope that only reacts with neutrons above a certain threshold energy. The particles or charged nuclei liberated in these reactions can be detected e.g., in plastic detectors (cellulose-nitrate or cellulose-acetate foils) by an etching technique, and evaluated under a microscope or with automatic pattern recognition methods (compare section 4.21). Table 6.1 lists several threshold reactions used for neutron detection.

To cover different energy ranges of neutrons in a single exposure, one uses stacks of plastic foils coated with different isotopes. From the counting rates in the individual carrier foils with different energy thresholds, a rough determination of the neutron energy spectrum can be made [65].

Table 6.1. *Threshold reactions for neutron energy measurements [65]*

Reaction	Threshold energy [MeV]
Fission of ^{234}U	0.3
Fission of ^{236}U	0.7
^{31}P (n,p) ^{31}Si	0.72
^{32}S (n,p) ^{32}P	0.95
Fission of ^{238}U	1.3
^{27}Al (n,p) ^{27}Mg	1.9
^{56}Fe (n,p) ^{56}Mn	3.0
^{27}Al (n,α) ^{24}Na	3.3
^{24}Mg (n,p) ^{24}Na	4.9
^{65}Cu $(n,2n)$ ^{64}Cu	10.1
^{58}Ni $(n,2n)$ ^{57}Ni	12.0

6.2   Neutrino detectors

Neutrino detectors have to be extremely large because the cross section for neutrino interactions is very small. Depending on the neutrino species the following reactions for neutrino detection can be considered

$$\begin{aligned} \nu_e + n &\rightarrow e^- + p \\ \bar{\nu}_e + p &\rightarrow e^+ + n \end{aligned} \qquad (6.10)$$

$$\begin{aligned} \nu_\mu + n &\rightarrow \mu^- + p \\ \bar{\nu}_\mu + p &\rightarrow \mu^+ + n \end{aligned} \qquad (6.11)$$

$$\begin{aligned} \nu_\tau + n &\rightarrow \tau^- + p \\ \bar{\nu}_\tau + p &\rightarrow \tau^+ + n \end{aligned} \qquad . \qquad (6.12)$$

For higher energies inelastic neutrino reactions on nucleons or nuclei can be used as well. Since the cross section for neutrinos with energies in the several MeV range for reaction (6.10) is of the order of 10^{-43} cm^2 per nucleon, the reaction probability and consequently the efficiency for neutrino detection for a detector with a length times density of 1000 g/cm^2 (for iron approximately 1.3 m) is only of the order of $6 \cdot 10^{-17}$. Neutrino detectors therefore require massive targets and high neutrino fluxes to yield significant reaction rates.

In experiments where neutrinos are produced (e.g., in electron-positron annihilation or proton-antiproton interactions) it is possible, however, to detect them indirectly. If the detector covers the full solid angle around the interaction point ('hermeticity'), if the total reaction energy (center of mass energy) is known, and if the energies and momenta of all produced particles except the neutrino are recorded, then the characteristics of the escaping neutrino can be inferred from the missing energy and the direction of the missing momentum. This 'missing energy' technique was employed in the discovery of the leptonic decays of the Ws, the charged carriers of the electroweak interaction [443, 444].

6.3 Time-of-flight counters

Particle identification with time-of-flight techniques requires excellent time resolution. The working principle for a time-of-flight counter is shown schematically in figure 6.5.

After a particle has passed through a first detector (e.g., scintillation counter), it starts a time-to-amplitude converter (TAC). This is stopped by a second counter when reached by the particle. The time information stored in the time-to-amplitude converter can be read out by a multichannel analyzer (MCA or PHA (=pulse height analyzer)) or processed by a computer.

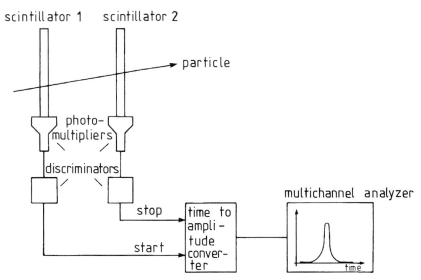

Fig. 6.5. Working principle of time-of-flight measurement.

If, for example, the momentum of a beam of particles of unknown composition has already been determined in a magnetic spectrometer, the time-of-flight measurement can be used to identify particles of different mass, because for the same momentum they have different velocities [1].

Two particles of mass m_1 and m_2 have for the same momentum and flight distance L the time-of-flight difference

$$\Delta t = L\left(\frac{1}{v_1} - \frac{1}{v_2}\right) = \frac{L}{c}\left(\frac{1}{\beta_1} - \frac{1}{\beta_2}\right). \tag{6.13}$$

Using $\gamma = 1/\sqrt{1-\beta^2}$ it follows that

$$\Delta t = \frac{L}{c}\left\{\sqrt{\frac{\gamma_1^2}{\gamma_1^2-1}} - \sqrt{\frac{\gamma_2^2}{\gamma_2^2-1}}\right\}. \tag{6.14}$$

Substituting $\gamma = E/m_0 c^2$ into equation (6.14) yields

$$\Delta t = \frac{L}{c}\left\{\sqrt{\frac{1}{1-\left(\frac{m_1 c^2}{E_1}\right)^2}} - \sqrt{\frac{1}{1-\left(\frac{m_2 c^2}{E_2}\right)^2}}\right\}. \tag{6.15}$$

For relativistic particles ($E \gg m_0 c^2$) one obtains

$$\Delta t = \frac{L}{c}\left\{\sqrt{1+\left(\frac{m_1 c^2}{E_1}\right)^2} - \sqrt{1+\left(\frac{m_2 c^2}{E_2}\right)^2}\right\}. \tag{6.16}$$

Since in this case $E \approx pc$, the expansion of the square roots leads to

$$\Delta t = \frac{Lc}{2p^2}(m_1^2 - m_2^2). \tag{6.17}$$

Suppose that for a mass separation a significance of $\Delta t = 4\sigma_t$ is demanded. That is, a time-of-flight difference four times the time resolution is required. In this case a pion–kaon separation can be achieved up to momenta of $1\,\mathrm{GeV}/c$ for a flight distance of $3\,\mathrm{m}$ and a time resolution of $\sigma_t = 300\,\mathrm{ps}$, which can be obtained with, e.g., scintillation counters [1]. For higher momenta the time-of-flight systems become increasingly long since $\Delta t \propto 1/p^2$.

Because of the excellent time resolution of spark counters ($\sigma_t \approx 30\,\mathrm{ps}$) time-of-flight systems with these detectors can be correspondingly shorter ($L = 30\,\mathrm{cm}$ for pion–kaon separation up to $p = 1\,\mathrm{GeV}/c$).

The time-of-flight differences for different pairs of charged particles for a flight distance of $1\,\mathrm{m}$ are plotted in figure 6.6. For highly relativistic particles the time-of-flight differences approach zero. This reduces the application of time-of-flight measurement techniques to particles whose velocities are not too close to the velocity of light.

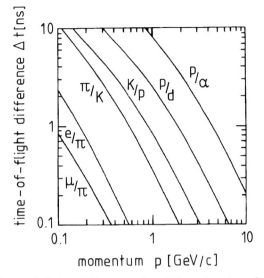

Fig. 6.6. Time-of-flight differences for different pairs of particles for a flight distance of 1 m (after [1]).

6.4 Cherenkov counters

A charged particle, traversing a medium with refractive index n with a velocity v exceeding the velocity of light c/n in that medium, emits a characteristic electromagnetic radiation, called Cherenkov radiation [445, 446]. Cherenkov radiation is emitted because the charged particle polarizes the atoms along its track so that they become electric dipoles. The time variation of the dipole field leads to the emission of electromagnetic radiation. As long as $v < c/n$, the dipoles are symmetrically arranged around the particle's path, so that the dipole field integrated over all dipoles vanishes and no radiation results. If, however, the particle moves with $v > c/n$, the symmetry is broken resulting in a non-vanishing dipole moment, which leads to the emission of radiation. Figure 6.7 illustrates the difference in polarization for the cases $v < c/n$ and $v > c/n$ [68].

The contribution of Cherenkov radiation to the energy loss is small compared to that from ionization and excitation (equation (1.12)) even for minimum-ionizing particles. For gases with $Z \geq 7$ the energy loss by Cherenkov radiation amounts to less than 1 % of the ionization loss of minimum-ionizing particles. For light gases (He, H) this fraction amounts to about 5 % [5, 6].

The angle between the emitted Cherenkov photons and the track of the charged particle can be obtained from a simple argument (figure 6.8).

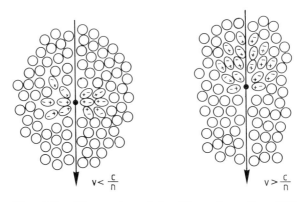

Fig. 6.7.   Illustration of the Cherenkov effect [68].

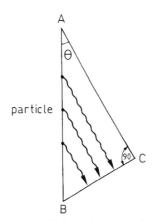

Fig. 6.8.   Simple geometric determination of the Cherenkov angle.

While the particle has travelled the distance $AB = t\beta c$, the photon has advanced by $AC = t \cdot c/n$. Therefore one obtains

$$\cos\theta_{\mathrm{c}} = \frac{c}{n\beta c} = \frac{1}{n\beta} \; . \qquad (6.18)$$

In principle, the emission of a Cherenkov photon leads to a recoil of the charged particle, which then slightly changes its direction. Taking this effect into account, the exact treatment of the kinematics leads to

$$\cos\theta_{\mathrm{c}} = \frac{1}{n\beta} + \frac{\hbar k}{2p}\left(1 - \frac{1}{n^2}\right) , \qquad (6.19)$$

where k is the wave vector of the photon ($k = 2\pi/\lambda$; λ – wavelength), $\hbar k$ the momentum of the photon and p the momentum of the charged particle. θ_{c} is the angle between the momentum vector of the incident particle and the direction of the emitted photon. Since $\hbar k \ll p$, equation (6.18) represents an excellent approximation for all practical cases.

For the emission of Cherenkov radiation there is a threshold effect. Cherenkov radiation is only emitted, if $\beta > \frac{1}{n}$. At threshold, Cherenkov radiation is emitted in the forward direction. The Cherenkov angle increases until it reaches a maximum for $\beta = 1$, namely

$$\theta_c = \arccos \frac{1}{n} . \tag{6.20}$$

Consequently, Cherenkov radiation is emitted only if the medium and the frequencies ν are such that $n(\nu) > 1$.

The threshold velocity for the emission of Cherenkov radiation corresponds to a threshold energy given by

$$E_{th} = \gamma_{th} m_0 c^2 , \tag{6.21}$$

where

$$\gamma_{th} = \frac{1}{\sqrt{1 - \beta_{th}^2}} = \frac{1}{\sqrt{1 - \frac{1}{n^2}}} = \frac{n}{\sqrt{n^2 - 1}} . \tag{6.22}$$

For fixed energy, the threshold Lorentz factor depends on the mass of the particle. Therefore, the measurement of Cherenkov radiation is ideally suited for particle identification purposes.

The dependence of the Cherenkov angle on the velocity of the particle is shown in figures 6.9a and 6.9b [71] for different indices of refraction.

The number of Cherenkov photons emitted per unit path length with wavelengths between λ_1 and λ_2 is given by

$$\frac{dN}{dx} = 2\pi\alpha z^2 \int_{\lambda_1}^{\lambda_2} \left(1 - \frac{1}{n^2\beta^2}\right) \frac{d\lambda}{\lambda^2} , \tag{6.23}$$

for $n(\lambda) > 1$, where z is the electric charge of the particle producing Cherenkov radiation and α is the fine structure constant.

Neglecting the dispersion of the medium (i.e., n independent of λ) leads to

$$\frac{dN}{dx} = 2\pi\alpha z^2 \cdot \sin^2 \theta_c \cdot \frac{\lambda_2 - \lambda_1}{\lambda_1 \lambda_2} . \tag{6.24}$$

For the optical range ($\lambda_1 = 400\,\text{nm}$ and $\lambda_2 = 700\,\text{nm}$) one obtains for singly charged particles ($z = 1$)

$$\frac{dN}{dx} = 490 \sin^2 \theta_c \ [\text{cm}^{-1}] . \tag{6.25}$$

Figures 6.10a and 6.10b show the number of Cherenkov photons emitted per unit path length for various materials as a function of the velocity of the particle [71].

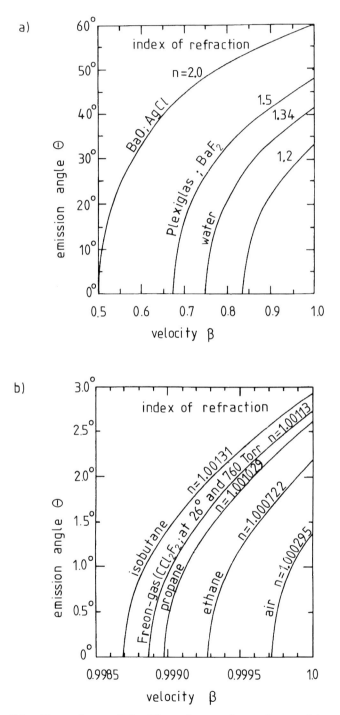

Fig. 6.9.   Dependence of the Cherenkov angle on the particle velocity for different indices of refraction [71].

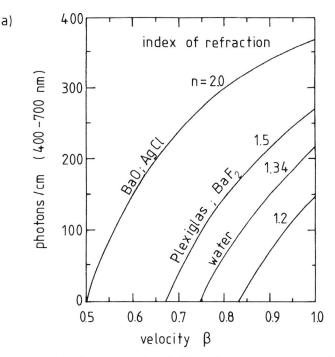

Fig. 6.10 a. Number of produced Cherenkov photons per unit path length for various materials as a function of the particle velocity [71].

The photon yield can be increased by up to a factor of two or three if the photons emitted in the ultraviolet range can also be detected. Although the number of emitted Cherenkov photons exhibits a $1/\lambda^2$-dependence (see equation (6.23)), Cherenkov photons are not emitted in the X-ray range because in this region the index of refraction is $n = 1$, and therefore the condition for Cherenkov emission cannot be fulfilled.

To obtain the correct number of photons produced in a Cherenkov counter, equation (6.23) must be integrated over the region for which $\beta \cdot n(\nu) \gg 1$. Also the response function of the light collection system must be taken into account to obtain the number of photons arriving at the photon detector.

All transparent materials are candidates for Cherenkov radiators. In particular, Cherenkov radiation is emitted in all scintillators and in the light guides which are used for the readout. The scintillation light, however, is approximately 100 times more intense than the Cherenkov light. A large range of indices of refraction can be covered by the use of solid, liquid or gaseous radiators (see table 6.2).

Ordinary liquids have indices of refraction greater than $\approx 1.33(\mathrm{H_2O})$ and gases have n less than about 1.002 (pentane). Although gas

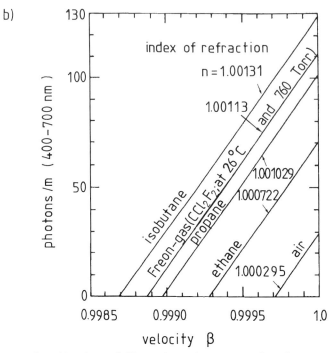

Fig. 6.10 b. Number of Cherenkov photons produced per unit path length in various gases as a function of the particle velocity [71].

Cherenkov counters can be operated at high pressure, thus increasing the index of refraction, the substantial gap between $n = 1.33$ and $n = 1.002$ cannot be bridged in this way.

By use of silica aerogels, however, it has become feasible to cover this missing range of the index of refraction. Aerogels are phase mixtures from m (SiO_2) and $2m$ (H_2O) where m is an integer. Silica aerogels form a porous structure with pockets of air. The diameter of the air bubbles in the aerogel is small compared to the wavelength of the light so that the light 'sees' an average index of refraction between the air and the solid forming the aerogel structure. Silica aerogels can be produced with densities between 0.1 to 0.3 g/cm^3 [1, 34, 35].

The required length of Cherenkov radiators for an effective particle identification can be derived from the threshold condition for the Cherenkov effect.

Consider two particles with different masses m_1 and m_2 with the same momentum [1]. To be able to distinguish these two particles in a threshold Cherenkov counter, it is required that the lighter particle with mass m_1 emits Cherenkov radiation; however, the heavier particle with mass m_2 is

Table 6.2. *Compilation of Cherenkov radiators [1, 34, 35, 122]. The index of refraction for gases is for 0 °C and 1 atm (STP). Solid sodium is transparent for wavelengths below 2000 Å [447, 448]*

material	$n - 1$	β-threshold	γ-threshold
solid sodium	3.22	0.24	1.029
lead sulfite	2.91	0.26	1.034
diamond	1.42	0.41	1.10
zinc sulfide (ZnS(Ag))	1.37	0.42	1.10
silver chloride	1.07	0.48	1.14
flint glass (SFS1)	0.92	0.52	1.17
lead fluoride	0.80	0.55	1.20
Clerici solution	0.69	0.59	1.24
lead glass	0.67	0.60	1.25
thallium formate solution	0.59	0.63	1.29
scintillator	0.58	0.63	1.29
Plexiglas (lucite)	0.48	0.66	1.33
boron silicate glass (Pyrex)	0.47	0.68	1.36
water	0.33	0.75	1.52
silica aerogel	0.025 - 0.075	0.93 - 0.976	4.5 - 2.7
pentane (STP)	$1.7 \cdot 10^{-3}$	0.9983	17.2
CO_2 (STP)	$4.3 \cdot 10^{-4}$	0.9996	34.1
air (STP)	$2.93 \cdot 10^{-4}$	0.9997	41.2
H_2 (STP)	$1.4 \cdot 10^{-4}$	0.99986	59.8
He (STP)	$3.3 \cdot 10^{-5}$	0.99997	123

supposed to be precisely at threshold and does not radiate. Under these circumstances one has:

$$\beta_2 = \frac{1}{n} \qquad (6.26)$$

or

$$\gamma_2 = \frac{1}{\sqrt{1 - \frac{1}{n^2}}} \,. \qquad (6.27)$$

Rewriting equation (6.27) yields:

$$n^2 = \frac{\gamma_2^2}{\gamma_2^2 - 1} \,. \qquad (6.28)$$

The lighter particle emits $490 \cdot \sin^2 \theta_c$ photons per centimeter of path length (see equation (6.25)), where

$$\sin^2 \theta_c = 1 - \cos^2 \theta_c = 1 - \frac{1}{(\beta_1 n)^2} \tag{6.29}$$

$$= 1 - \frac{1}{\beta_1^2 \frac{\gamma_2^2}{\gamma_2^2 - 1}} = 1 - \frac{1}{\frac{\gamma_1^2 - 1}{\gamma_1^2} \frac{\gamma_2^2}{\gamma_2^2 - 1}} \tag{6.30}$$

$$= \frac{\gamma_1^2 - \gamma_2^2}{(\gamma_1^2 - 1)\gamma_2^2} . \tag{6.31}$$

Since normally $\gamma_1^2 \gg 1$, it follows

$$\sin^2 \theta_c = \frac{1}{\gamma_2^2} - \frac{1}{\gamma_1^2} = \frac{m_2^2 c^4}{E_2^2} - \frac{m_1^2 c^4}{E_1^2} . \tag{6.32}$$

If $\gamma_1^2 \gg 1$, also $E_1 \approx p_1 c$ holds; since $p_1 = p_2 = p$ and assuming that also for the heavier particle with mass m_2 the energy can be approximated by $E_2 \approx p_2 c$, it follows that:

$$\frac{dN}{dx} = 490 \cdot \sin^2 \theta_c \, [\mathrm{cm}^{-1}] = 490 \cdot \frac{c^2}{p^2} (m_2^2 - m_1^2) \, [\mathrm{cm}^{-1}] . \tag{6.33}$$

For a radiator length L (in cm) and a quantum efficiency of the photomultiplier for Cherenkov photons of q, the number of photoelectrons for complete light collection is

$$N = 490 \frac{c^2}{p^2} (m_2^2 - m_1^2) \cdot L \cdot q . \tag{6.34}$$

If the efficient detection of the fast particle requires N_0 photoelectrons, the necessary radiator length can be calculated from

$$L = \frac{N_0 p^2}{490 \cdot c^2 (m_2^2 - m_1^2) \cdot q} \, [\mathrm{cm}] . \tag{6.35}$$

To achieve a kaon-proton separation ($m_{kaon} = 494 \, \mathrm{MeV}/c^2$, $m_{proton} = 938 \, \mathrm{MeV}/c^2$) at momenta of 10 GeV/$c$ assuming $N_0 = 10$ and a quantum efficiency of $q = 0.25$, a minimum radiator length of $L = 12.8 \, \mathrm{cm}$ is required [1]. Figure 6.11 shows the necessary detector length for the separation of particle pairs as a function of the momentum under the conditions mentioned ($q = 0.25$, $N_0 = 10$).

The index of refraction, however, has to be precisely chosen so that the particle with the higher mass does not radiate. This necessitates, e.g., for a kaon-pion separation at $p = 10 \, \mathrm{GeV}/c$ an index of refraction of $n = 1.005$, which, for example, can be obtained with pentane under increased pressure.

It is common practice to use a combination of several threshold Cherenkov counters. In this way a pion, kaon, and proton separation

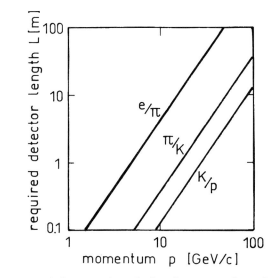

Fig. 6.11. Required detector length for the separation of particle pairs with a threshold Cherenkov counter versus the momentum ($N_0 = 10$, $q = 0.25$) [1, 449].

can be achieved in a momentum selected beam (see figure 6.12).

At $p = 10\,\mathrm{GeV}/c$ the charged pion ($m_\pi \approx 0.14\,\mathrm{GeV}/c^2$) is above Cherenkov threshold in all radiators. A charged kaon produces a signal in the aerogel and neopentane counters, but not in the argon-neon counter, while a proton is above threshold only in the aerogel counter. The coincidence requirements $C1 \cdot C2 \cdot C3$ and $C1 \cdot C2 \cdot \overline{C3}$ and $C1 \cdot \overline{C2} \cdot \overline{C3}$ therefore select pions, kaons and protons, respectively. By varying the gas pressure in the gas Cherenkov counters the Cherenkov thresholds can be continuously adjusted. Up to about $20 - 30\,\mathrm{GeV}/c$ such a combination of threshold Cherenkov counters allows a precise particle identification.

Additional information can be obtained by measuring the Cherenkov angle. These differential Cherenkov counters supply in fact a direct measurement of the particle velocity. The working principle of a differential Cherenkov counter which accepts only particles in a certain velocity interval is shown in figure 6.13 [2].

All particles with velocities above $\beta_{\min} = \frac{1}{n}$ are accepted. With increasing velocity the Cherenkov angle increases and finally reaches the critical angle for internal reflection θ_t in the radiator so that no light can escape into an air light guide. The critical angle for internal reflection can be computed from Snell's law of refraction to be

$$\sin \theta_t = \frac{1}{n} \,. \tag{6.36}$$

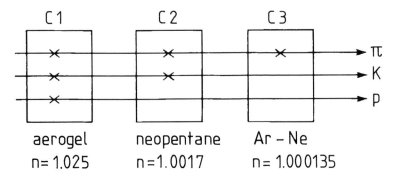

Fig. 6.12.   Principle of particle identification by threshold Cherenkov counters (x represents production of Cherenkov photons).

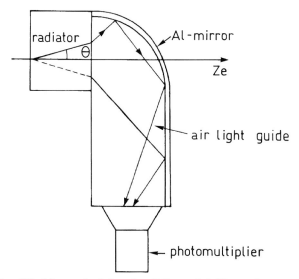

Fig. 6.13.   Working principle of a differential Cherenkov counter [2].

Because

$$\cos\theta = \sqrt{1 - \sin^2\theta} = \frac{1}{n\beta} \tag{6.37}$$

the maximum detectable velocity is

$$\beta_{\mathrm{max}} = \frac{1}{\sqrt{n^2 - 1}} \, . \tag{6.38}$$

For diamond ($n = 2.42$) β_{min} is 0.413 and β_{max} is equal to 0.454. In this way, such a differential Cherenkov counter selects a velocity window of $\Delta\beta = 0.04$. If the optical system of a differential Cherenkov counter is optimized, so that chromatic aberrations are corrected (DISC counter, DIScriminating Cherenkov counter), a velocity resolution of $\Delta\beta/\beta = 10^{-7}$ can be achieved. With these DISC counters pion-kaon separation up to momenta of several hundred GeV/c can be obtained [1, 450].

Differential Cherenkov counters, however, can only be used if the incident particles are parallel to the optical axis, which means their direction of incidence has to be exactly known. This is only true for fixed target experiments at an accelerator. In storage ring experiments where particles can be produced in the full solid angle, differential Cherenkov counters cannot be used. This is the domain of RICH counters (Ring Imaging Cherenkov counters) [451, 452]. In RICH counters a spherical mirror of radius R_S, whose center of curvature coincides with the interaction point, projects the cone of Cherenkov light produced in the radiator onto a ring on the surface of a spherical detector of radius R_D (see figure 6.14, [450]).

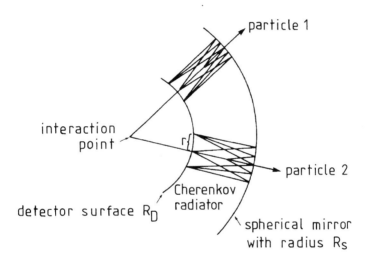

Fig. 6.14. Working principle of a RICH counter [450].

The radiator fills the volume between the spherical surfaces with radii R_S and R_D. In general, one takes $R_D = R_S/2$. The focal length f of a spherical mirror is $R_S/2$. The Cherenkov photons are emitted under an angle θ_c. From this one can calculate the radius of the Cherenkov ring on the detector surface

$$r = f \cdot \theta_c = \frac{R_S}{2} \cdot \theta_c \; . \tag{6.39}$$

The measurement of r allows one to determine the particle velocity via

$$\cos \theta_c = \frac{1}{n\beta} \Longrightarrow \beta = \frac{1}{n \cos \left(\frac{2r}{R_S} \right)} \; . \tag{6.40}$$

The error of the velocity measurement $\Delta\beta$ originates mainly from the experimental uncertainties in the determination of the radius r of the Cherenkov ring. An uncertainty $\Delta\beta$ corresponds to an uncertainty in the Lorentz factor of

$$\Delta\gamma = \beta\gamma^3\Delta\beta \ . \tag{6.41}$$

If the mass, i.e., identity of the particle is known, equation (6.41) permits the determination of the particle's momentum $p = \gamma m_0 \beta c$. Because of

$$\gamma = \frac{1}{\sqrt{1 - \beta^2}} \tag{6.42}$$

or

$$\beta\gamma = \sqrt{\gamma^2 - 1} \tag{6.43}$$

the momentum error is calculated to be

$$\Delta p = \frac{m_0 c \gamma}{\sqrt{\gamma^2 - 1}}\Delta\gamma = \frac{m_0 c}{\beta}\Delta\gamma \ , \tag{6.44}$$

and the relative momentum resolution

$$\frac{\Delta p}{p} = \frac{\Delta\gamma}{\beta^2\gamma} = \gamma^2\frac{\Delta\beta}{\beta} \ , \tag{6.45}$$

which means for fast particles ($\beta \approx 1$):

$$\frac{\Delta p}{p} = \frac{\Delta\gamma}{\gamma} \ . \tag{6.46}$$

For high energy particles ($E \approx pc$) this result could have been derived more easily by

$$\frac{\Delta p}{p} = \frac{\Delta(pc)}{pc} = \frac{\Delta E}{E} = \frac{\Delta(\gamma m_0 c^2)}{\gamma m_0 c^2} = \frac{\Delta\gamma}{\gamma} \ . \tag{6.47}$$

If, however, the momentum of the charged particle is already known, e.g., by magnetic deflection, then the particle can be identified (i.e., its mass m_0 determined) from the size of the Cherenkov ring r. The measurement of r yields, using equation (6.40), the particle velocity β, and by use of the relation

$$p = \gamma m_0 \beta c = \frac{m_0 c \beta}{\sqrt{1 - \beta^2}} \tag{6.48}$$

the mass m_0 can be determined if the momentum is known.

The most crucial aspect of RICH counters is the detection of Cherenkov photons with high efficiency on the detector surface. Since one is not only interested in detecting the photons, but also in measuring their coordinates, a position sensitive detector is necessary. Normally, multiwire proportional chambers are used where a photosensitive vapor is mixed into the chamber gas. Vapor additions such as triethylamine (TEA: $(C_2H_5)_3N$) with an ionization energy of 7.5 eV and tetrakis-dimethylaminoethylene (TMAE: $[(CH_3)_2N]_2C = C_5H_{12}N_2$; $E_{ion} = 5.4$ eV) are the most interesting candidates. Additional problems arise from the fact that the Cherenkov ring is normally defined by only very few photoelectrons. For usual Cherenkov radiators singly charged particles only produce between 3 and 5 photoelectrons. Figure 6.15 shows the pion-kaon separation in a RICH counter at 200 GeV/c. For the same momentum kaons are slower compared to pions, and consequently produce (see equations (6.39, 6.40)) Cherenkov rings with smaller radii [453].

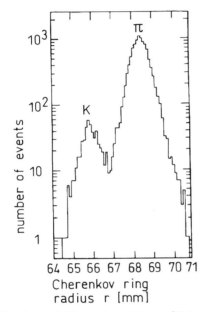

Fig. 6.15. Distribution of Cherenkov ring radii in a pion-kaon beam at 200 GeV/c. The Cherenkov photons have been detected in a multiwire proportional chamber filled with helium (83 %), methane (14 %) and TEA (3 %). Calcium fluoride crystals (CaF_2-crystal), having a high transparency in the ultraviolet region, were used for the entrance window [453].

Further problems arise if jets of particles with overlapping Cherenkov rings — which is normally the case in storage ring experiments — have to be disentangled. Better Cherenkov rings are obtained from fast heavy ions, because the number of produced photons is proportional to the square of the projectile charge. Figure 6.16 [454] shows a Cherenkov ring produced by a relativistic heavy ion. The center of the ring is also shown since the ionization loss in the photon detector leads to a high energy deposit at the center of the ring (see figure 6.14). Spurious signals,

Fig. 6.16.   Cherenkov ring of a relativistic heavy ion in a RICH counter [454].

normally not lying on the Cherenkov ring, are caused by δ-rays, which are produced in interactions of the heavy ions with the chamber gas.

Figure 6.17 [455] shows an example of Cherenkov rings obtained by superimposing 100 collinear events from a monoenergetic collinear particle beam. The four square contours show the size of the calcium fluoride crystals ($10 \times 10\,cm^2$ each), which served as the entrance window for the photon detector. The ionization loss of the particles is also seen at the center of the Cherenkov rings.

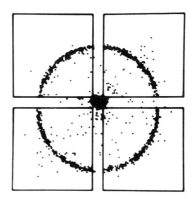

Fig. 6.17.   Superposition of Cherenkov rings of 100 collinear events in a RICH counter. The square contours indicate the calcium-fluoride entrance windows of the photon detector [455].

Figure 6.18 shows the computer reconstructed Cherenkov ring of a single charged particle (right) and the result of the superposition of 10 collinear events (left). The Cherenkov rings were detected with a microchannel plate of 40 mm diameter after being amplified with an image intensifier. The rings, increased this way in brightness, were read out by a photodiode matrix consisting of 80 000 pixels [456].

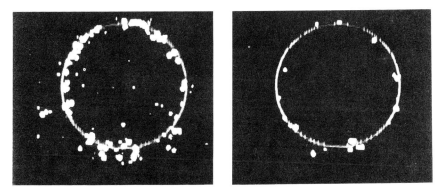

Fig. 6.18. Computer reconstruction of the Cherenkov ring of an individual charged particle (right) and superposition of 10 collinear events (left). The rings were recorded with a multichannel plate [456].

It is even possible to obtain Cherenkov rings from electromagnetic cascades initiated by high energy electrons or photons. The secondary particles produced during cascade development in the radiator follow closely the direction of the incident particle. They are all together highly relativistic and therefore produce concentric rings of Cherenkov light with equal radii lying on top of one another. Figure 6.19 shows a distinct Cherenkov ring produced by a 5 GeV electron [457]. The large number of produced Cherenkov photons can be detected via the photoelectric effect in a position sensitive detector.

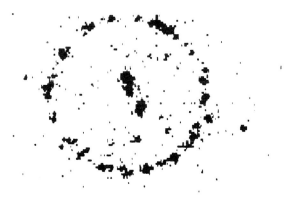

Fig. 6.19. Cherenkov ring produced by a high energy (5 GeV) electron [457].

The shape and position of such Cherenkov rings (elliptically distorted for inclined angle of incidence) can be used to determine the incoming direction of high energy gamma rays in the field of gamma-ray astronomy.

6.5   Transition radiation detectors (TRD)

Also below Cherenkov threshold, charged particles may emit electromagnetic radiation. This radiation is emitted in those cases where charged particles traverse the boundary between media with different dielectric properties [458]. This occurs, for example, when a charged particle enters a dielectric through a boundary from the vacuum or from air, respectively. The energy loss by transition radiation represents only a negligibly small contribution to the total energy loss of charged particles.

A charged particle moving towards a boundary forms together with its mirror charge an electric dipole, whose field strength varies in time, i.e., with the movement of the particle (see figure 6.20). The field strength vanishes when the particle enters the medium. The time dependent dipole electric field causes the emission of electromagnetic radiation.

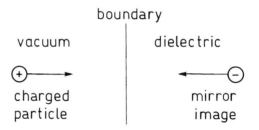

Fig. 6.20.   Illustration of the production of transition radiation at boundaries.

The emission at boundaries can be understood in such a way that although the electric displacement $\vec{D} = \varepsilon\varepsilon_0\vec{E}$ varies continuously in passing through the boundary, the electric field strength does not [459, 460, 461].

The number of transition radiation photons produced can be increased if the charged particle traverses a large number of boundaries, e.g., in porous media or periodic arrangements of foils and air gaps.

The attractive feature of transition radiation is that the radiated energy by transition radiation photons increases with the Lorentz factor γ, (i.e., the energy) of the particle, and not only proportional to its velocity [462, 463]. Since most processes used for particle identification (energy loss by ionization, time-of-flight, Cherenkov radiation, ...) depend on the velocity, thereby representing only very moderate identification possibilities for relativistic particles ($\beta \to 1$), the γ-dependent effect of transition radiation is extremely valuable for particle identification at high energies.

An additional advantage is the fact that transition radiation photons are emitted in the X-ray range [464]. The increase of the radiated energy in transition radiation proportional to the Lorentz factor originates mainly from the increase of the average energy of X-ray photons and much less from the increase of the radiation intensity. In figure 6.21 the average energy of transition radiation photons is shown in their dependence on the electron momentum for a typical radiator [450].

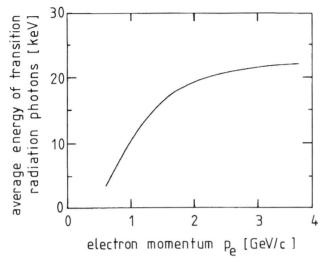

Fig. 6.21. Typical dependence of the average energy of transition radiation photons on the electron momentum for standard radiator arrangements [450].

The angle of emission of transition radiation photons is inversely proportional to the Lorentz factor

$$\theta = \frac{1}{\gamma_{\text{particle}}} . \tag{6.49}$$

For periodical arrangements of foils and gaps interference effects occur, which produce an effective threshold behaviour at a value of $\gamma \approx 1000$ [465, 466], i.e., for particles with $\gamma < 1000$ almost no transition radiation photons are emitted.

The typical arrangement of a transition radiation detector (TRD) is shown in figure 6.22. The TRD is formed by a set of foils consisting of a material with an atomic number Z as low as possible. Because of the strong dependence of the photoabsorption cross section on Z ($\sigma_{\text{photo}} \propto Z^5$) the transition radiation photons would otherwise not be able to escape from the radiator. The transition radiation photons have to be recorded in a detector with a high efficiency for X-ray photons. This requirement is fulfilled by a multiwire proportional chamber filled with krypton or xenon, i.e., gases with high atomic number for an effective absorption of X-rays.

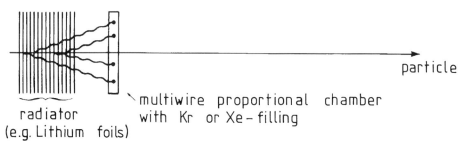

radiator
(e.g. Lithium foils)

Fig. 6.22.   Working principle of a transition radiation detector.

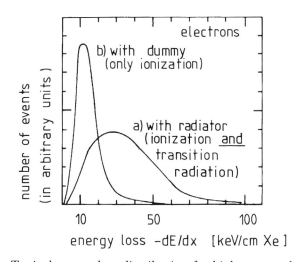

Fig. 6.23.   Typical energy-loss distribution for high energy electrons in a transition radiation detector with radiator and with dummy radiator [450].

In the set-up sketched in figure 6.22 the charged particle also traverses the photon detector, leading to an additional energy deposit by ionization and excitation. This energy loss is superimposed onto the energy deposit by transition radiation. Figure 6.23 shows the energy-loss distribution in a transition radiation detector for highly relativistic electrons for the case that a) the radiator has gaps and b) the radiator has no gaps (dummy). In both cases the amount of material in the radiators is the same. In the first case, because of the gaps, transition radiation photons are emitted, leading to an increased average energy loss of the electrons, while in the second case only the ionization loss of electrons is measured [450].

For basic detector studies concerning production of transition radiation the inconvenient effects of the ionization loss can be excluded by deflecting the charged particles in a magnetic field, thus preventing them from reaching the photon detector (see figure 6.24). For applications of

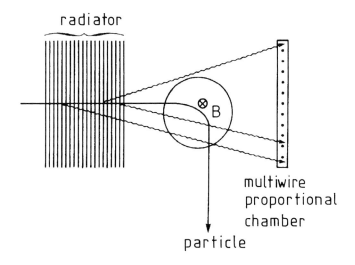

radiator

⊗ B

multiwire
proportional
chamber

particle

Fig. 6.24. Investigation of the effect of transition radiation production without disturbance from the incoming particle by means of magnetic deflection.

transition radiation in a particle physics experiment, however, this is not feasible.

Since the effective threshold for the production of transition radiation in periodic structures amounts to approximately $\gamma \approx 1000$, pions with energies below about 140 GeV would not produce transition radiation photons, but only deposit the energy loss by ionization and excitation in the detector. This provides a means to distinguish electrons from pions. Figure 6.25 [467] shows the energy-loss distribution of 15 GeV electrons ($\gamma_e \approx 30\,000$) and 15 GeV pions ($\gamma_\pi \approx 110$) in a transition radiation detector. This figure indicates that a cut in the measured energy loss allows one to separate to a certain extent electrons from pions. The effective separation of the transition radiation signal from the Landau tail of the ionization loss, however, presents large problems. These can be reduced if one considers the different nature of energy loss by ionization on the one hand and by transition radiation on the other. The idea of an effective separation consists of not only measuring the total deposited charge in a multiwire proportional chamber, but also its spatial distribution.

The total energy loss by ionization and excitation originates from a large number of low energy transfers to electrons. Only very rarely are somewhat more energetic δ-electrons produced. Quite in contrast to this, the energy loss by transition radiation is made up of very few local strong energy depositions which originate from the absorbed transition radiation photons. This feature is sketched in figure 6.26. If only the local energy deposits above a certain preselected charge $Q_{\text{threshold}}$ are measured, the

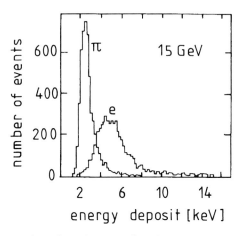

Fig. 6.25. Energy-loss distribution of 15 GeV electrons and pions in a transition radiation detector [467].

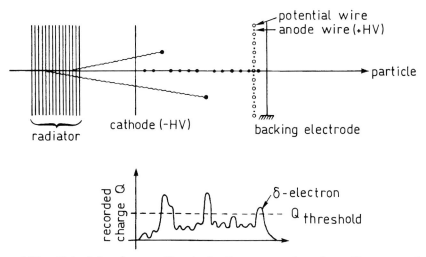

Fig. 6.26. Principle of separating ionization energy loss from the energy loss from emission of transition radiation photons.

more or less continuous energy loss by ionization can be effectively suppressed.

The method of cluster counting was applied for the data, for which the total energy loss is shown in figure 6.25. The result (figure 6.27, [467]) shows a much better electron-pion separation. To quantify this effect, a cut in the number of recorded clusters is applied. This, of course, reduces the efficiency of the electron detection. A certain number of pions exceeding the cut value would, however, be misinterpreted as electrons.

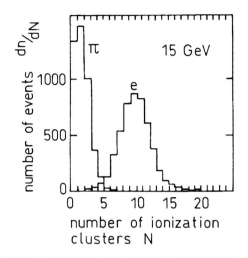

Fig. 6.27. Cluster-frequency distribution for pions and electrons in a transition radiation detector [467].

One has always to find a compromise between an electron efficiency as high as possible for a low pion contamination. Figure 6.28 shows the pion contamination as a function of the electron acceptance for both separation methods either on the basis of the total energy loss, or using the cluster counting method [467]. The latter allows a pion suppression by a factor 10^3 for an electron acceptance of 90 %, which is clearly superior to the first method. The data in figure 6.28 have been obtained for 15 GeV electrons and pions in a lithium-foil radiator. Instead of lithium foils, radiators from small diameter carbon fibers ($\phi < 20\,\mu$m), thin mylar foils or porous foams can also be used [468].

   Just as electrons can be separated from pions, pions can be distinguished from kaons at correspondingly higher energies. For momenta in excess of 140 GeV/c ($\gamma_\pi = 1000$) pions will produce transition radiation; kaons ($\gamma_K = 280$), however, do not. With a long lithium radiator kaons can be suppressed to the level of 10 % for a pion acceptance of 99 % [467, 1]. If, conversely, kaons are to be selected, the energetic pions can be suppressed by a factor of 10^2 by requiring the emission of transition radiation, while kaons, which do not emit transition radiation photons, are identified with an efficiency of 90 % via their low energy loss from ionization only.

   In the TeV-region even a $\pi/K/p$ separation with transition radiation detectors appears feasible. Monte Carlo calculations have indicated that X-ray transition radiation detectors can be extended to Lorentz factors of 10^4 to 10^5 with possible applications at the Tevatron (Fermilab), UNK (Russia), LHC and CLIC (= CERN Linear Collider) [469].

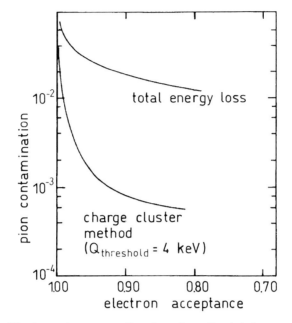

Fig. 6.28. Electron-pion separation based on the total energy loss and on the cluster counting method [467]. The electron detection efficiency can in principle be made very large. However, to reduce the pion contamination, not all electrons are accepted. Therefore the detection efficiency for electrons is called electron acceptance.

For a radiator of 1 m length made from 100 μm lithium foils (gaps also 100 μm) 5000 foils would be required. The production of extremely thin lithium foils is rather complicated.

The actual set-up of a transition radiation detector with such a large number of foils would be made using chambers for photon measurement inserted into the stack of foils at regular intervals. This would reduce the effect of the absorption of photons in the radiator foils themselves to as low a level as possible.

6.6 Energy-loss sampling

In particle physics experiments it is often desirable to cover a wide energy range with different methods of particle identification. Time-of-flight measurements have to be abandoned if the velocity differences for particles to be distinguished become too small. This method allows a pion-kaon separation up to momenta of 2 GeV/c ($\gamma_\pi = 14$). Threshold Cherenkov counters achieve a pion-kaon separation up to $p = 20$ GeV/c ($\gamma_\pi = 140$). Differential Cherenkov counters can only be incorporated in special set-ups, which allow a pion-kaon identification up to $200\,\mathrm{GeV}/c$ ($\gamma_\pi = 1400$). RICH counters cover a comparable momentum range, but require, however, a sophisticated construction and readout. Methods based on transition radiation emission can only be used for very energetic particles ($\gamma = 1000$). If one discards for the moment DISC counters because of their restricted use and the complicated RICH counters, there is still a gap for particle identification in the energy range $100 \leq \gamma \leq 1000$. This gap can be bridged by the measurement of the relativistic rise in the ionization loss of charged particles. The measurement of the energy loss has to be quite accurate to allow a reliable particle identification in this energy range.

The average energy loss of electrons, muons, pions, kaons, and protons in the momentum range between 0.1 and $100\,\mathrm{GeV}/c$ in a 1 cm-layer of argon-methane ($80\,\% - 20\,\%$) is shown in figure 6.29 [1, 470]. It is immediately clear that a muon-pion separation on the basis of an energy-loss measurement is practically impossible. These two particles are too close in mass ($m_\mu = 105.7\,\mathrm{MeV}/c^2$; $m_\pi = 139.6\,\mathrm{MeV}/c^2$) corresponding to almost equal β for equal momentum. However, a $\pi/K/p$ separation should be achievable. The logarithmic rise of the energy loss in gases ($\propto \ln \gamma$, see equation (1.12)) amounts to 50 to $60\,\%$ compared to the energy loss of minimum-ionizing particles at a pressure of 1 atm [5, 6].

The measurement of the average energy loss $\mathrm{d}E/\mathrm{d}x$ can be used to identify particles in a particle beam of known momentum and unknown composition. One problem will arise from the fact that the curves in figure 6.29 only show the average energy loss; the energy loss of an individual measurement, however, fluctuates according to a Landau distribution. The long tails of the asymmetric energy-loss distributions actually render the particle separation considerably more difficult compared to a symmetric distribution about the mean.

Normally the energy-loss measurements are made in gaseous detectors, but they can also be performed in, e.g., solid state counters. Figure 6.30 shows the energy-loss spectra of $600\,\mathrm{MeV}/c$ pions and protons in a 3 mm lithium-drifted silicon counter in an unseparated particle beam [471].

The asymmetric energy-loss distributions with tails to large energy transfers, making particle identification more difficult, are also clearly visible here. In this example, the energy-loss differences at momenta below the minimum of ionization are used for particle separation. Pions at $600\,\mathrm{MeV}/c$ are already minimum ionizing, while protons of the same momentum, because of their lower velocity and due to the $1/\beta^2$ dependence of the energy loss, exhibit larger signals (compare figure 6.29).

The Landau fluctuations can be effectively suppressed by making multiple measurements of the energy loss and using only those measurements in which the measured value falls below a certain cut (typically includ-

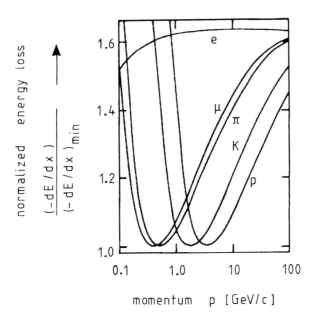

Fig. 6.29.   Average energy loss of electrons, muons, pions, kaons and protons, normalized to minimum-ionizing value [1, 470].

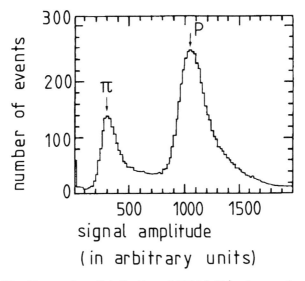

Fig. 6.30.   Energy-loss distribution of 600 MeV/c pions and protons in a lithium-drifted silicon semiconductor counter with a thickness of 3 mm [471].

ing $40-60\%$ of all individual energy-loss measurements). This method excludes high energy transfers which are caused by the occasional production of δ-electrons. The restricted energy-loss sample is used to form a 'truncated mean' which is used for particle identification. With about 100 dE/dx measurements energy-loss resolutions of

$$\frac{\sigma(dE/dx)}{(dE/dx)} = 2\% \tag{6.50}$$

for pions, kaons and protons of 50 GeV can be achieved [1]. The resolution can be improved by increasing the number N of individual measurements according to $1/\sqrt{N}$; which means, to improve the dE/dx-resolution by a factor of two, one has to take four times as many dE/dx-measurements. For a fixed total length of a detector, however, there exists an optimum number of measurements. If the detector is subdivided in too many dE/dx-layers, the energy loss per layer will eventually become too small, thereby increasing its fluctuation.

The resolution should also improve with increasing gas pressure in the detector like $1/\sqrt{p}$. One must, however, be careful not to increase the pressure too much, otherwise the logarithmic rise of the energy loss, which is the basis for the particle identification, will be reduced by the onset of the density effect. The increase of the energy loss compared to the minimum of ionization at 1 atm amounts to about 55 %. For 7 atm it is reduced to 30 %. In total, although a pressure rise leads to an improvement of the accuracy of the energy-loss measurement, the discriminating power between two mass hypotheses does not improve.

An alternative, more sophisticated method compared to the use of the truncated mean of a large number of energy-loss samples, which also provides more accurate results, will be discussed in the following.

Figures 6.31 (a) and (b) show the energy-loss distribution of 50 GeV pions and kaons in a 1 cm layer of argon/methane ($80\% - 20\%$) in a linear and logarithmic scale. The Landau distributions may be interpreted as probability distributions, that a pion or kaon produces a signal of given size. Let $P_\pi^i(x)$ be the probability that the pion produces a signal of size x in the detector. Each particle yields a set of $x_i (i = 1, 2, \ldots, N)$ signals. The probability that this set was produced by a pion is

$$P_1 = \prod_{i=1}^{N} P_\pi^i(x_i) . \tag{6.51}$$

Correspondingly, a kaon will produce the same set of signals with the probability of

$$P_2 = \prod_{i=1}^{N} P_K^i(x_i) . \tag{6.52}$$

Consequently, in a beam of pions and kaons of constant (known) momentum the probability to identify a particle as pion is

$$P = \frac{P_1}{P_1 + P_2} .$$
(6.53)

As an example, let us consider a five-fold measurement of the energy loss as shown in figure 6.32. The kaon hypothesis for five dE/dx measurements leads to a set of probabilities (0.124; 0.061; 0.025; 0.013; 0.006) with $P_2 = 1.5 \cdot 10^{-8}$; correspondingly, the pion hypothesis (0.031; 0.236; 0.192; 0.108; 0.047) yields a value of $P_1 = 7.1 \cdot 10^{-6}$, so that the particle that produced this set of energy-loss measurements is a pion with a probability of

$$P = \frac{P_1}{P_1 + P_2} = 99.8\% .$$
(6.54)

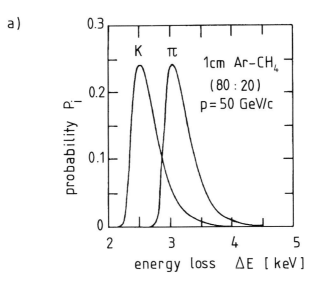

Fig. 6.31 a.   Energy-loss distribution of 50 GeV/c pions and kaons in a layer of 1 cm argon methane shown with a linear scale.

This method of interpreting the energy-loss distributions as probability distributions is rather time consuming, but yields, however, much better results compared to using the truncated mean only.

It is, however, also possible to combine both methods of particle identification based on the logarithmic rise of the energy loss by applying the method of probability distributions to the restricted set of energy-loss samples. The gain in particle identification reliability that can be obtained with this combined method depends on the cut in the energy-loss distribution but is in most cases only marginal.

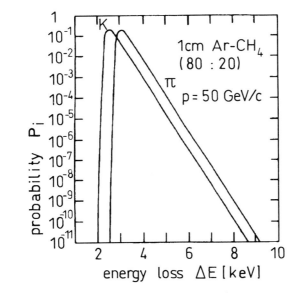

Fig. 6.31 b. Energy-loss distribution of 50 GeV/c pions and kaons in a layer of 1 cm argon methane shown with a semi-logarithmic scale.

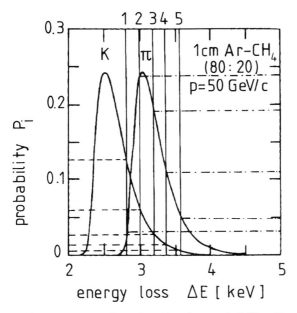

Fig. 6.32. π/K-separation based on Landau probability distributions illustrated by a five-fold measurement of the energy loss. The five dE/dx measurements with their corresponding probabilities P_π^i and P_K^i are indicated.

Figure 6.33 shows the results of energy-loss measurements in a mixed particle beam [455, 472, 473]. This figure very clearly shows that the method of particle separation by multiple dE/dx sampling only works either below the minimum of ionization ($p < 1\,\mathrm{GeV}/c$) or in the relativistic rise region.

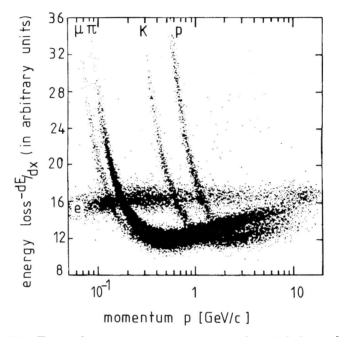

Fig. 6.33. Energy-loss measurements in a mixed particle beam [455, 472, 473].

6.7 Comparison of particle identification methods

On the basis of the methods described, such as ionization measurement, dE/dx sampling, time-of-flight measurement and the detection of Cherenkov and transition radiation, a successful particle identification can be achieved at practically all momenta. The momentum ranges that allow a pion-kaon separation with a realistic detector are sketched in figure 6.34.

A detector specialized in particle identification is sketched as an example in figure 6.35. In a real experiment, however, such a detector would become very bulky, because in general the various subdetectors must be large in order to yield reliable results.

The particle identification methods described so far are essentially non-destructive, which means that the energy loss of charged particles is rela-

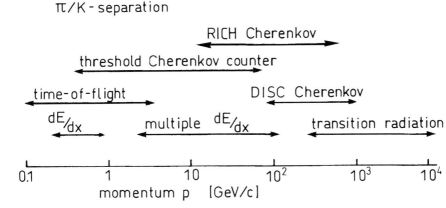

Fig. 6.34. Illustration of various particle identification methods for π/K-separation along with characteristic momentum ranges.

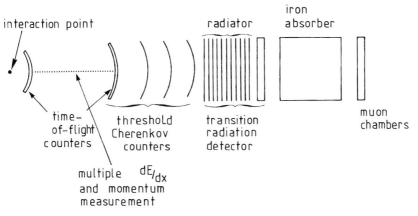

Fig. 6.35. Sketch of a detector specialized in particle identification.

tively low compared to their energy. Many identification methods rely on the knowledge of the particle momentum. For this reason magnet spectrometers (see chapter 8) are an important tool for particle identification.

There are further methods, however, to identify, for example, electrons at high energies very reliably. Electrons initiate cascades in electromagnetic calorimeters (see chapter 7), whose development is characteristically different from those of other charged particles. The problem of muon-pion separation can be solved by taking advantage of the high penetration probability of muons through massive absorbers, in which pions are stopped by the development of hadronic cascades. The calorimetric particle-identification method, however, totally absorbs the particles (except muons) so that no further measurements on these particles can be made.

Particle identification with calorimeters is described in section 7.4.

The methods of particle identification presented here can only be used if the particles are sufficiently long-lived to allow the application of the described identification techniques. For short-lived particles other means of particle identification must be used.

The standard technique to identify short-lived particles is to measure their decay products with the described methods and to determine the invariant mass of the system of decay particles. This requires an accurate momentum and energy measurement of the decay products. Combinatorial background in high multiplicity events due to wrong particle combinations normally leads to a smooth mass distribution. Superimposed on this background one finds the signal distribution. The width of the signal is given by the experimental resolution of the invariant mass (resulting from the finite momentum and energy resolution) and the lifetime (width) of the unstable particle.

Looking for the invariant mass of two oppositely charged pions in multihadronic decays of the Z^0-particle one will find, e.g., enhancements at masses of $498 \, \mathrm{MeV}/c^2$ and $768 \, \mathrm{MeV}/c^2$ corresponding to K_s^0 and ϱ^0 decays. The width of the ϱ^0-signal distribution will be dominated by the short lifetime of the ϱ^0 (width $\Gamma = 152 \, \mathrm{MeV}$), while the width of the relatively long-lived K_s^0 (lifetime $\tau = 8.9 \cdot 10^{-11} \, \mathrm{s}$) will be dominated by the experimental resolution of the invariant mass. In this method the short-lived particles are identified by their reconstructed invariant mass.

7

Energy measurement

Many detectors can be used for energy measurements. For example, the energies of X-ray photons can be determined in proportional counters. Every device that totally absorbs an incoming particle also measures its energy if it is deposited in the sensitive volume of the detector. At high energies ($\geq 1\,\mathrm{GeV}$) the determination of energy is done with calorimetric methods; depending on the particle species it is done in electromagnetic calorimeters for photons and electrons and in hadron calorimeters for strongly interacting particles [474, 475]. At these energies one has $E \approx p \cdot c$, so that momentum spectrometers (see chapter 8) at high momenta provide energy information at the same time.

For lower energies (MeV-range) solid state counters allow a precise energy determination. These detectors can also be used for track measurements with high accuracy [124, 476, 477]. Electron-hadron calorimeters, on the other hand, may provide tracking information as well.

Modern methods of energy measurement of elementary particles have to cover a large dynamical range of more than 20 orders of magnitude in energy. The detection of extremely small energies (milli-electronvolts) is of great importance in astrophysics if one searches for the remnants of the Big Bang. At the other end of the spectrum, one measures in cosmic rays particles with energies of up to $10^{20}\,\mathrm{eV}$, which are presumably of extra-galactic origin [478].

7.1 Solid state detectors

Solid state detectors are essentially ionization chambers with solids as the counting medium. Because of their high density compared to gaseous detectors they can absorb particles of correspondingly higher energy.

Charged particles or photons produce electron-hole pairs in a semiconducting material. An electric field is applied across the semiconductor

223

crystal which allows the produced charge carriers to be collected. The main advantage of solid state detectors is that the average energy required for the production of an electron-hole pair compared to that in gases is rather small. For silicon (germanium) one needs 3.6 eV (2.8 eV) for the production of an electron-hole pair in contrast to approximately 30 eV in gases. For comparison, the energy gap between valence and conduction bands is 1.14 eV (0.67 eV) in silicon (germanium) at room temperature. Because of the low band gap in germanium such detectors generally have to be cooled to reduce thermal noise. Because of the low W-value (see section 1.1.2) solid state counters provide excellent energy resolution. Solid state detectors (SSD) can be compared to scintillation counters (SC) where an energy between 400 and 1000 eV is required to produce one photoelectron. A rough idea for the ratio of energy resolutions in these two detector types is obtained from

$$\frac{\sigma_{\mathrm{SSD}}(E)/E}{\sigma_{\mathrm{SC}}(E)/E} = \frac{\sqrt{N_{\mathrm{SC}}}}{\sqrt{N_{\mathrm{SSD}}}} = \frac{\sqrt{E/700\,\mathrm{eV}}}{\sqrt{E/3\,\mathrm{eV}}} = 6 \cdot 10^{-2} \;, \qquad (7.1)$$

where $N_{\mathrm{SSD}}(N_{\mathrm{SC}})$ is the number of produced charge carriers in a solid state detector (scintillator-photomultiplier system). Of course, care must be taken not to spoil the excellent resolution of solid state detectors by noisy preamplifiers.

The energy resolution of solid state detectors is typically superior by a factor of 10–50 over that of scintillators. Figure 7.1 [479] shows the result of the photon energy measurement from a ^{60}Co-source in a germanium solid state detector compared to the measurement in a NaI(Tl)-scintillation counter. The γ-ray lines at 1.17 MeV and 1.33 MeV are resolved in the solid state detector with a full width at half maximum of 1.9 keV ($\sigma(E) = 0.80$ keV) and in the NaI(Tl)-scintillator with 90 keV ($\sigma(E) = 38$ keV).

The operating principle of solid state detectors can best be understood from the band structure of solids. A germanium or silicon crystal becomes n-conducting (i.e., conducting for negative charge carriers, electrons), when electron donor impurities are introduced into the lattice. Germanium and silicon have four electrons in the outer shell. Atoms with five electrons in the outer shell act as electron donor impurities. Similarly, germanium and silicon become p-conducting (i.e., conducting for holes) if trivalent atoms which act as electron acceptor impurities are introduced into the crystal lattice.

Figure 7.2 shows the band structure of a solid state detector. Phosphor and arsenic act as electron donors. The neighbouring silicon (germanium) atoms can only bind four electrons. The fifth electron of the electron donor impurity is only weakly bound and can easily reach the conduction band. The donor levels are situated approximately 0.05 eV below the edge of the

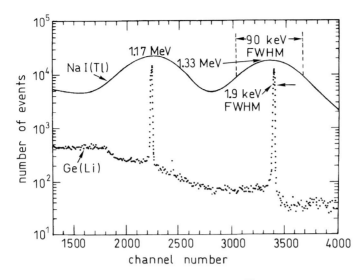

Fig. 7.1. Comparison of measurements of the ^{60}Co-gamma ray spectrum, obtained with a NaI(Tl)-scintillator and a Ge(Li)-solid state detector [479].

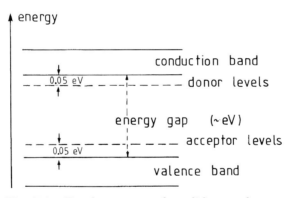

Fig. 7.2. Band structure of a solid state detector.

conduction band.

If trivalent electron acceptor impurities like boron or indium are added to the lattice, one of the silicon bonds remains incomplete. This acceptor level, which is about 0.05 eV above the edge of the valence band, tries to attract one electron from a neighbouring silicon atom. As a consequence, the state of the missing electron (the hole) migrates through the crystal. The addition of small amounts of energy can cause the holes to be transferred to the valence band, thereby causing a hole current. Lithium acts as an electron donor because it has only one weakly bound electron in the outer shell.

If a charged particle traverses an n or p-conducting crystal, it will produce along its track electron-hole pairs. The primary electrons can produce further secondary electron pairs or excite lattice vibrations (phonons). A plasma channel along the particle track with charge carrier concentrations from 10^{15} up to $10^{17}/cm^3$ is produced [1]. The working principle of a solid state detector now consists of collecting the free charge carriers in an external drift field before they can recombine with the holes. If this is successful, the measured charge signal is proportional to the energy loss of the particle or, if the particle deposits its total energy in the sensitive volume of the detector, it is proportional to the particle energy.

Semiconductor counters must be operated with a reverse bias voltage to produce a sufficiently large electric field in the crystal to collect the electrons. Normally one uses diodes with pn junctions, surface barrier detectors, or p-i-n structures (see figures 7.3 to 7.6) [480, 481, 482].

To produce a pn junction two semiconductors of p and n-type are joined together. The electrons of the n-type semiconductor diffuse into the p-type, and the holes from the p-type to the n-type region. This leads to a recombination of the charge carriers near the boundary. In this region a depletion layer without free charge carriers is produced, even if no external voltage is applied to the semiconductor. Figure 7.3 shows the working principle of a pn semiconductor. Shown in each case are the free charge carriers in the n or p-regions. The immobile ions are positive in the n-region (they are electron donors) and negative in the p-region (they are electron acceptors). The charged ions produce a voltage difference across the depletion layer.

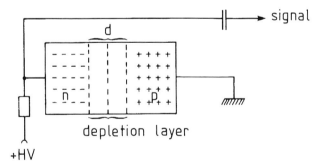

Fig. 7.3. Working principle of a pn semiconductor counter (the free charge carriers only are shown).

The pn junction has the properties of a diode. An external positive voltage applied to the n-conducting region increases the depletion layer. Depletion layers of up to 1 mm thickness can be obtained in this way. Since the regions of the crystal next to the depletion layer have a relatively good conductivity, the applied reverse voltage effectively produces a field across the depletion layer. Free electrons, which are produced in this volume by charged particles or photons, can be collected in this field before they recombine.

Typical widths of depletion layers are $d = 300\,\mu$m. For field strengths of $E = 10^3$ V/cm and charge carrier mobilities of $\mu = 10^3$ cm^2/Vs collection times of

$$t_{\mathrm{s}} = \frac{d}{\mu E} \approx 3 \cdot 10^{-8}\,\mathrm{s} \tag{7.2}$$

are obtained. For α-particle and electron spectroscopy it is necessary that the depletion layers in semiconductor counters are near the surface. These surface barrier detectors are made of an n-conducting silicon crystal. At its surface a depletion layer is produced by contact with a p-conducting silicon crystal. (In practice it is of course also possible to produce a depletion layer by doping the silicon crystal differently on each side.) A thin evaporated gold layer of several μm thickness serves as a high voltage contact. This side also is used as the entrance window for charged particles (see figure 7.4). In this way depletion layers of thickness

$$d = 0.309\sqrt{U \cdot \varrho_{\mathrm{p}}}\ \ [\mu\mathrm{m}] \tag{7.3}$$

for p-doped silicon and

$$d = 0.505\sqrt{U \cdot \varrho_{\mathrm{n}}}\ \ [\mu\mathrm{m}] \tag{7.4}$$

for n-doped silicon can be obtained [1, 2]. U is the reverse bias voltage (in volts); ϱ_{p} and ϱ_{n} the specific resistivity in the p and n-doped silicon in Ω cm; and d the thickness of the depletion layer in μm. For typical values of $\varrho_{\mathrm{p}} = 3 \cdot 10^3\,\Omega$ cm, depletion layer thicknesses of $170\,\mu$m are obtained for bias voltages of 100 V.

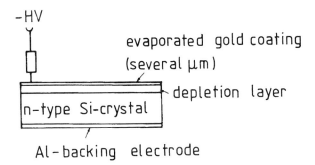

Fig. 7.4. Working principle of a surface barrier detector.

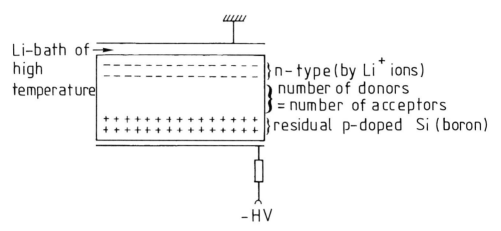

Fig. 7.5. Principle of construction of a *p-i-n* solid state detector. The zone where the number of donors equals the number of acceptors corresponds to the region of intrinsic conductivity.

Thicker depletion layers can be obtained with *p-i-n* structures. These are devices in which a region of intrinsic conductivity is produced between a *p* and an *n*-conducting layer. Such *p-i-n* structures can be made by drifting lithium into *p*-conducting, e.g., boron-doped, silicon (see figure 7.5). Lithium has three electrons in the outer shell and is therefore an electron donor, since its outer electron is only weakly bound. Lithium atoms are allowed to diffuse into the *p*-conducting crystal at a temperature of about 400 °C. Because of their small size, reasonable diffusion velocities are obtained with lithium atoms. After the diffusion process has been completed, an external electric field is used to drift the positive lithium ions further into the crystal. A region is formed in which the number of lithium ions is compensated by the remaining boron ions. In this way a specific resistivity of $3 \cdot 10^5 \, \Omega$ cm is produced in the depletion layer, which is approximately equal to the intrinsic conductivity of silicon without any impurities. An applied external bias voltage transforms the depletion layer into a barrier layer which serves as detector volume. In this way, *p-i-n* structures can be produced with relatively thin *p* and *n*-regions and with *i*-zones up to 5 mm.

   The space charge (produced by positive and negative ions), the electric field strength and potential across a solid state detector with *p-i-n* structure are shown in figure 7.6 [2, 63]. The space charges confined to the relatively thin *p* and *n*-regions cause a constant electric field strength to be produced across the intrinsic region. The potential difference between the *n* and *p*-conducting zone is composed of the external bias voltage V_0 and the diffusion voltage V_D.

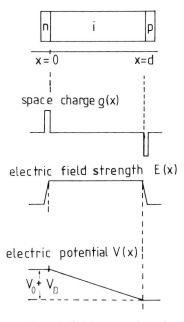

Fig. 7.6. The space charge, electric field strength and and potential as a function of position across a semiconducting *p-i-n* structure (V_D is the diffusion voltage) [2].

Such Si(Li) or Ge(Li) solid state detectors are well suited for gamma and electron spectroscopy in the MeV-range where they have excellent resolution. To avoid thermally caused dark currents, germanium detectors have to be cooled because of their low energy gap (0.67 eV).

Since it has become possible to grow large germanium crystals without impurities, Ge(Li) detectors are now frequently replaced by ultra-pure germanium detectors. These have the additional advantage that they only have to be cooled during operation, while Ge(Li) detectors must be permanently cooled to prevent the lithium ions from diffusing out of the intrinsically conducting region.

Gallium-arsenide (GaAs) solid state devices are also excellent candidates for particle detectors in the fields of nuclear and elementary particle physics [483]-[486].

Figures 7.7 and 7.8 show the ranges of electrons, protons, deuterons, α-particles and some heavy ions in silicon [487]. A silicon lithium-drifted semiconductor counter of 5 mm thickness will stop α-particles up to 120 MeV, protons up to 30 MeV and electrons up to about 3 MeV for normal incidence.

Figure 7.9 shows part of the conversion line spectrum of ^{207}Bi, which decays by electron capture into lead, recorded with a Si(Li) detector.

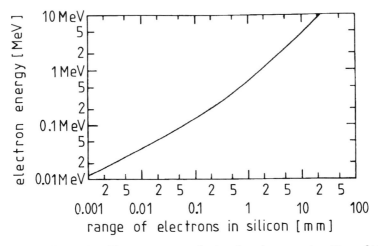

Fig. 7.7.   Energy-range relation for electrons in silicon [487].

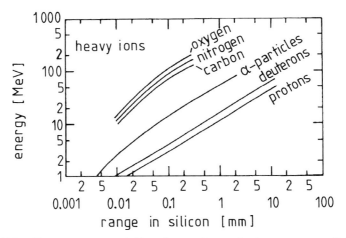

Fig. 7.8.   Energy-range relation for protons, deuterons, α-particles and some heavy ions in silicon [487].

Clearly visible are two K and L line pairs corresponding to the nuclear level transitions of 570 keV and 1064 keV. The 976 keV-conversion line is resolved with a relative width of 1.4 % fwhm. The best Si(Li) semi-conductor counters even allow a separation of the M-electrons from the K and L-electrons. Figure 7.10 shows part of the ^{207}Bi-conversion line spectrum in the range of the 570 keV transition. A distinct line that is visible can even be traced back to the production of a K-shell conversion electron with subsequent absorption of the characteristic L_α line $(K+L_\alpha)$ [488].

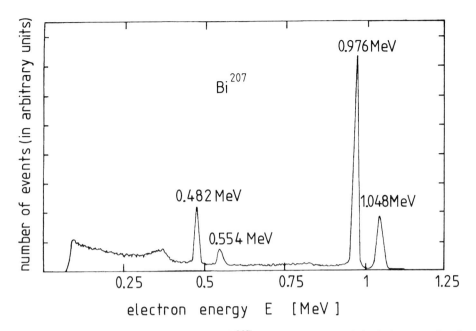

Fig. 7.9. Conversion line spectrum of ^{207}Bi, recorded in a Si(Li) detector [489].

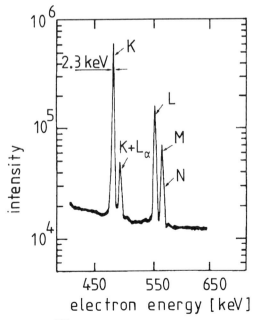

Fig. 7.10. Part of the ^{207}Bi-conversion line spectrum in the region of the 570 keV transition [488].

To compare the properties of solid state detectors with other counters, figures 7.11 and 7.12 show the energy spectra of the isotope ^{207}Bi, recorded with liquid-argon and liquid-xenon ionization chambers. The liquid-argon chamber separates the K and L-electrons relatively well and achieves a resolution of $\sigma_E = 11$ keV. Because of the high atomic number of xenon ($Z = 54$), it also records the photons of 570 keV and 1064 keV energies, emitted by ^{207}Bi with high probability. The K and L-conversion lines can no longer be separated with this chamber. In both counters, Compton scattered photons cause a significant background [421].

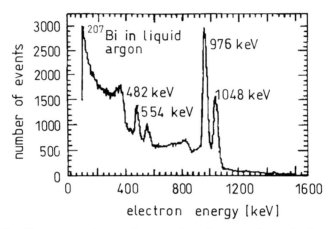

Fig. 7.11.   Energy spectrum of conversion electrons from the isotope ^{207}Bi in a liquid-argon chamber [421].

In a liquid-xenon detector doped with 20 ppm TEA ((C_2H_5)$_3$N) a resolution of 30 keV fwhm for the 570 keV photon line has been obtained [490].

For solid state counters, just as with gaseous detectors, the statistical fluctuation of the number of produced charge carriers is smaller than Poissonian fluctuations, $\sqrt{n}$. The shape of the monoenergetic peak is somewhat asymmetric and narrower than a Gaussian distribution. The Fano factor F (for silicon 0.16, see section 1.1.2) modifies the Gaussian variance σ^{*2} to $\sigma^2 = F\sigma^{*2}$, so that the energy resolution — because E is proportional to n — can be represented by

$$\frac{\sigma(E)}{E} \propto \frac{\sqrt{F\sigma^{*2}}}{n} = \frac{\sqrt{n}\sqrt{F}}{n} = \frac{\sqrt{F}}{\sqrt{n}} \ . \tag{7.5}$$

Using $n = E/W$, where W is the average energy required for the produc-

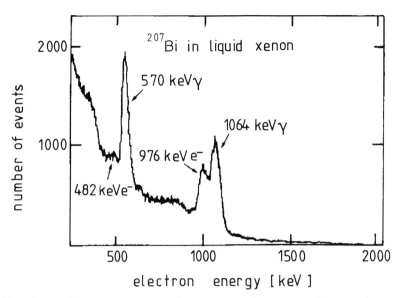

Fig. 7.12. Energy spectrum of conversion electrons and γ-ray photons from the isotope ^{207}Bi in a liquid-xenon chamber [421].

tion of one charge carrier pair, one obtains

$$\frac{\sigma(E)}{E} = \frac{\sqrt{F \cdot W}}{\sqrt{E}} \ . \tag{7.6}$$

Consequently, the energy resolution of the 976 keV-line of the ^{207}Bi-conversion spectrum recorded with a Si(Li) counter could reach a theoretical value of $8 \cdot 10^{-4}$. Similarly, the 1.33 MeV γ-ray peak of the ^{60}Co spectrum could in principle be resolved with a relative precision of $9 \cdot 10^{-4}$ (the Fano factor for germanium is $F = 0.4$), which in fact can be achieved with the best detectors.

The processing of signals from solid state detectors requires the use of low-noise charge sensitive amplifiers.

Germanium and silicon solid state detectors are preferentially used for alpha, beta and gamma-ray spectroscopy. These semiconductor counters are characterized by quantum transitions in the range of several electron volts. The energy resolution could be further improved if the energy absorption were done in even finer steps, such as by the break-up of Cooper pairs in superconductors. Figure 7.13 shows the amplitude distribution of current pulses, caused by manganese K_α and K_β X-ray photons in an Sn/SO$_x$/Sn-tunnel junction layer at $T = 400$ mK. The obtainable resolutions are in this case already significantly better than the results of the best Si(Li) semiconductor counters [491].

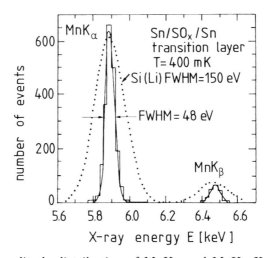

Fig. 7.13. Amplitude distribution of MnK$_\alpha$ and MnK$_\beta$ X-ray photons in an Sn/SO$_x$/Sn-tunnel junction layer. The dotted line shows the best obtainable resolution with a Si(Li) semiconductor detector for comparison [491].

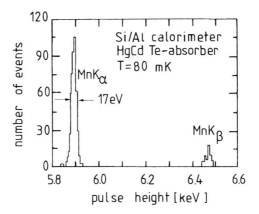

Fig. 7.14. Amplitude distribution of 5.9 keV and 6.47 keV X-rays from the MnK$_\alpha$ and K$_\beta$-lines in a bolometer consisting of an HgCdTe-absorber and a Si/Al-calorimeter. The K$_\alpha$-line corresponds to a transition from the L into the K-shell, the K$_\beta$-line a transition from the M into the K-shell [493].

For even lower temperatures ($T = 80$ mK) resolutions of 17 eV fwhm for the manganese K$_\alpha$ line have been obtained with a bolometer made from an Hg Cd Te-absorber in conjunction with a Si/Al-calorimeter (see figure 7.14) [492, 493].

With the bolometer, a deposited energy of 5.9 keV from K$_\alpha$ X-rays is registered by means of a temperature rise. These microcalorimeters must have an extremely low heat capacity, and they have to be operated at cryo-

genic temperatures (see also section 7.6). In most cases they consist of an
absorber with a relatively large surface (some millimeters in diameter),
which is coupled to a semiconductor thermistor. The deposited energy is
collected in the absorber part, which forms together with the thermistor
readout a totally absorbing calorimeter. Such two-component bolometers
allow one to obtain excellent energy resolution, but they cannot, at the
moment, process high rates of particles since the decay time of the thermal signals is of the order of $20\,\mu$s. Compared to standard calorimetric
techniques, which are based on the production and collection of ionization
electrons, bolometers have the large advantage that they can in principle also detect weakly or non-ionizing particles such as slow magnetic
monopoles or astrophysical neutrinos, for example neutrino radiation as
a remnant from the Big Bang with energies around $0.2\,$meV ($\approx 1.9\,$K),
corresponding to the $2.7\,$K microwave background radiation (see section
7.6). Excellent energy resolution for X-rays also has been obtained with
large area superconducting Nb/Al–AlO$_x$/Al/Nb-tunnel junctions [494].

In high energy physics experiments, solid state detectors can also be
used as a 'living target', by triggering a more complex detector system.
Figure 7.15 shows the schematic set-up of such an experiment. The complex detector system is only read out if the incoming particle undergoes
an interaction in the target, which in this case consists of a solid state
detector. The charge recorded in the target allows one to decide whether
an interaction has occurred or not. If a trigger threshold $Q \geq k \cdot Q_{\min}$ with
$k > 1$ is required, where $Q_{\min}$ is the charge deposited by a minimum ionizing beam particle in the solid state counter, the readout of background
events can be suppressed.

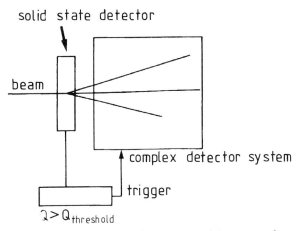

Fig. 7.15. Solid state detector as 'living target' in a complex experiment.

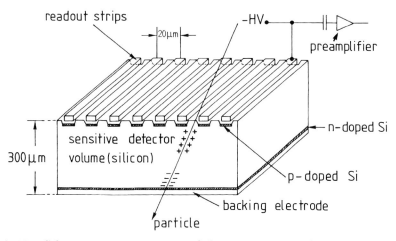

Fig. 7.16. Schematic representation of the construction of a silicon microstrip detector. Each readout strip is at negative potential. The strips are capacitively coupled (not to scale, from [481]).

Solid state detectors can also be used for track measurements if the electrodes are segmented in the form of strips or pads [477]. The charge distribution on the readout strips allows a spatial resolution of the order of 10 μm or even better [476, 495]. Such silicon microstrip counters are frequently used in storage ring experiments as vertex detectors, in particular to determine the lifetimes of unstable hadrons in the picosecond range and to tag shortlived mesons in complicated final states (see section 11.12). This technique of using silicon microstrip detectors in the vicinity of interaction points mimics the ability of high resolution bubble chambers (see section 4.11, figure 4.67) or nuclear emulsions (see section 4.16, figure 4.95) but uses a purely electronic readout. Because of the high spatial resolution of microstrip detectors, secondary vertices can be reconstructed and separated from the primary interaction relatively easily.

Silicon microstrip detectors can also be operated as solid state drift chambers [480, 496]. They only provide, however, projections of an event. A two-dimensional readout can be obtained if the detector is read out on both sides of the silicon chip by orthogonal strips.

Figure 7.16 shows the operation principle of the silicon microstrip detector with sequential cathode readout [481]. In figure 7.17 a readout with orthogonally segmented anodes and cathodes is shown schematically. The double-sided readout allows one to reconstruct three-dimensional spatial coordinates with only *one* detector.

If a silicon chip is subdivided in a matrix-like fashion into many pads that are electronically shielded by potential wells with respect to one

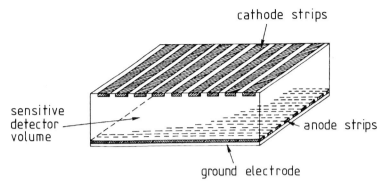

Fig. 7.17. Schematic representation of a double-sided readout of a silicon microstrip detector (the lower layer of anode strips is indicated with dashed lines).

another, the energy depositions produced by complex events which are stored in the cathode pads can be read out strip by strip. The readout time is rather long because of the sequential data processing. It supplies, however, two-dimensional pictures in a plane perpendicular to the beam direction. For a pixel size of $20 \times 20 \,\mu m^2$ spatial resolutions of $5 \,\mu m$ can be obtained. Because of the charge coupling of the pads, this type of silicon detector is also called a 'Charge Coupled Device'. Commercially available CCD-detectors with external dimensions of $1 \times 1 \, cm^2$ have about 10^5 pixels [1, 497, 498, 499].

Microstrip detectors for charged particles can also be made from gallium arsenide instead of silicon [483]-[486].

All semiconductor counters exhibit ageing effects in a high radiation environment which results in an increased leakage current [500, 501]. For example, the leakage current in a typical silicon microstrip detector is increased by a factor of ten for an absorbed dose of $1 \, kGy$ ($= 100 \, krad$) [502]. Solid state detectors with their sensitive highly integrated preamplifiers can therefore only be operated with a limited lifetime in a high radiation environment. Radiation hard detectors are required, e.g., for experiments at the LHC (Large Hadron Collider at CERN) or HERA (Electron-Proton Storage Ring at DESY, Hamburg), and also for experiments in outer space.

7.2 Electron-photon calorimeters

The dominating interaction processes for spectroscopy in the MeV energy range are the photoelectric and Compton effects for photons and ionization and excitation for charged particles. At high energies ($>$ several GeV) electrons lose their energy almost exclusively by bremsstrahlung

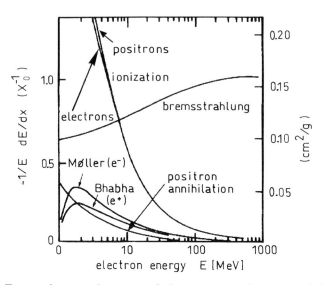

Fig. 7.18. Energy loss mechanisms of electrons as a function of the electron energy (Møller scattering: $e^-e^- \rightarrow e^-e^-$; Bhabha scattering: $e^+e^- \rightarrow e^+e^-$; annihilation: $e^+e^- \rightarrow \gamma\,\gamma$) [27, 34, 35, 503].

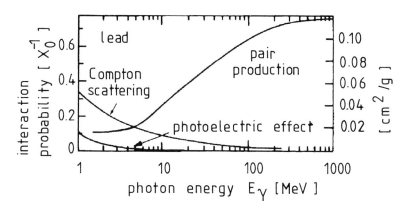

Fig. 7.19. Cross sections for photon interactions in lead as a function of the photon energy [27, 34, 35, 503].

and photons their energy by electron-positron pair production (see figures 7.18 and 7.19, [34, 35]). An electromagnetic cascade is produced. The development of such an electromagnetic shower is sketched in figure 7.20.

The longitudinal and lateral development of electron or photon initiated cascades can be described either with analytical or with Monte Carlo methods. The total track length T, which is the summed length of all individual tracks of charged particles in the shower is proportional to the

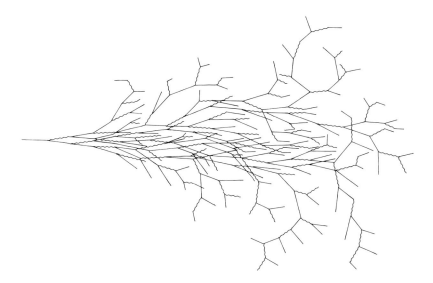

Fig. 7.20. Schematic representation of an electromagnetic cascade. The wavy
lines are photons and the solid lines electrons or positrons.

energy E_0 of the incident particle.

$$T \propto \frac{E_0}{E_c} ,\qquad (7.7)$$

where E_c is the critical energy of the material in which the shower develops
(see section 1.1.4).

If $\eta(\xi)$ is the minimum energy of individual shower particles that can
be detected in a calorimeter, the measurable track length is given by [58]

$$T_m = F(\xi) \cdot \frac{E_0}{E_c} \cdot X_0 \,[\mathrm{g/cm^2}] ,\qquad (7.8)$$

where $T_m \leq T$ and $\xi = \xi(\eta)$. The total number of particles in the shower,
of course, depends on the cut-off parameter η. However, this dependence
is not very pronounced if η is chosen to be sufficiently small ($\approx$ MeV). The
function $F(\xi)$ takes into account the effect of the cut-off parameter on the
total measurable track length for completely contained electromagnetic
cascades in a calorimeter. $F(\xi)$ can be parametrized as [58]:

$$F(\xi) = \{1 + \xi \ln(\xi/1.53)\}e^\xi ,\qquad (7.9)$$

where

$$\xi = 2.29 \cdot \frac{\eta}{E_c} .\qquad (7.10)$$

The longitudinal distribution of the energy loss can be approximated by

$$\frac{\mathrm{d}E}{\mathrm{d}t} = \mathrm{const} \cdot t^a e^{-b \cdot t} , \qquad (7.11)$$

where $t = x/X_0$ is the shower depth in units of the radiation length X_0, and a and b are adjustable parameters [504, 505].

Such a parametrization is motivated by the physical processes of shower formation. For low shower depths t the number of secondary particles increases like t^a. The increase of the number of shower particles goes along with a decrease in their average energy. The total number of particles finally reaches a maximum value. Beyond the maximum of shower development absorption processes dominate which are described by the exponential function e^{-bt} (compare also figure 7.21).

A more accurate treatment of the longitudinal profile of electromagnetic cascades based on the Monte Carlo program EGS [34, 35, 506, 507] yields the parametrization

$$\frac{\mathrm{d}E}{\mathrm{d}t} = E_0 \cdot f \cdot \frac{(f \cdot t)^{g-1} e^{-ft}}{\Gamma(g)} , \qquad (7.12)$$

where $\Gamma(g)$ is Euler's Γ-function, defined by

$$\Gamma(g) = \int_0^\infty e^{-x} x^{g-1} \mathrm{d}x . \qquad (7.13)$$

The gamma function has the property

$$\Gamma(g+1) = g\Gamma(g) . \qquad (7.14)$$

The parameters g and f must be obtained from fits to data, and E_0 is the energy of the incident particle. In this parametrization the maximum of shower development is reached at

$$t_{\mathrm{max}} = \frac{g-1}{f} . \qquad (7.15)$$

For the design of a calorimeter the longitudinal and lateral shower development is of great importance; 98 % of the shower energy is contained in a length of

$$L(98\,\%) = 2.5 \, t_{\mathrm{max}}[X_0] \qquad (7.16)$$

for incident energies between 10 and 1000 GeV [508]. The depth in radiation lengths, t_{max}, describes the position of the maximum number of particles in the shower development. This is at [34, 35]

$$t_{\mathrm{max}}^{\mathrm{e}} = \ln\left(\frac{E_0}{E_{\mathrm{c}}}\right) - 0.5 \qquad (7.17)$$

for electrons and

$$t_{\max}^{\gamma} = \ln\left(\frac{E_0}{E_c}\right) + 0.5 \qquad (7.18)$$

for incident photons, which (using equation 7.15), leads to

$$t_{\max} = \frac{g-1}{f} = \ln\left(\frac{E_0}{E_c}\right) + C_i \qquad (7.19)$$

with $C_\gamma = +0.5$ and $C_e = -0.5$ for gamma and electron induced cascades.

The longitudinal development of electron cascades in matter is shown in figure 7.21 [509, 503] for various incident energies. The longitudinal shower profiles depend in general on the absorber material. This dependence is taken into account in figure 7.21 by measuring the incident energy in units of the critical energy of the material, and the shower depths in units of radiation lengths.

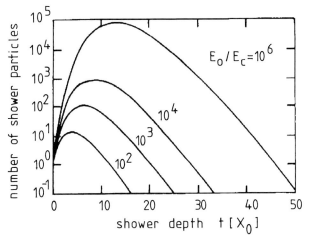

Fig. 7.21. Longitudinal shower development of electromagnetic cascades (E_c - critical energy) [509, 503].

At very high energies the development of electromagnetic cascades in dense media is influenced by the Landau-Pomeranchuk-Migdal (LPM) effect [510, 511]. This effect predicts that the production of low energy photons by high energy electrons is suppressed in dense media. When an electron interacts with a nucleus producing a bremsstrahlung photon the longitudinal momentum transfer between the electron and nucleus is very small. Heisenberg's uncertainty principle therefore requires that the interaction must take place over a long distance, which is called the formation zone. If the electron is disturbed while travelling this distance, the photon emission can be disrupted. This can occur for very dense media, where the distance between scattering centers is small compared to the spatial extent of the wave function. The Landau-Pomeranchuk-Migdal effect predicts that in dense media multiple scattering of electrons is enough to suppress photon production at the low energy end of the bremsstrahlung spectrum. The validity of this effect has been demonstrated by a recent experiment at SLAC with 25 GeV electrons on various targets. The magnitude of the photon suppression is consistent with the LPM prediction [512, 513].

The LPM effect is relevant for experiments with ultra high energy cosmic rays and should be taken into account for the design of calorimeters at high energy accelerators and storage rings such as the LHC.

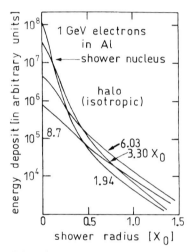

Fig. 7.22. Lateral widths of electromagnetic showers as a function of the shower depths [505, 508, 514].

The lateral width of an electromagnetic cascade is mainly caused by multiple scattering and can best be characterized by the Molière radius

$$R_{\rm m} = \frac{21\,{\rm MeV}}{E_{\rm c}} X_0 \, [{\rm g/cm}^2] \, . \tag{7.20}$$

With increasing longitudinal shower depth the lateral width of an electromagnetic shower increases. Figure 7.22 [505, 508, 514] shows the radial shower profiles of 1 GeV electrons in an aluminium absorber for various longitudinal shower depths. The largest part of the energy is deposited in a relatively narrow shower core. Generally speaking, one can say that 95 % of the shower energy is contained in a cylinder around the shower axis whose radius is

$$R(95\,\%) = 2R_{\mathrm{m}} = \frac{42\,\mathrm{MeV}}{E_{\mathrm{c}}} X_0 \; . \tag{7.21}$$

The dependence of the containment radius on the material is taken into account by the critical energy and radiation length appearing in equation 7.21.

Figure 7.23 shows the longitudinal and lateral development of a 6 GeV electron cascade in a lead calorimeter (based on [504, 505]).

The most important properties of electron cascades can be understood in a very simplified model [1, 515, 516]. Let E_0 be the energy of a photon incident on a total absorption calorimeter (figure 7.24).

After one radiation length the photon produces an e^+e^- pair; electrons and positrons emit after a further radiation length one bremsstrahlung photon each, which again are transformed into electron-positron pairs. Let us assume that the energy is symmetrically shared between the particles at each step of the multiplication. The number of shower particles (electrons, positrons and photons together) at the depth t is

$$N(t) = 2^t \; , \tag{7.22}$$

where the energy of the individual particles in generation t is given by

$$E(t) = E_0 \cdot 2^{-t} \; . \tag{7.23}$$

The multiplication of the shower particles continues as long as $E_0/N > E_{\mathrm{c}}$. If the shower particles fall below the critical energy, absorption processes like ionization for electrons, and Compton and photoelectric effects for photons start to dominate. The shower slowly dies out.

The position of the shower maximum is reached in the last step of multiplication when the energy of shower particles equals the critical energy

$$E_{\mathrm{c}} = E_0 \cdot 2^{-t_{\max}} \; . \tag{7.24}$$

This leads to

$$t_{\max} = \frac{\ln E_0/E_{\mathrm{c}}}{\ln 2} \propto \ln E_0/E_{\mathrm{c}} \tag{7.25}$$

in qualitative agreement with equation (7.17). The total number of shower particles is therefore given by

$$S = \sum_{t=0}^{t_{\max}} N(t) = \sum_{t=0}^{t_{\max}} 2^t = 2^{t_{\max}+1} - 1 \approx 2^{t_{\max}+1} \tag{7.26}$$

or

$$S = 2 \cdot 2^{t_{\max}} = 2 \cdot \frac{E_0}{E_{\mathrm{c}}} \propto E_0 \; . \tag{7.27}$$

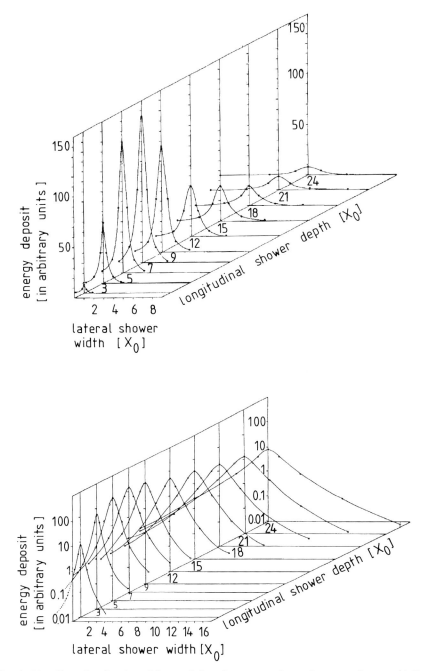

Fig. 7.23.   Longitudinal and lateral development of an electron shower (6 GeV) in lead shown with linear and logarithmic scales (based on [504, 505]).

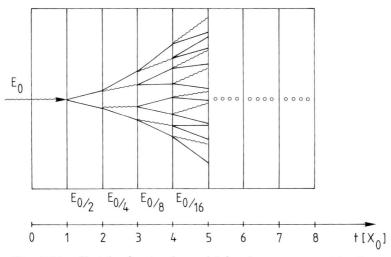

Fig. 7.24. Sketch of a simple model for shower parametrization.

The particle track segments counted in equidistant steps t measured in units of X_0 yield the total track length in a shower

$$S^* = 2 \cdot \frac{E_0}{E_c} \cdot \frac{1}{t} \; . \tag{7.28}$$

For a sampling calorimeter, in which the track segments are counted only in fixed distances t in active detector layers, one would expect to obtain an energy resolution of

$$\frac{\sigma}{E_0} = \frac{\sqrt{S^*}}{S^*} = \frac{1}{\sqrt{S^*}} = \frac{\sqrt{t}}{\sqrt{2E_0/E_c}} \propto \frac{\sqrt{t}}{\sqrt{E_0}} \; . \tag{7.29}$$

This very simple model already correctly describes the most important qualitative characteristics of electromagnetic cascades.

   The lateral width of electromagnetic showers as given by equation (7.21) is only valid for total absorption calorimeters (i.e., totally absorbing in the active medium). If sampling calorimeters are used which consist of alternating detector and absorber layers, the lateral width is increased. For example, this is the case for gas-sampling calorimeters, depending on the space used by the gaseous detectors. If $y = \sum y_i$ is the sum of the thicknesses of gaseous detectors and $x = \sum x_i$ the total thickness of the absorber in which the shower almost exclusively develops, the lateral width is increased by a factor $\frac{x+y}{x}$ compared to a homogeneous calorimeter, leading to:

$$R(95\,\%) = 2R_{\mathrm{m}} \cdot \frac{x+y}{x} \; . \tag{7.30}$$

This equation says that the effective radius for a 95 % lateral containment is given by averaging over the thicknesses of detectors and absorbers in a sampling calorimeter.

In the following the various fluctuations that limit the energy resolution of sampling calorimeters will be discussed in somewhat more detail [1, 504, 505, 517, 518, 519]. In a sampling calorimeter only that part of the energy of an electromagnetic cascade is recorded which is sampled in the detection layers. The energy loss in the absorbers and detector layers, of course, varies from event to event, leading to the so-called 'sampling fluctuations' which have a considerable influence on the energy resolution. If the energy is determined by detectors in which only track segments of shower particles are registered, the number of intersection points with the detector layers is given by

$$N = \frac{T_{\mathrm{m}}}{d} \,, \tag{7.31}$$

when T_{m} is the total measurable track length (see equation (7.8)) and d the distance between two detector layers. Using the measurable track length defined by equation (7.8) the number of track segments is then

$$N = F(\xi) \cdot \frac{E_0}{E_{\mathrm{c}}} \cdot \frac{X_0}{d} \,. \tag{7.32}$$

Using Poisson statistics the sampling fluctuations limit the energy resolution to

$$\frac{\sigma(E)}{E} = \frac{\sqrt{N}}{N} = \sqrt{\frac{E_{\mathrm{c}} \cdot d}{F(\xi) \cdot E_0 \cdot X_0}} \,. \tag{7.33}$$

In this case it has not been considered that because of multiple scattering, the shower particles have a certain angle θ with respect to the shower axis. The effective sampling thickness is therefore not d, but rather $d/\cos\theta$. This leads to a modification of equation (7.33) to

$$\left(\frac{\sigma(E)}{E}\right)_{\mathrm{Sampling}} = \sqrt{\frac{E_{\mathrm{c}} \cdot d}{F(\xi) \cdot E_0 \cdot X_0 \cdot \cos\theta}} \,. \tag{7.34}$$

Naturally, the scattering angles θ vary for the shower particles, so one has to average over $\cos\theta$. However, for electromagnetic cascades, the scattering angles are rather small (see figures 4.70 and 4.84). The average value of $\cos\theta$ in equation (7.34) depends on the energy of the incident particle E_0. It can be determined by Monte Carlo simulation or by calibration.

As can be seen from equation (7.34), the energy resolution of a sampling calorimeter for a fixed given material improves with $\sqrt{d/E_0}$.

Energy depositions in the detector from large energy transfers in ionization processes can further deteriorate the energy resolution. These

Landau fluctuations are of particular importance for thin detector layers. If δ is the average energy loss per detector layer, the Landau fluctuations of the ionization loss yield a contribution to the energy resolution of [58, 508, 520]

$$\left(\frac{\sigma(E)}{E}\right)_{\text{Landau fluctuations}} \propto \frac{1}{\sqrt{N}\ln(k \cdot \delta)}, \qquad (7.35)$$

where k is a constant and δ is proportional to the matter density per detector layer. It can happen that a calorimeter is not big enough to contain the shower completely, and longitudinal and lateral leakages may occur. The influence of such leakages is shown in figure 7.25 [521]. While lateral leakage has only a minor influence on the energy resolution, longitudinal leakage deteriorates the energy resolution considerably. The latter grows for fixed calorimeter size logarithmically with the energy, i.e.,

$$\left(\frac{\sigma(E)}{E}\right)_{\text{leakage}} \propto \ln E \qquad (7.36)$$

[517]. The length of a calorimeter should also not be oversized, because unused detector layers will deteriorate the energy resolution by additional noise contributions.

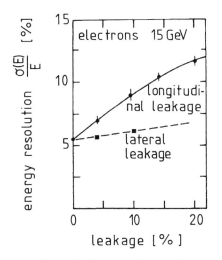

Fig. 7.25. Influence of longitudinal and lateral leakage on the energy resolution [521, 9].

It is very important to calibrate the calorimeter as a whole and also the individual detector layers with respect to each other. Relative calibrations can be performed with, e.g., minimum-ionizing muons, which penetrate the whole calorimeter. For absolute calibrations one has to consider, for example, that electrons and muons, for the same total energy absorption, do not necessarily lead to the same signal in the calorimeter, because in electron cascades many low energy photons are produced which do not contribute to the 'visible energy'.

The measured longitudinal shower profiles in a streamer-tube calorimeter with $0.8\,X_0$ copper sampling for various electron energies are shown in figure 7.26 [518, 522]. The experimentally measured influence of longitudinal leakage on the energy resolution for this calorimeter can be seen in figure 7.27 [518, 522].

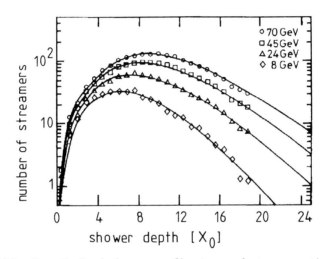

Fig. 7.26. Longitudinal shower profiles in an electromagnetic streamer-tube calorimeter with $0.8\,X_0$ copper sampling [518, 522].

In streamer-tube calorimeters tracks are essentially counted, at least as long as the particles are not incident under too large an angle with respect to the shower axis, which is assumed to be perpendicular to the detector planes. For each ionization track exactly one streamer is formed — independent of the ionization produced along the track. For this reason Landau fluctuations have practically no effect on the energy resolution for this type of detector.

The measured amplitude distributions for electrons of 5, 25, 50, and 100 GeV in an iron streamer-tube calorimeter with 2 cm iron sampling are shown in figure 7.28 [523].

As for all calorimeters, the energy resolution can be represented by a constant and an energy dependent term in the following fashion:

$$\frac{\sigma(E)}{E} = a \oplus \frac{b}{\sqrt{E}} \,, \tag{7.37}$$

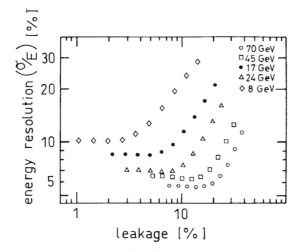

Fig. 7.27. Influence of the longitudinal leakage on the energy resolution in an electromagnetic calorimeter [518, 522].

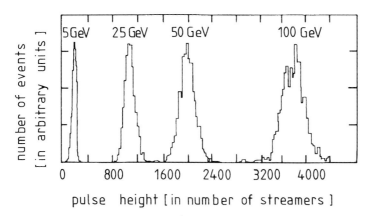

Fig. 7.28. Amplitude distributions of 5, 25, 50, and 100 GeV electrons in an iron streamer-tube calorimeter [523].

where $\oplus$ means that both contributions have to be added in quadrature. The constant term does not originate from fluctuations in shower development but is caused by systematic errors, for example in the readout system of the calorimeter. It can also be caused by noise contributions from the sampling detectors or by a time dependent drift of calibration parameters.

Figure 7.29 [522] shows the energy resolution for electrons in a copper streamer-tube calorimeter, both for an anode wire readout, and for the measurement of signals induced on cathode pads.

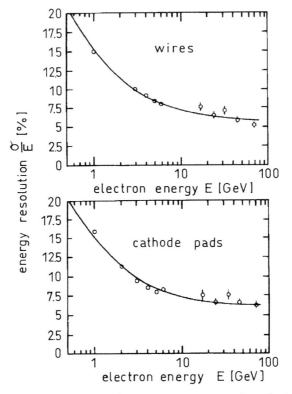

Fig. 7.29.   Energy resolution of a copper streamer-tube calorimeter for electrons [522].

The energy resolution of scintillator or liquid-argon sampling calorimeters is superior to that achievable with gaseous detectors [524, 525]. For a 1 mm sampling thickness, lead scintillator or lead liquid-argon calorimeters allow energy resolutions of

$$\frac{\sigma(E)}{E} = \frac{7\,\%}{\sqrt{E}} \oplus 1\,\% ; \quad E \text{ in GeV} . \tag{7.38}$$

With lead liquid-argon or lead liquid-xenon calorimeters values as good as $\sigma(E)/E \approx 3\,\%/\sqrt{E}$ have been obtained [526]. Liquid-krypton calorimeters also show excellent performance [527, 528].

Similar and even better results can be achieved with total absorption (homogeneous) calorimeters [529]. In large NaI(Tl) calorimeter hodoscopes typical values of

$$\frac{\sigma(E)}{E} = 2.5\,\%/\sqrt{E} ; \quad E \text{ in GeV} \tag{7.39}$$

have been obtained. Best values of individual calorimeter elements have given $\sigma(E)/E = 0.9\,\%/\sqrt{E}$ [1, 530]. Also bismuth germanate (BGO) gives a comparable energy resolution.

Total absorption lead-glass counters, which register the Cherenkov light of shower particles, show somewhat worse performance characteristics, because Cherenkov light is much less intense than scintillation light [531]. Using lead-glass Cherenkov counters, energy resolutions of [1, 532]

$$\frac{\sigma(E)}{E} = 5.3\,\%/\sqrt{E} \oplus 1.2\,\% \;; \quad E \text{ in GeV} \tag{7.40}$$

have been obtained. For lead-fluoride radiators comparable resolutions have been achieved [533].

A very important criterion for the application of Cherenkov shower counters in high-rate experiments is their radiation hardness. For example, $BaYb_2F_8$-radiators exhibit an excellent resistance against radiation [534].

Heavy liquids can also be used for total absorption Cherenkov shower counters. Aqueous solutions of thallium formate (HCOOTl, density $\varrho = 3.3\,\text{g/cm}^3$, index of refraction $n = 1.59$) or a mixture of thallium formate and thallium malonate ($CH_2(COOTl)_2$, 'Clerici' solution, $\varrho = 4.21\,\text{g/cm}^3$, $n = 1.69$) [535, 536] are excellent candidates for heavy liquids. The Clerici solution is also known from mining applications, where minerals of different density are separated by gravimetric methods.

If calorimeters are constructed from submodules or if the readout is segmented, they also allow one to determine the spatial coordinates of incident electrons or photons.

In an arrangement of $3 \times 3\,\text{cm}^2$ calorimeter modules, spatial resolutions of the order of magnitude of 1 mm for electron energies exceeding 20 GeV are obtained. For this the measurement is based on the partition of the shower energy in various calorimeter modules [270].

Already relatively large calorimeter modules with scintillator sampling allow spatial resolutions of approximately 10 % of the calorimeter width by comparing the light sharing on two sides of the calorimeter. Because of light attenuation and solid angle effects, this light sharing depends on the particle's point of incidence.

Details on the readout of calorimeters with scintillator sampling and also for spaghetti-type calorimeters have already been described in section 5.2. The readout of calorimeters with scintillating fibers has the additional advantage of providing accurate tracking information ([537, 538, 539] and references in section 5.2).

7.3   Hadron calorimeters

In principle, hadron calorimeters work along the same lines as electron-photon calorimeters, the only difference being that for hadron calorimeters the longitudinal development is determined by the average nuclear absorption length. In normal detector materials this is much larger than the radiation length X_0, which describes the behaviour of electron cascades. This is the reason why hadron calorimeters have to be much larger than electron shower counters. Frequently, electron and hadron calorimeters are integrated to a single detector. For example, figure 7.30 [540] shows an iron-scintillator calorimeter with separate wavelength shifter readout for electrons and hadrons. The electron part has a depth of 14 radiation lengths, and the hadron section corresponds to 3.2 absorption lengths.

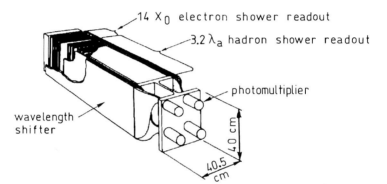

Fig. 7.30. Typical set-up of an iron-scintillator calorimeter with wavelength shifter readout [540].

In figure 7.31 an electron-hadron calorimeter with streamer-tube sampling is shown [523]. In this case the information of the individual detector layers is read out by means of anode wires and cathode pads.

Apart from the larger longitudinal development of hadron cascades, their lateral width is also sizably increased compared to electron cascades. While the lateral structure of electron showers is mainly determined by multiple scattering, in hadron cascades it is caused by large transverse momentum transfers in nuclear interactions. Typical processes in a hadron cascade are shown in figure 7.32. Figure 7.33 shows a neutrino-induced hadron shower in a sampling calorimeter [517, 541].

The different structures of 250 GeV photon and proton induced cascades in the earth's atmosphere are clearly visible from figure 7.34 [542]. The results shown in this case were obtained from a Monte Carlo simulation.

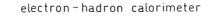

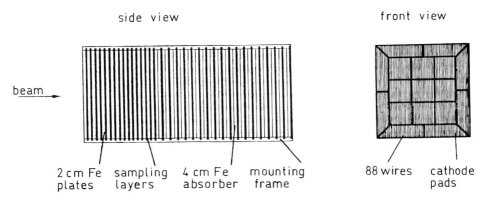

Fig. 7.31. Typical set-up of an electron-hadron calorimeter with streamer-tube sampling [523].

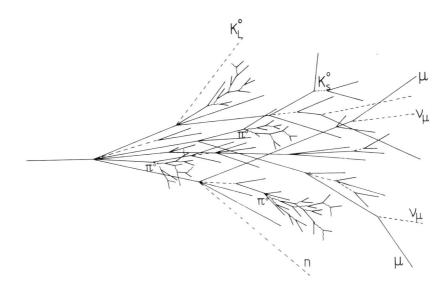

Fig. 7.32. Sketch of a hadron cascade in an absorber.

The production of secondary particles in a hadron cascade is caused by inelastic hadronic processes. Mainly charged and neutral pions, but with lower multiplicities also kaons, nucleons and other hadrons are produced. The particle multiplicity per interaction varies only weakly with energy ($\propto \ln E$). The average transverse momentum of secondary particles can

Fig. 7.33. Digital pattern of a neutrino-induced hadron cascade in a flash-chamber calorimeter [517, 541], (©1981 IEEE).

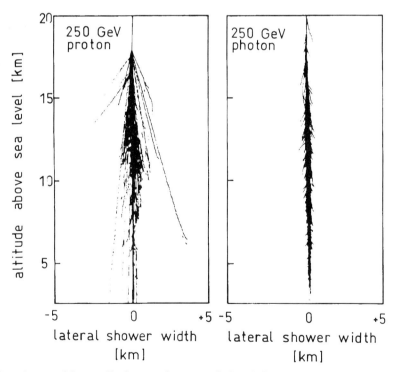

Fig. 7.34. Monte Carlo simulations of the different development of hadronic and electromagnetic cascades in the earth's atmosphere, induced by 250 GeV protons and photons [542].

be characterized by

$$\langle p_\mathrm{T} \rangle \approx 0.35\,\mathrm{GeV}/c\,. \tag{7.41}$$

The average inelasticity, that is, the fraction of energy which is transferred to secondary particles in the interaction, is around 50 %.

A large component of the secondary particles in hadron cascades are neutral pions, which represent approximately one third of the pions produced. The fraction of neutral pions increases with increasing primary energy of the incident hadron. Neutral pions decay rather quickly ($\approx 10^{-16}$ s) into two energetic photons, thereby initiating electromagnetic subcascades in a hadron shower. The π^0-production, however, is subject to large fluctuations, which are determined essentially by the properties of the first inelastic interaction.

In contrast to electrons and photons, whose electromagnetic energy is completely recorded in the detector, a substantial fraction of the energy in hadron cascades remains 'invisible'. This is related to the fact that some part of the hadron energy is used to break up nuclear bonds. This nuclear binding energy must be provided by the primary and secondary hadrons and does not contribute to the visible energy. The invisible energy fraction is of the order of 20 % of the total energy. Furthermore, extremely short range nuclear fragments are produced in the break-up of nuclear bonds. In sampling calorimeters, these fragments do not contribute to the signal since they are absorbed before reaching the detection layers. In addition, long lived or stable neutral particles like neutrons, K_L^0 or neutrinos can escape from the calorimeter, thereby reducing the visible energy. Muons created as decay products of pions and kaons deposit in most cases only a very small fraction of their energy in the calorimeter. All these effects cause the energy resolution for hadrons to be significantly inferior to that of electrons because of the different interaction and particle production properties.

Figure 7.35 shows for various simulations [58], [543]-[546] how the primary hadron energy is shared between the different energy-loss processes [517]. It is clear from the figure that the variety of simulations leads to quite different results for the relative energy fractions.

It is important to remember that only the electromagnetic energy and the energy loss of charged particles can be recorded in a calorimeter. Consequently, a hadron signal for the same particle energy is normally smaller than an electron signal.

The different response of calorimeters to hadrons and electrons is a substantial drawback if one is interested in the measurement of the total energy in an event which contains hadrons and electrons in unknown composition. It is, however, possible to regain some part of the 'invisible' energy in hadron cascades, thereby equalizing the response to electrons

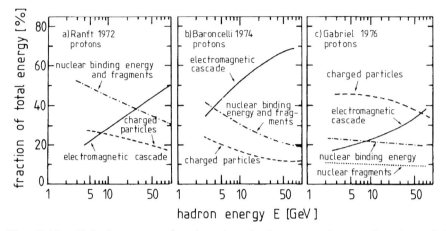

Fig. 7.35. Relative energy fractions in a hadron cascade as a function of the hadron energy [58], [543]-[546].

and hadrons. This hadron calorimeter compensation is based on the following physical principles [520, 547, 548].

If uranium is used as an absorber material, neutrons will also be produced in nuclear interactions. These neutrons may induce fission of other target nuclei which produces more neutrons, and also energetic γ-rays as a consequence of nuclear transitions. These neutrons and γ-rays can enhance the amplitude of the hadron shower signal if their energy is recorded. Also for absorber materials other than uranium where fission processes are endotherm, neutrons and γ-rays may be produced. The γ-rays can contribute to the visible energy by a suitable choice of sampling detectors, and neutrons can produce low energy recoil protons in (n,p)-reactions in detector layers containing hydrogen.

These recoil protons also increase the hadron signal. The regaining of nuclear binding energy or parts thereof depends decisively on parameters like density, atomic number and thickness of the active detector layers. For gaseous detectors these parameters can be optimized relatively easily by a suitable choice of gas mixture and pressure. Of course, the material of the passive absorber (uranium, lead, iron) also plays an important rôle. The sampling time can have a considerable influence on the visible energy as well [520, 548]. This is because photons and neutrons, being produced in nuclear processes, can be delayed by a time characteristic for the absorber material used.

Figure 7.36 shows the amplitude ratio of electron and hadron-induced showers in various calorimeter structures as a function of the particle energy ([549, 550] and references therein). For energies below 1 GeV even in uranium-sampling calorimeters, the lost energy in hadron cascades cannot be regained. By suitable means (uranium/liquid-argon; uranium/copper-scintillator) compensation can be achieved for energies exceeding several GeV. For very high energies ($\geq$ 100 GeV) even overcompensation can occur. Such overcompensation can be avoided by limiting the sampling time. Overcompensation can also be caused by a reduction of the electron signal due to saturation effects in the detector layers. Because of the different lateral structure of electron and hadron cascades saturation effects affect the electron and hadron signals differently.

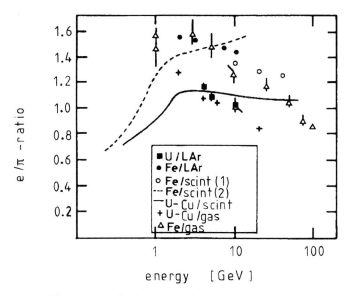

Fig. 7.36. Electron to hadron signal ratio in various calorimeter structures as a function of the incident particle energy (based on [549, 550] and references therein).

It has also been attempted to achieve compensation in Fe/TMP and Pb/TMP sampling calorimeters [551].

Another way to achieve the compensation condition ($e/\pi = 1$, i.e., equal response to pions and electrons) in hadron calorimeters is presented by the filtering effect [552, 553]. The value of the electron energy below which the ionization loss starts to dominate over the energy lost by radiation is the critical energy E_c. Therefore, the lower the critical energy of the absorber, the softer the energy spectrum of particles in electromagnetic subshowers in a hadron cascade when exiting from a high Z absorber and entering a low Z absorber. As a result, the energy distribution of the shower is transformed and the response of the calorimeter to the incoming shower can be substantially modified. The degree of the effect obviously depends on the fractions of low and high Z materials in the absorber.

Although the radiation lengths X_0 will be quite different for low and high Z absorbers, the interaction or absorption lengths may be comparable. If this is the case, even if the passive medium of a hadron calorimeter consists of a low Z absorber with a thickness smaller than (or comparable to) that of the high Z material in units of X_0, the low Z material may very well be the thickest in terms of interaction lengths.

Figures 7.37 and 7.38 show the longitudinal shower developments of pions of different energies in iron and tungsten calorimeters [505], [554]-[558]. The lateral shower profiles of $10\,\mathrm{GeV}/c$ pions in iron and $10\,\mathrm{GeV}/c$ protons in aluminum are shown in figures 7.39 and 7.40 [559].

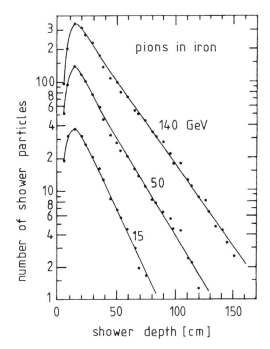

Fig. 7.37.   Longitudinal shower development of pions in iron [555].

The so-called length of a hadron cascade depends on exactly how this is defined. Regardless of the definition, the length increases with the energy of the incident particle. Figure 7.41 shows the shower length for two different definitions [555]. One possible definition is given by the requirement that the shower length is reached if, on average, only one particle or less is registered at the depth t. According to this definition a $50\,\mathrm{GeV}$ pion shower in an iron-scintillator calorimeter is approximately

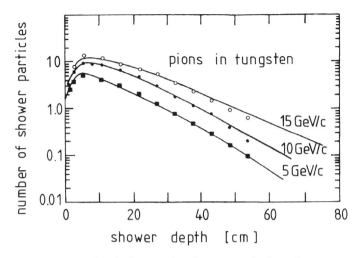

Fig. 7.38. Longitudinal shower development of pions in tungsten [556, 557]. The solid lines are from Monte Carlo simulations [558].

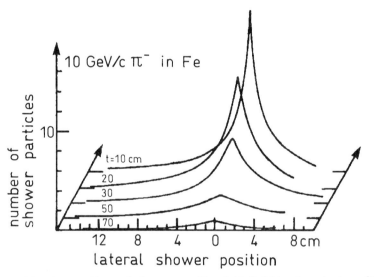

Fig. 7.39. Lateral shower profile of $10\,\mathrm{GeV}/c$ pions in iron [559].

$120\,\mathrm{cm}$ Fe 'long'. An alternative definition is given by the depth before which a certain fraction of the primary energy (e.g., 95 %) is contained. A 95 % energy containment would lead to a length of $70\,\mathrm{cm}$ iron for a $50\,\mathrm{GeV}$ pion shower. The longitudinal center of gravity of the shower only increases logarithmically with the energy. The position of the center of gravity of the shower is also shown in figure 7.41.

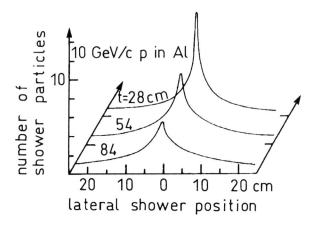

Fig. 7.40. Lateral shower profile of $10\,\mathrm{GeV}/c$ protons in aluminum [559].

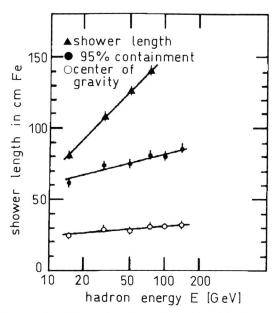

Fig. 7.41. Shower lengths and center of gravity of hadron cascades for various definitions [555].

The lateral distribution of a hadron shower is initially very narrow but becomes wider with increasing calorimeter depth (see figures 7.39 and 7.40).

Figure 7.42 a) shows the development of 20 GeV and 30 GeV hadron cascades, which on the one hand demonstrate the difficulty of the length definition and on the other hand illustrate the large fluctuations for individual showers [560, 561]. For higher energies (figure 7.42b) saturation effects in a flash-tube calorimeter become clearly visible [560]. Looking at the flash-tube patterns of this figure, a discrimination between even 50 GeV and 300 GeV hadronic cascades appears difficult.

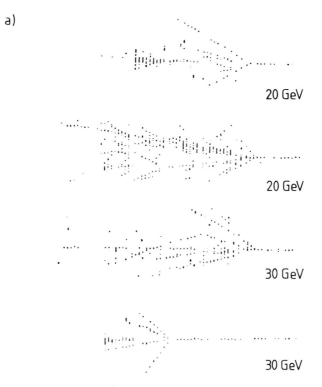

Fig. 7.42 a. Digital pattern of individual 20 and 30 GeV hadron cascades in a flash-chamber calorimeter [560, 561].

To be able to absorb hadrons of several hundred GeV, approximately 2 m iron with a lateral extension of $60 \times 60 \, \text{cm}^2$ is required. The 95 % longitudinal containment in iron can be approximated by

$$L(95 \, \%) = (9.4 \ln E + 39) \, [\text{cm}] \tag{7.42}$$

[1], where E is measured in GeV. Similarly the lateral distribution of cascades can be characterized by a radial width.

The required lateral calorimeter radius for a 95 % containment as a function of the longitudinal shower depth is shown in figure 7.43 for pions in iron of two different energies [555].

In experiments where muons are to be distinguished from hadrons, the punch-through probability is an important quantity. Energetic muons

b)

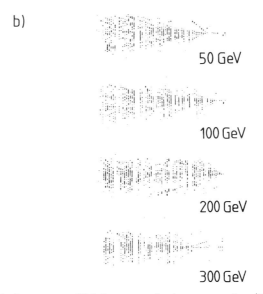

50 GeV

100 GeV

200 GeV

300 GeV

Fig. 7.42 b.   Digital pattern of high energy hadron cascades (50 - 300 GeV) in a flash-chamber calorimeter [560].

will penetrate through massive shielding (e.g., the flux-return yoke of a magnet) quite easily. Hadrons incident onto the shielding will normally develop hadronic cascades. If the shielding is not instrumented, a shower particle leaking from the shielding may fake a muon. The frequency of this effect is characterized by the punch-through probability. The punch through has, in fact, two components: a hadronic component and a tail composed mainly of penetrating muons from pion and kaon decay. Punch-through probabilities for pions of different momenta for iron absorbers of different thicknesses are shown in table 7.1 [562]. The punch-through probability for 50 GeV/c pions through 2 m of iron, for example, is 3 %.

In addition to punch through, hadrons may also penetrate through shielding without interaction ('sail through'). Sail-through hadrons cannot be easily distinguished from muons. The sail-through probability is simply given by $\exp\left(-x/\lambda_{\mathrm{a}}\right)$ where the absorber thickness x is measured in units of the absorption lengths λ_{a}. For example, an energetic pion may sail through an absorber of 1 m iron with a probability of 0.3 %. For thick absorbers ($\geq$ 2 m iron equivalent) sail through is negligible compared to punch through.

The linear relation between the produced number of particles and the incident energy is shown in figure 7.44 for an iron streamer-tube calorimeter [523]. The slight deviation from linearity at high energies originates from the increased π^{0} production. These decay into photon pairs result-

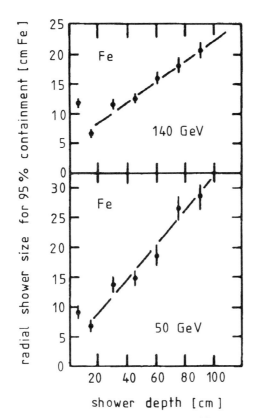

Fig. 7.43. Radius of hadronic showers for 95 % containment as a function of the depth in iron [555]. The corresponding total width of the hadron shower is twice the radius.

ing in higher signal amplitudes. Saturation effects are not important, in this case, because the π^0 are well separated in space.

The energy resolution for hadrons is significantly worse compared to electrons because of the large fluctuations in the hadron shower development. High local energy depositions caused by π^0 production, leading to a significant degradation of the energy resolution, can be suppressed by an appropriate weighting technique. If w_i are the measured energy depositions in the various detector layers, a weighting according to

$$w_i^* = w_i(1 - c \cdot w_i) \qquad (7.43)$$

with the weighting factor c reduces the strong fluctuations, thereby improving the energy resolution [1, 523]. Figure 7.45 [518] shows the amplitude distributions of 100 GeV pions in an iron streamer-tube calorimeter before and after weighting according to equation (7.43). The energy reso-

Table 7.1. *Punch-through probabilities for pions of different momentum for various iron absorber thicknesses [562]*

thickness of iron [m]	pion momentum [GeV/c]	punch-through probability
1	50	0.53
2	50	0.03
3	50	0.004
4	50	0.0016
5	50	0.0008
1	300	0.96
2	300	0.24
3	300	0.028
4	300	0.008
5	300	0.004

lution in this case is improved from $\frac{\sigma(E)}{E} \approx 80\,\%/\sqrt{E}$ to $\frac{\sigma(E)}{E} = 50\,\%/\sqrt{E}$ (compare figure 7.46 [518, 523]). It should be mentioned, however, that in this particular case the weighting factor c has been optimized using the knowledge of the pions' energy. Therefore, the improvement of the energy resolution given above represents the optimum that can be achieved by such a weighting procedure.

The best hadron sampling calorimeters (e.g., uranium scintillator; uranium liquid-argon) can achieve energy resolutions of

$$\frac{\sigma(E)}{E} = \frac{35\,\%}{\sqrt{E}} \; ; \quad E \text{ in GeV} \tag{7.44}$$

[563]. A possible constant term in the parametrization of the energy resolution analogous to equation (7.37) can safely be neglected for hadronic cascades because the large sampling fluctuations dominate the energy resolution. Only for extremely high energies ($\approx 1000\,\text{GeV}$) will a constant term limit the energy resolution.

The energy resolution attainable in hadron calorimeters varies with the number of detector layers (sampling planes) in the calorimeter N as

$$\frac{\sigma(E)}{E} \propto \frac{1}{\sqrt{N}} \; . \tag{7.45}$$

The number of sampling planes is related to the calorimeter length L and

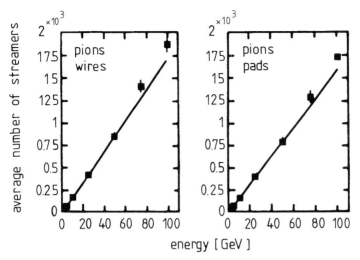

Fig. 7.44. Linear relation between the number of produced particles and the incident energy in an iron streamer-tube calorimeter. Shown are results for wire and pad readout [523].

the thickness of the absorber plates d by

$$N = \frac{L}{d} \, , \qquad (7.46)$$

so that

$$\frac{\sigma(E)}{E} \propto \sqrt{d} \, , \qquad (7.47)$$

exactly as with electromagnetic calorimeters. Experimentally one finds that absorber thicknesses $d < 2\,\mathrm{cm}$ iron do not lead to an improvement of the energy resolution [1]. Depending on the application and also on the available financial resources, a large variety of sampling detectors can be considered. Possible candidates for sampling elements in calorimeters are scintillators, liquid-argon or liquid-xenon layers, multiwire proportional chambers, layers of proportional tubes, flash chambers, streamer tubes, Geiger-Müller tubes (with local limitation of the discharge – 'limited Geiger-mode'), parallel plate chambers and layers of 'warm' (i.e., room temperature) liquids like tetramethylsilane (TMS), tetramethylpentane (TMP) or tetramethylgermane (TMG). Ionization chambers under high pressure also can be used [564]. For absorber materials, uranium, copper, tungsten and iron are most commonly used, although aluminum and marble calorimeters have also been constructed and operated.

An attractive alternative to longitudinally segmented calorimeters are 'spaghetti calorimeters', which allow at the same time an excellent track

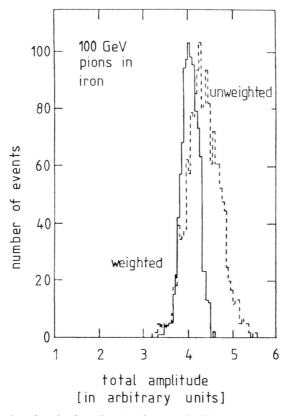

Fig. 7.45. Amplitude distribution for 100 GeV pions in an iron streamer-tube calorimeter with and without weighting factors [518]. The weighting factor was optimized for an incident energy of 100 GeV.

reconstruction (see section 5.2). Completely different, specially designed absorber geometries, however, can also be used ('accordion calorimeter') [565, 566].

In the course of time very imaginative and creative calorimeter designs have been realized for various applications. In high energy physics experiments, electrons, photons and hadrons are often detected with the same calorimeter. For these applications a hybrid set-up of a sampling calorimeter can be recommended, whose first part essentially measures the electron-photon component (about $25\,X_0$ lead), followed by a high quality uranium hadron calorimeter (about six absorption lengths of uranium). To catch the tails of hadron cascades, frequently an 'inexpensive' iron calorimeter ($\approx 4\lambda_a$) can be employed, which can serve at the same time as a flux return yoke for a magnetic coil. Such an arrangement is sketched in figure 7.47.

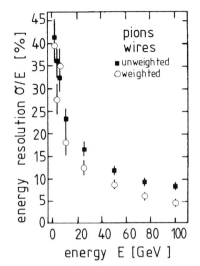

Fig. 7.46. Energy resolution of an iron streamer-tube hadron calorimeter for unweighted and weighted energy depositions [518, 523].

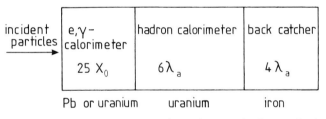

Fig. 7.47. Hybrid arrangement of an electron-hadron calorimeter.

Higher resolutions compared to sampling calorimeters can of course be obtained with total absorption hadron calorimeters (TANC – Total Absorption Nuclear Cascade). However, these detectors are in most cases too large and above all too expensive. Furthermore, the intrinsic losses for hadron cascades, e.g., by escaping neutrinos, muons and neutrons cannot be totally avoided. Therefore, in nearly all experiments one employs sampling calorimeters, at least as far as hadron calorimetry is concerned.

A prominent feature of calorimeters is that their energy resolution $\sigma(E)/E$ improves with increasing energy like $1/\sqrt{E}$, quite in contrast to momentum spectrometers, whose resolution deteriorates linearly with increasing momentum. In addition, calorimeters are rather compact even for high energies, because the shower length only increases logarithmically with the particle energy.

In cosmic ray experiments concerned with the energy determination of protons, heavy nuclei and also photons of the primary cosmic radiation in the energy range $> 10^{14}$ eV, alternative calorimetric measurement methods are needed to account for the low particle intensities. Cosmic ray particles initiate in the earth's atmosphere hadronic or electromagnetic cascades (see figure 7.34) which can be detected by quite different techniques. The energy of extensive air showers is normally determined by sampling their lateral distribution at sea level. This classical method quite obviously suffers from a relatively inaccurate energy determination [567]. Better results are obtained if the scintillation or Cherenkov light of the shower particles produced in the atmosphere is recorded (compare section 11.8). The latter technique, however, requires — because of the low light yield — clear and moonless nights.

An alternative method of energy determination can be considered for energy determination of high energy cosmic neutrinos or muons. These particles penetrate the earth's atmosphere without difficulty, so that one can also take advantage of the clear highly transparent water of the ocean, deep lakes or even polar ice as a Cherenkov medium. Muons undergo energy loss at high energies (> 1 TeV) mainly by bremsstrahlung and direct electron-pair production (see figure 1.6). These two energy-loss processes are both proportional to the muon energy. A measurement of the energy loss using a three-dimensional matrix of photomultipliers in deep water shielded from sun light allows a determination of the muon's energy. Similarly, the energy of electron or muon neutrinos can be roughly determined, if these particles produce electrons or muons in inelastic interactions in water, that, for the electron case, induce electromagnetic cascades, and for the muon case produce a signal proportional to the energy loss. The deep ocean, lake water or polar ice in this case are both interaction targets and detectors for the Cherenkov light produced by the interaction products. The produced electrons or muons in neutrino interactions closely resemble the direction of incidence of the neutrinos. Therefore, these deep water neutrino detectors are at the same time neutrino telescopes allowing one to enter the domain of neutrino astronomy in the TeV energy range [568, 569, 570].

7.4 Particle identification with calorimeters

In addition to energy determination, calorimeters are also capable of separating electrons from hadrons. The longitudinal and lateral shower development of electromagnetic cascades is determined by the radiation length X_0, and that of hadronic cascades by the much larger interaction length λ_{w} or absorption length λ_{a}. Calorimetric electron-hadron separation is based on these characteristic differences of shower development.

Figure 7.48 [518] shows the longitudinal development of 100 GeV electron and pion showers in a streamer-tube calorimeter. Since sampling calorimeters are usually longitudinally segmented into detector layers, various criteria for electron identification can be constructed. Essentially these separation methods are based on the following characteristics [1, 519, 549].

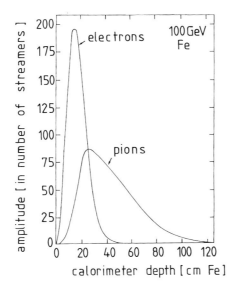

Fig. 7.48. Comparison of the longitudinal development of 100 GeV pions and electrons in a streamer-tube calorimeter [518].

1. Electrons deposit the largest fraction of their energy in the front part of a calorimeter. If the calorimeter is subdivided into a front and rear part, the amplitude ratio of the front divided by the rear part will more likely be large for electrons and small for hadrons.

2. Since for all materials normally used in calorimeters the absorption length λ_a is much larger than the radiation length X_0, electrons interact earlier in the calorimeter compared to hadrons. For example, 95 % of electrons have already initiated a shower after three radiation lengths, while a large fraction of hadrons (72 % in iron) have not undergone any interaction. The starting point of the shower development therefore is an additional separation criterion.

3. Hadronic cascades are much wider compared to electromagnetic showers (see figures 7.22 and 7.43). In a compact iron calorimeter 95 % of the electromagnetic energy is contained in a cylinder of 3.5 cm radius. For hadron cascades the 95 % lateral containment

radius is larger by a factor of about five, depending on the energy. From the different lateral behaviour of electromagnetic and hadronic cascades a typical characteristic compactness parameter can be derived.

4. Finally, the longitudinal center of gravity of the shower can also be used as an electron-hadron separation criterion.

Each separation parameter can be used to define a certain probability to the electron or pion hypothesis in an unseparated electron-pion beam. A multiplication of these probabilities allows to obtain an excellent electron-pion separation in calorimeters. One must take into account, however, that the separation criteria may be strongly correlated. Figure 7.49 [519, 549] shows such combined probability distributions exhibiting only a small overlap between the electron and pion hypothesis. The resulting e/π-misidentification probability for a given electron efficiency is shown in figure 7.50 [519, 549]. For a 95 % electron acceptance one obtains in this example a 1 % pion contamination for a particle energy of 75 GeV. With more sophisticated calorimeters a pion contamination as low as 0.1 % can be reached with calorimetric methods.

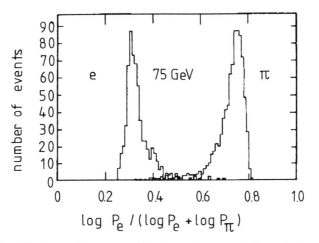

Fig. 7.49.   Electron-pion separation in a streamer-tube calorimeter [519, 549].

Muons can be distinguished not only from pions but also from electrons by their low energy deposition in calorimeters. Figure 7.51 [518] shows the amplitude distributions of 50 GeV electrons and muons. The possibility of an excellent electron-muon separation is already evident from this diagram.

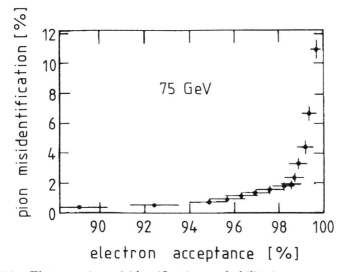

Fig. 7.50. Electron-pion misidentification probability in a streamer-tube calorimeter [519, 549]. The electron acceptance is the fraction of electrons accepted by a cut in the probability distribution. Correspondingly, the pion misidentification represents the fraction of accepted electron candidates that are really pions.

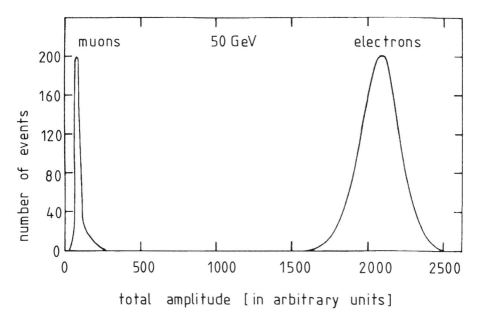

Fig. 7.51. Amplitude distribution of 50 GeV electrons and muons in a streamer-tube calorimeter [518].

For higher beam energies the interaction probability of muons for processes with higher energy transfers, e.g., by muon bremsstrahlung, increases [47, 59, 61, 571, 572, 573]. Although these processes are still quite rare, they can nevertheless lead to a small μ/e-misidentification probability in purely calorimetric measurements.

Figure 7.52 [571] shows the small overlap of the muon energy loss signal of 192 GeV muons in an electron calorimeter with the amplitude distribution of electrons of the same energy. Copper was used in this case as the absorber material. The overlap leads here to a purely calorimetric μ/e-misidentification probability of $1.7 \cdot 10^{-5}$ ($2.8 \cdot 10^{-5}$) for an electron acceptance of 95% (99%). This means that in an unseparated beam of electrons and muons of equal intensity a cut in the energy distribution alone, corresponding to an electron acceptance of 95%, also accepts a fraction $1.7 \cdot 10^{-5}$ of muons, i.e., misidentifies them as electrons.

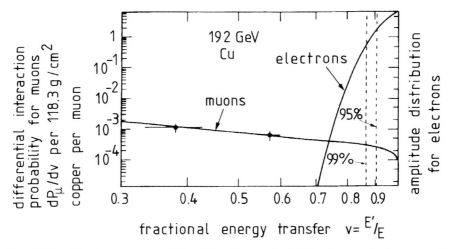

Fig. 7.52. Electron-muon misidentification probabilities in an electromagnetic streamer-tube calorimeter. The differential interaction probability for muons corresponds to the energy-loss distribution of the muons in the calorimeter [571].

The digital hit pattern of a 10 GeV pion, muon and electron in a streamer-tube calorimeter is shown in figure 7.53 [270].

Since the energy loss of high energy muons (> 500 GeV) in matter is dominated by processes with large energy transfers (bremsstrahlung, direct electron pair production, nuclear interactions), and these energy losses are proportional to the muon energy (see equation (1.67)), one can even build muon calorimeters for high energies in which the measurement of the muon energy loss allows an energy determination. This possibility of muon calorimetry will certainly be applied in proton-proton collision experiments at the highest energies (LHC – Large Hadron Collider,

$\sqrt{s} = 16\,\mathrm{TeV}$; ELOISATRON, $\sqrt{s} = 200\,\mathrm{TeV}$ [574]). The calorimetric method of muon energy determination can also be employed in deep water experiments used as neutrino telescopes.

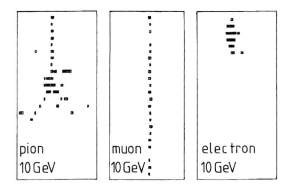

Fig. 7.53. Digital patterns (i.e., dot = fired tube) of 10 GeV pions, muons and electrons in a streamer-tube hadron calorimeter [270].

7.5 Calibration and monitoring of calorimeters

The calibration of individual calorimeters is usually done in accelerator test beams which supply identified particles of known momenta. By varying the beam energy the linearity of the calorimeter can be tested and characteristic shower parameters can be recorded. For the calibration of calorimeters designed for low energies, e.g., semiconductor detectors, radioactive sources are normally used. Preferentially used are K-line emitters, like ^{207}Bi with well defined monoenergetic electrons or gamma ray lines, which allow a calibration via the total absorption peaks.

In addition to energy calibration, the dependence of the calorimeter signal on the point of particle impact, the angle of incidence and the behaviour in magnetic fields is of great importance. In particular, for calorimeters with gas sampling, magnetic field effects can cause spiraling electrons, which can significantly modify the calibration. This is demonstrated in figure 7.54 [522] which shows the amplitude distribution of 6 GeV electrons in a copper streamer-tube calorimeter without and with a magnetic field parallel to the anode wires, for a field strength of 1.1 Tesla. The deterioration of the energy resolution caused by low energy electrons which, because of the magnetic field, have longer path lengths in the detector layers, is clearly visible.

In gas-sampling calorimeters the particle rate can have an influence on the signal amplitude because of dead time or recovery time effects.

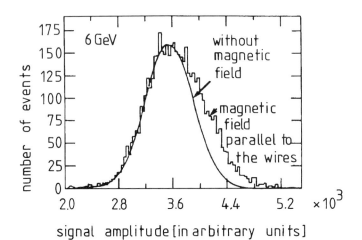

Fig. 7.54. Magnetic field dependence of signal amplitudes of 6 GeV electrons in a streamer-tube calorimeter [522].

A thorough calibration of a calorimeter therefore requires an extensive knowledge of the various parameter dependent characteristics.

Big experiments can contain a large number of calorimeter modules, not all of which can be calibrated in test beams. If some of the modules are calibrated in a test beam, the rest can be adjusted relative to them. This relative calibration can be done by using minimum-ionizing muons that penetrate many calorimeter modules. In uranium calorimeters, the constant noise caused by the natural radioactivity of the uranium can be used for a relative calibration. If one uses non-radioactive absorber materials in gas-sampling calorimeters, a test and relative calibration can also be performed with radioactive noble gases like ^{85}Kr.

Scintillator calorimeters can best be calibrated by feeding defined light signals, e.g., via light-emitting diodes (LEDs), into the detector layers and recording the output signals from the photomultipliers. To avoid variations in the injected light intensity, which may be caused by different light yields of the individual light diodes, a single light source can be used (e.g., a laser), which distributes its light via a manifold of light fibers to the scintillation counters [1].

Once a complex calorimeter system has been calibrated, one has to ensure that the calibration constants do not vary or, if they do, the drift of the calibration parameters must be monitored. The time stability of the calibration can be checked with, e.g., cosmic ray muons. In some cases some calorimeter modules may be positioned unfavorably so that the rate of cosmic ray muons is insufficient for accurate stability control. Therefore, reference measurements have to be performed periodically by injecting calibrated reference signals into the various detector layers or into the inputs of the readout electronics.

In gas-sampling calorimeters the output signal can in principle only vary because of a change of gas parameters and high voltage. In this case, an identical test chamber can be supplied with the detector gas and the characteristic X-rays of a radioactive source can be permanently monitored. A change in the measured X-ray energy in this test chamber indicates a time dependent calibration which can be compensated by an adjustment of the high voltage.

In some experiments there are always particles available that can be used for calibration and monitoring. For example, elastic Bhabha scattering ($e^+e^- \rightarrow e^+e^-$) can be used to calibrate the electromagnetic calorimeters in an e^+e^- scattering experiment, since the final state particles — if one neglects radiative effects — have known beam energy. In the same way, the reaction $e^+e^- \rightarrow q\bar{q}$ with subsequent hadronization of the quarks can be used to check the performance of a hadron calorimeter. Finally, muon pair production ($e^+e^- \rightarrow \mu^+\mu^-$) supplies final state muons with known momentum (= beam momentum at high energies), which can reach all detector modules because of their nearly flat angular distribution ($\mathrm{d}\sigma/\mathrm{d}\Omega \sim 1 + \cos^2\theta$, where θ = angle between e^- and μ^-).

7.6 Cryogenic calorimeters

The calorimeters described so far can be used for the spectroscopy of particles from the MeV-range up to the highest energies. For many investigations, e.g., in astrophysics, the detection of particles of extremely low energy in the range between 1 and 1000 eV is of great interest. Calorimeters for such low energy particles are used for the detection of and search for low energy cosmic neutrinos, weakly interacting massive particles (WIMPs) or other candidates of dark, non-luminous matter [575]-[588]. To reduce the detection threshold and improve at the same time a calorimeter's energy resolution, it is only natural to replace the ionization or electron-hole pair production by quantum transitions requiring lower energies (see section 7.1).

Typical energies to break the bond of a Cooper pair in a superconductor are of the order of 1 meV ($= 10^{-3}$ eV). Cooper pairs are bound states of two electrons with opposite spin which behave like bosons and will form at sufficiently low temperatures a Bose condensate. Phonons in solid state materials have energies around 10^{-5} eV for temperatures around 100 mK. If it is possible to efficiently detect broken Cooper pairs ('quasi particles') or phonons, the low threshold energies for these processes would present an advantageous working principle for the detection of low energy particles. To avoid thermal excitations of these quantum processes, such calorimeters, however, would have to be operated at extremely low temperatures, typically in the milli-Kelvin range. For this reason, such calorimeters are called cryogenic detectors. Cryogenic calorimeters can be subdivided in two main categories: firstly, detectors for quasi particles in superconducting crystals, and secondly, phonon detectors in insulators [589, 590, 591].

Cooper pairs in superconductors have binding energies in the range between $4 \cdot 10^{-5}$ eV (Ir) and $3 \cdot 10^{-3}$ eV (Nb). Even extremely low energy depositions would break up a large number of Cooper pairs. The main purpose of these detectors is to detect these 'quasi particles', which is difficult. One detection method is based on the fact that the superconductivity of a substance is destroyed by the energy deposition if the detector element is sufficiently small. This is the working principle of superheated superconducting granules [590]. In this case the cryogenic calorimeter is made of a large number of superconducting spheres with diameters in the μm-range. If these granules are embedded in a magnetic field, and the energy deposition of a low energy particle transfers one particular granule from the superconducting to the normal conducting state, this transition can be detected by the suppression of the Meissner effect. This is where the magnetic field, which does not enter the granule in the superconducting state, now again passes through the normal conducting granule. The transition from the superconducting to the normal conducting state can be detected by pick-up coils coupled to very sensitive preamplifiers or by SQUIDs (Superconducting Quantum Interference Devices) [591]. These quantum interferometers are extremely sensitive detection devices for magnetic effects. The operation principle of a SQUID is based on the Josephson effect, which is a tunnel effect operating between two superconductors separated by thin insulating layers. In contrast to the normal one-electron tunnel effect, known, e.g., from alpha decay, the Josephson effect involves the tunneling of Cooper pairs. In Josephson junctions, interference effects of the tunnel current occur which can be influenced by magnetic fields. The structure of these inference effects is related to the size of the magnetic flux quanta [448, 592, 593].

An alternative method to detect quasi particles is to let them directly tunnel through an insulating foil between two superconductors (SIS – Superconducting-Insulating-Superconducting transition) [589]. In this case the problem arises of keeping undesired leakage currents at an extremely low level.

In contrast to quasi particles, phonons, which can be exited by energy deposition in insulators, can be detected with the methods of classical calorimetry. If ΔE is the absorbed energy, this results in a temperature rise of

$$\Delta T = \Delta E/mc , \qquad (7.48)$$

where c is the specific heat and m the mass of the calorimeter. If these calorimetric measurements are performed at very low temperatures, where c can be very small (the lattice contribution to the specific heat is proportional to T^3 at low temperatures), this method can in principle also be used to detect single individual particles. In a practical experiment, the temperature change is recorded with a thermistor, which is basically an NTC resistor (negative temperature coefficient), embedded into or fixed to an ultra-pure crystal. The crystal represents the absorber, i.e., the detector for the radiation that is to be measured. Because of the discrete energy of phonons, one would expect discontinuous thermal energy fluctuations which can be detected with electronic filter techniques.

In this way α-particles and γ-rays have been detected in a small TeO_2-crystal at 15 mK in a purely thermal detector with thermistor readout with an energy resolution of 5 keV fwhm [594].

Special bolometers have also been developed in which heat and ionization signals are measured simultaneously [595].

Thermal detectors provide promise for improvements of energy resolutions. For example, a 1 mm cubic crystal of silicon kept at 20 mK would have a heat capacity of $5 \cdot 10^{-15}$ J/K and a fwhm energy resolution of 0.1 eV (corresponding to $\sigma = 42$ meV) [582]. There are, however, still various important problems to be solved, before these values can be reached.

1. Non-uniformity of phonon collection, especially in large detectors.

2. Spatial non-uniformity of the recombination of electron-hole pairs trapped by various impurities.

3. Noise due to electromagnetic sources, especially microphonics (generation of noise due to mechanical, acoustic and electromagnetic excitation).

4. Problems with keeping the temperature of the bolometer constant, and consequently its gain.

Joint efforts in the fields of cryogenics, particle physics and astrophysics are required, which may lead to exciting and unexpected results. One interesting goal would be to detect relic neutrinos of the Big Bang with energies around 200 μeV [582].

The development of cryogenic calorimeters for the measurement of extremely low energies has only recently begun. The set-up of a cryogenic detector, based on the energy absorption in superheated superconducting granules, is shown in figure 7.55 [596]. The system of granules and

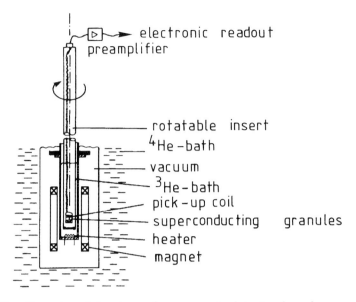

Fig. 7.55. Experimental set-up of a cryogenic detector based on superheated superconducting granules (SSG) [596].

pick-up coil was rotatable by 360° around an axis perpendicular to the magnetic field. This was used to investigate the dependence of the critical field strength for reaching the superconducting state on the orientation of the granules with respect to the magnetic field. This system succeeded in detecting quantum transitions in tin, zinc and aluminum granules at ^{4}He and ^{3}He-temperatures. Figure 7.56 shows a microphotograph of tin granules [590, 597]. At present it is already possible to manufacture tin granules with diameters as small as $5\,\mu$m.

With a detector consisting of superheated superconducting granules, it has already been shown that one can detect minimum-ionizing particles unambiguously [598].

The detection of transitions from the superconducting into the normal conducting state with signal amplitudes of about $100\,\mu$V and recovery times of 10 to 50 ns already indicates that superconducting strip counters are a possible candidate for microvertex detectors in particle physics experiments [599].

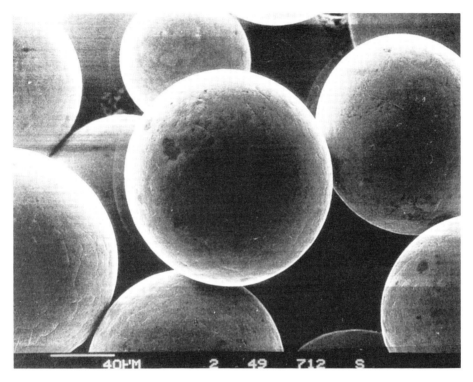

Fig. 7.56. Tin granules (diameter = 130 μm) as a cryogenic calorimeter. Low energy absorption can warm the granules by an amount sufficient to cause a change from the superconducting state to the normal conducting state, thereby providing a detectable signal [590, 597].

8

Momentum measurement

The momenta of charged particles are usually determined in magnetic spectrometers.

The Lorentz force causes the particles to follow circular or helical trajectories around the direction of the magnetic field. The bending radius of particle tracks is related to the magnetic field strength and the momentum component of the particle perpendicular to the magnetic field. Depending on the experimental situation or the particle species to be investigated, different magnetic spectrometers are used.

8.1 Magnetic spectrometers for fixed target experiments

The basic set-up of a magnetic spectrometer for fixed target experiments (in contrast to storage ring experiments) is sketched in figure 8.1. Particles of known identity and also, in general, of known energy are incident on a target thereby producing secondary particles in an interaction. The purpose of the spectrometer is to measure the momenta of the charged secondary particles.

Let the magnetic field B be oriented along the y-axis, $\vec{B} = (0, B_y, 0)$, whereas the direction of incidence of the primary particles is taken to be parallel to the z-axis. In hadronic interactions typical transverse momenta of

$$p_{\mathrm{T}} \approx 350 \,\mathrm{MeV}/c \tag{8.1}$$

are transferred to secondary particles, where

$$p_{\mathrm{T}} = \sqrt{p_x^2 + p_y^2} \,. \tag{8.2}$$

Normally, $p_x, p_y \ll p_z$, where the momenta of outgoing particles are described by $\vec{p} = (p_x, p_y, p_z)$ [1]. The trajectories of particles incident into the spectrometer are determined in the most simple case by track

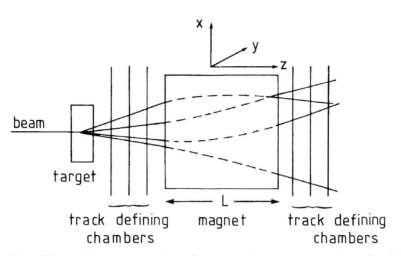

Fig. 8.1. Schematic representation of a magnetic spectrometer in a fixed target experiment with a stationary target.

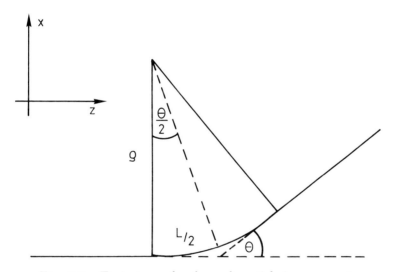

Fig. 8.2. Trajectory of a charged particle in a magnet.

detectors before they enter and after they have left the magnet. Since the magnetic field is oriented along the y-axis, the deflection of charged particles is in the xz plane. Figure 8.2 sketches the track of a charged particle in this plane.

The Lorentz force is balanced by the centrifugal force. We choose our coordinate system in such a way that the particles incident into the spectrometer are parallel to the z-axis, i.e., $|\vec{p}| = p_z = p$, where $\vec{p}$ is the momentum of the particle to be measured. One then has (for $\vec{p} \perp \vec{B}$, where m – mass, v – velocity and ϱ – bending radius of the track in the magnetic field):

$$\frac{mv^2}{\varrho} = e\,v\,B_y \; . \tag{8.3}$$

The bending radius ϱ itself is obtained from equation (8.3) by

$$\varrho = \frac{p}{eB_y} \; . \tag{8.4}$$

The particles pass through the magnet following a circular trajectory, where the bending radius ϱ, however, is normally very large compared to the magnet length L. Therefore, the deflection angle θ can be approximated by

$$\theta = \frac{L}{\varrho} = \frac{L}{p}eB_y \; . \tag{8.5}$$

Because of the magnetic deflection, the charged particles obtain an additional transverse momentum of

$$\Delta p_x = p \cdot \sin\theta \approx p \cdot \theta = L\,e\,B_y \; . \tag{8.6}$$

If the magnetic field varies along L, equation (8.6) is generalized to

$$\Delta p_x = e \int_0^L B_y(l)\mathrm{d}l \; . \tag{8.7}$$

The accuracy of the momentum determination is influenced by a number of different effects. Let us first consider the influence of the finite track resolution of the detector on the momentum determination. Using equations (8.4) and (8.5), we obtain

$$\begin{aligned} p &= e\,B_y \cdot \varrho \\ &= e\,B_y \cdot \frac{L}{\theta} \; . \end{aligned} \tag{8.8}$$

Since the tracks of ingoing and outgoing particles are straight, the deflection angle θ is the actual quantity to be measured. Because of

$$\left|\frac{\mathrm{d}p}{\mathrm{d}\theta}\right| = e\,B_y\,L \cdot \frac{1}{\theta^2} = \frac{p}{\theta} \; , \tag{8.9}$$

one has

$$\frac{\mathrm{d}p}{p} = \frac{\mathrm{d}\theta}{\theta} \; , \tag{8.10}$$

and

$$\frac{\sigma(p)}{p} = \frac{\sigma(\theta)}{\theta} \; . \tag{8.11}$$

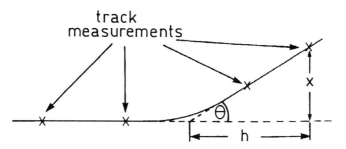

Fig. 8.3.   Sketch illustrating the determination of the track measurement error.

To determine the deflection angle, at least four track coordinates are required, i.e., two before and two after the magnet. If all track measurements have the same measurement error $\sigma(x)$, the variance of the deflection angle is obtained to be (see figure 8.3)

$$\sigma^2(\theta) \propto \sum_{i=1}^{4} \sigma_i^2(x) = 4\sigma^2(x) \tag{8.12}$$

$$\sigma(\theta) \propto 2\sigma(x) \ . \tag{8.13}$$

Because of

$$\theta = \frac{x}{h} \ , \tag{8.14}$$

where h is the lever arm for the angular measurement (see figure 8.3), one obtains

$$\sigma(\theta) = \frac{2\sigma(x)}{h} \ . \tag{8.15}$$

Using equation (8.11), this leads to

$$\frac{\sigma(p)}{p} = \frac{2\sigma(x)/h}{eB_yL} \cdot p = \frac{2\sigma(x)}{h} \cdot \frac{p}{\Delta p_x} \ . \tag{8.16}$$

From equation (8.16) one sees that the momentum resolution $\sigma(p)$ is proportional to p^2. Depending on the quality of the track detectors, one may obtain [1]

$$\frac{\sigma(p)}{p} = (10^{-3} \text{ to } 10^{-4}) \cdot p \ [\text{GeV}/c] \ . \tag{8.17}$$

In cosmic ray experiments, it has become usual practice to define a maximum detectable momentum (mdm). This is defined by

$$\frac{\sigma(p_{\text{mdm}})}{p_{\text{mdm}}} = 1 \ . \tag{8.18}$$

For a magnetic spectrometer with a momentum resolution given by equation (8.17), the maximum detectable momentum would be

$$p_{\mathrm{mdm}} = 1\,\mathrm{TeV}/c \;\; \text{to} \;\; 10\,\mathrm{TeV}/c \;. \tag{8.19}$$

The momentum measurement is normally performed in an air-gap magnet. The effect of multiple scattering is low in this case and influences the measurement accuracy only at low momenta. Because of the high penetrating power of muons, their momenta can also be analyzed in solid iron magnets. For this kind of application, however, the influence of multiple scattering cannot be neglected.

A muon penetrating a solid iron magnet of thickness L obtains a transverse momentum $\Delta p_{\mathrm{T}}^{\mathrm{MS}}$ due to multiple scattering according to

$$\Delta p_{\mathrm{T}}^{\mathrm{MS}} = p \cdot \sin\theta_{\mathrm{rms}} \approx p \cdot \theta_{\mathrm{rms}} = 19.2\sqrt{\frac{L}{X_0}} \; [\mathrm{MeV}/c] \tag{8.20}$$

(see equation (1.48) with $\beta = 1$, and figure 8.4).

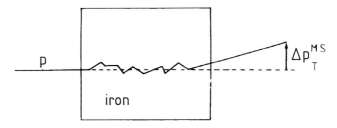

Fig. 8.4.   Illustration of the multiple scattering error.

Since the magnetic deflection is in the x-direction, only the multiple scattering error projected onto this direction is of importance:

$$
\begin{aligned}
\Delta p_x^{\mathrm{MS}} &= \frac{19.2}{\sqrt{2}}\sqrt{\frac{L}{X_0}} \; [\mathrm{MeV}/c] \\[2mm]
&= 13.6\sqrt{\frac{L}{X_0}} \; [\mathrm{MeV}/c] \;.
\end{aligned} \tag{8.21}
$$

The momentum resolution limited by the effect of multiple scattering is given by the ratio of the deflection by multiple scattering to the magnetic deflection according to [1]

$$\left.\frac{\sigma(p)}{p}\right|^{\mathrm{MS}} = \frac{\Delta p_x^{\mathrm{MS}}}{\Delta p_x^{\mathrm{magn}}} = \frac{13.6\,\sqrt{L/X_0}\,[\mathrm{MeV}/c]}{e\int_0^L B_y(l)\mathrm{d}l} \;. \tag{8.22}$$

Both the deflection angle θ caused by the Lorentz force and the multiple scattering angle are inversely proportional to the momentum. Therefore,

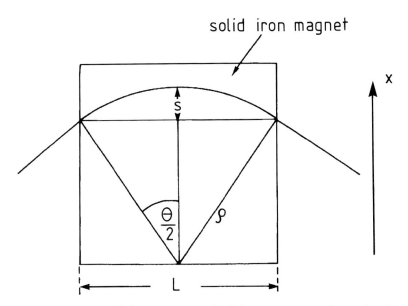

Fig. 8.5. Illustration of the sagitta method for momentum determination [1].

the momentum resolution in this case does not depend on the momentum of the particle.

For solid iron magnetic spectrometers ($X_0 = 1.76\,\text{cm}$) typical values of $B = 1.8\,\text{Tesla}$ are used, leading to a momentum resolution of (see equation (8.22))

$$\left.\frac{\sigma(p)}{p}\right|^{\text{MS}} = 0.19 \cdot \frac{1}{\sqrt{L}} , \tag{8.23}$$

with L in meters. This gives for $L = 3\,\text{m}$

$$\left.\frac{\sigma(p)}{p}\right|^{\text{MS}} = 11\,\% . \tag{8.24}$$

This equation only contains the effect of multiple scattering on the momentum resolution. In addition, one has to consider the momentum measurement error from the uncertainty of the position measurement. This error can be obtained from equation (8.16) or from the determination of the sagitta (see figure 8.5) [1]. The sagitta s is related to the magnetic bending radius ϱ and the magnetic deflection angle θ by

$$s = \varrho - \varrho \cos\frac{\theta}{2} = \varrho\left(1 - \cos\frac{\theta}{2}\right) . \tag{8.25}$$

Because of $1 - \cos\frac{\theta}{2} = 2\sin^2\frac{\theta}{4}$, one obtains

$$s = 2\varrho \sin^2\frac{\theta}{4} . \tag{8.26}$$

Since $\theta \ll 1$, the sagitta can be approximated by (θ in radians)

$$s = \frac{\varrho \theta^2}{8} \ . \tag{8.27}$$

In the following we will replace B_y by B for simplicity. Using equations (8.8) and (8.4) for θ and ϱ the sagitta can be expressed by

$$s = \frac{\varrho}{8} \cdot \left(\frac{eBL}{p}\right)^2 = \frac{eBL^2}{8p} \ . \tag{8.28}$$

If B is measured in Tesla, ϱ in meters and p in GeV/c, the sagitta is given by

$$s = 0.3 \, BL^2/(8p) \ . \tag{8.29}$$

The determination of the sagitta requires at least three position measurements x_i ($i = 1, 2, 3$). These can be obtained from three tracking detectors positioned at the entrance (x_1) and at the exit (x_3) of the magnet, while one chamber could be placed in the center of the magnet (x_2). Because of

$$s = x_2 - \frac{x_1 + x_3}{2} \tag{8.30}$$

and under the assumption that the track measurement errors $\sigma(x)$ are the same for all chambers, it follows that

$$\sigma(s) = \sqrt{\frac{3}{2}} \, \sigma(x) \ . \tag{8.31}$$

This leads to a momentum resolution from track measurement errors of

$$\left.\frac{\sigma(p)}{p}\right|^{\text{track resolution}} = \frac{\sigma(s)}{s} = \frac{\sqrt{\frac{3}{2}}\sigma(x) \cdot 8p}{0.3 \, BL^2} \ . \tag{8.32}$$

If the track is measured not only at three but at N points equally distributed over the magnet length L, it can be shown that the momentum resolution due to the finite track measurement error is given by [600]

$$\left.\frac{\sigma(p)}{p}\right|^{\text{track resolution}} = \frac{\sigma(x)}{0.3 \, BL^2} \sqrt{720/(N + 4)} \cdot p \ . \tag{8.33}$$

For $B = 1.8$ T, $L = 3$ m, $N = 4$ and $\sigma(x) = 0.5$ mm equation (8.33) leads to

$$\left.\frac{\sigma(p)}{p}\right|^{\text{track resolution}} \approx 10^{-3} \cdot p \ [\text{GeV}/c] \ . \tag{8.34}$$

If the N measurements are distributed over L in k constant intervals, one has

$$L = k \cdot N \tag{8.35}$$

and thereby (if $N \gg 4$):

$$\left. \frac{\sigma(p)}{p} \right|^{\text{track resolution}} \propto L^{-5/2} \cdot B^{-1} \cdot p \,. \tag{8.36}$$

To obtain the total error on the momentum determination, the multiple scattering and track resolution error have to be combined. Both contributions according to equations (8.24) and (8.34) are plotted in figure 8.6 for the above mentioned parameters of a solid iron magnetic spectrometer. At low momenta multiple scattering dominates the error and at high momenta it is limited by the track measurement error.

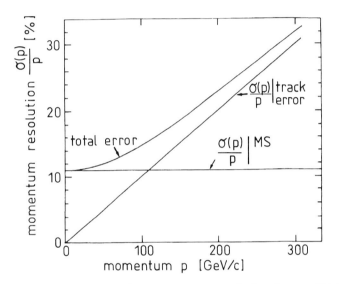

Fig. 8.6. Contributions to the momentum resolution for a solid iron magnetic spectrometer.

For an air gap magnet the error contribution due to multiple scattering is naturally much smaller. If equation (8.22) is applied to an air gap magnet, $(X_0 = 304 \,\text{m})$, one obtains

$$\left. \frac{\sigma(p)}{p} \right|^{\text{MS}} = 1.4 \cdot 10^{-3}/\sqrt{L} \,, \tag{8.37}$$

which means for $L = 3 \,\text{m}$:

$$\left. \frac{\sigma(p)}{p} \right|^{\text{MS}} = 0.08 \,\% \,. \tag{8.38}$$

8.2   Magnetic spectrometers for special applications

Fixed target experiments have the advantage that secondary beams can be produced from a primary target. These secondary beams can consist of many types of different particles so that one can perform experiments with, e.g., neutrino, muon, photon or K_L^0 beams. The disadvantage with fixed target experiments, however, is that the available center of mass energy is relatively small. Therefore, investigations in the field of high energy physics are frequently done at storage rings. In storage ring experiments, the center of mass system is identical with the laboratory system (for a crossing angle of zero), if the colliding beams have the same energy and are opposite in momentum. The event rates are in general rather low because the target density — one beam represents the target for the other and vice versa — is low compared to fixed target experiments. Because of the low luminosity and also in order to detect all interaction products, storage ring experiments normally cover the full solid angle of 4π, surrounding the interaction point. Such a hermeticity allows a complete reconstruction of individual events.

Depending on the type of storage ring different magnetic field configurations can be considered.

For proton-proton (or $p\bar{p}$) storage rings dipole magnets can be used, where the magnetic field is perpendicular to the beam direction. Since such a dipole also bends the stored beam, its influence must be corrected by compensation coils. The compensation coils are also dipoles, but with opposite field gradient, so that

$$\int \vec{B}(l) \cdot \mathrm{d}\vec{l} = 0 \qquad (8.39)$$

holds across the whole experiment (figure 8.7). Such a configuration is unsuitable for electron-positron storage rings, because the strong dipole field causes the emission of intense synchrotron radiation, which cannot be tolerated for the storage ring operation and the safe running of the detectors.

A dipole magnet can be made self-compensating if two dipoles with opposite field gradient on both sides of the interaction point are used instead of only one dipole. Equation (8.39) is automatically fulfilled in this case, but at the expense of strongly inhomogeneous magnetic fields at the interaction point which complicate the track reconstruction considerably (see figure 8.8 [1]). If, on the other hand, toroidal magnets (figure 8.9 [1]) are employed, one can achieve that the beams traverse the spectrometer in a region of zero field. Multiple scattering, however, on the inner cylinder of the toroidal magnet limits the momentum resolution.

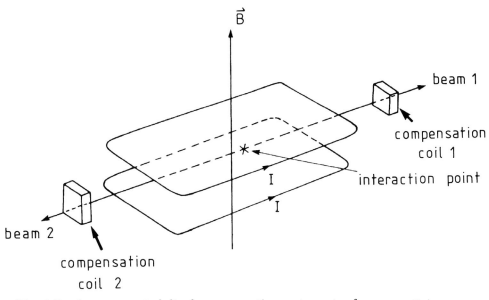

Fig. 8.7. A compensated dipole as magnetic spectrometer for pp or $p\bar{p}$ storage-ring experiments [1].

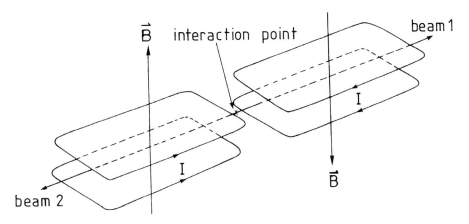

Fig. 8.8. A self-compensating (split field magnet) dipole arrangement (after [1]).

In most cases a solenoidal magnetic field is chosen, in which the stored beams run parallel to the magnetic field (figure 8.10). Therefore, the detector magnet has no influence on the beams, and also no synchrotron radiation is produced.

The track detectors are mounted inside the magnetic coil and are therefore also cylindrical. The longitudinal magnetic field acts only on the

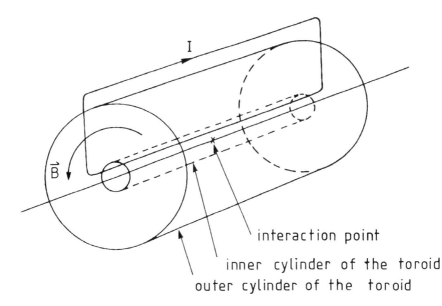

Fig. 8.9.   Sketch of a magnetic toroid. Only one winding of the coil is indicated (after [1]).

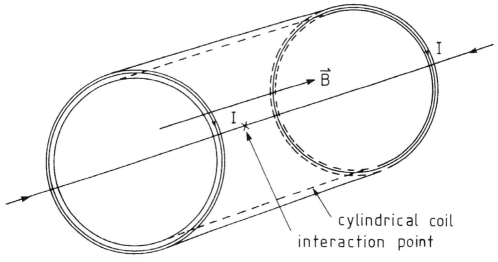

Fig. 8.10.   Sketch of a solenoid for the production of an axial magnetic field. The solenoid is a long cylindrical coil.

transverse momentum component of the produced particles and leads to a momentum resolution given by equation (8.33), where $\sigma(x)$ is the coordinate resolution in the plane perpendicular to the beam axis. Figure 8.11 shows schematically two tracks originating from the interaction point in a projection perpendicular to the beam ('$r\varphi$ plane') and parallel to the

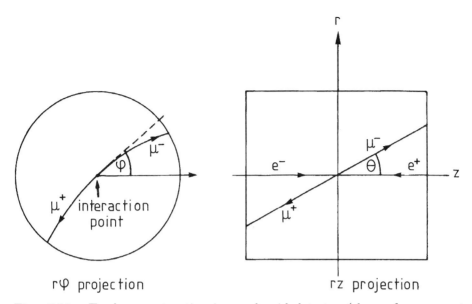

rφ projection   rz projection

Fig. 8.11. Track reconstruction in a solenoid detector (shown for an event $e^+e^- \to \mu^+\mu^-$).

beam ('rz plane'). The characteristic track parameters are given by the polar angle θ, the azimuthal angle φ and the radial coordinate r. If N coordinates are measured along a track of total length L (in m) with an accuracy of $\sigma_{r\varphi}$ (in m) in a magnetic field B (in Tesla) the transverse momentum resolution caused by the track measurement error is found to be [600] (see equation (8.33))

$$\left.\frac{\sigma(p)}{p_{\mathrm{T}}}\right|^{\text{track error}} = \frac{\sigma_{r\varphi}}{0.3BL^2}\sqrt{\frac{720}{N+4}} \cdot p_{\mathrm{T}} \ [\text{GeV}/c] \ . \qquad (8.40)$$

In addition to the track error one has to consider the multiple scattering error. This is obtained from equation (8.22) as

$$\left.\frac{\sigma(p)}{p_{\mathrm{T}}}\right|^{\text{MS}} = 0.045\frac{1}{B\sqrt{LX_0}} \ , \qquad (8.41)$$

where X_0 (in m) is the average radiation length of the material traversed by the particle.

The total momentum of the particle is obtained from p_{T} and the polar angle θ to be

$$p = \frac{p_{\mathrm{T}}}{\sin\theta} \ . \qquad (8.42)$$

As in the transverse plane, the measurement of the polar angle contains a track error and multiple scattering error.

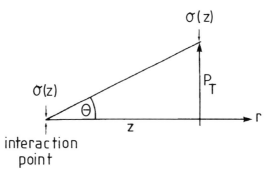

Fig. 8.12.   Illustration of the polar angle measurement error for the case of only two coordinates, defining a track.

If the z-coordinate in the track detector is determined with an accuracy $\sigma(z)$, the error on the measurement of the polar angle is given in the case of only two track coordinates by

$$\sigma(\theta)|^{\text{track error}} = \frac{\sigma(z)}{z} \cdot \sqrt{2}\,. \tag{8.43}$$

(For high energy particles the particle track in the rz plane is a straight line, see figure (8.12).) If the particle track is measured in N equidistant steps, equation (8.43) is generalized to [1, 600]

$$\sigma(\theta)|^{\text{track error}} = \frac{\sigma(z)}{z} \sqrt{\frac{12(N-1)}{N(N+1)}}\,. \tag{8.44}$$

In this formula z is the projected track length in the z-direction which is normally of the same order of magnitude as the transverse length of a track. Equation (8.44) describes only the track measurement error. In addition, one has to consider the multiple scattering error which can be derived from equation (1.47) to be

$$\sigma(\theta)|^{\text{MS}} = \frac{0.0136}{\sqrt{3}} \cdot \frac{1}{p} \cdot \sqrt{\frac{l}{X_0}}\,, \tag{8.45}$$

where p is measured in GeV/c, l is the track length (in units of radiation lengths), and $\beta = 1$ is assumed [34, 35].

The factor $\frac{1}{\sqrt{3}}$ requires some explanation. The multiple scattering angle θ, which is relevant for the polar angle measurement, has to be understood in this case as the ratio of the track displacement Δr by multiple scattering divided by the track length l (see figure 8.13). In contrast, the rms planar multiple scattering angle θ_{plane} is described by equation (1.47).

Gaseous detectors with extremely low transverse mass are generally used in solenoids. Therefore, the momentum measurement error due to

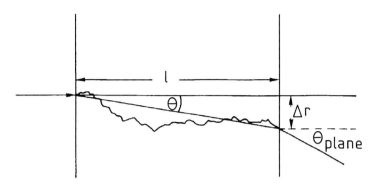

Fig. 8.13. Explanation of the definition of the polar angle measurement error. The uncertainty in the polar angle θ originates from the displacement of the particle track by Δr along the track length l. $\Delta r/l$ is related to the planar Coulomb multiple scattering angle θ_{plane} by $\Delta r/l = \frac{1}{\sqrt{3}}\theta_{\text{plane}}$ [34, 35].

multiple scattering plays only a minor rôle. Equation (8.40) shows that the momentum resolution improves with the product BL^2. It also improves for a fixed track length with the number of track measurement points although only approximately like $\frac{1}{\sqrt{N}}$.

Momentum measurements with magnetic spectrometers in the low-energy region, e.g., for beta spectroscopy, frequently require different magnet geometries. Because of their low momenta, these particles must be kept in a vacuum to avoid ionization losses and multiple scattering in the air. Double focusing spectrometers [63] allow one to obtain relative momentum resolutions of 10^{-5}, albeit with a low transmission probability from the source to the detector [601, 602]. Also semiconductor counters like Ge(Li) and Si(Li) detectors as well as high purity germanium crystals are used for alpha, beta, and gamma-ray spectroscopy (see section 7.1) with excellent energy resolution ($< 0.1\,\%$). Since solid state detectors are much easier to handle, magnetic spectrometers for applications in the MeV-range are less frequently employed.

At very low energies the momenta of particles can also be determined by evaluating in detail their multiple scattering behaviour. This method has been used, e.g., in the analysis of cosmic ray experiments with nuclear emulsions (see section 4.16).

9

Electronics

Multipurpose detectors for experiments in high energy physics, astrophysics or cosmic radiation often involve a large number of subdetectors. The purpose of these detector elements is to obtain information on arrival time, direction of incidence, energy, momentum, and other characteristics of the particle to be detected. For modern detection techniques the detector signals are provided in the form of electronic pulses.

The aim of the first section of this chapter is to give a brief overview of the variety of electronic techniques used to process the analog information coming from the subdetectors. The detector output is usually converted to a digital signal which can be reliably transmitted for further processing.

The second section describes techniques and architectures for trigger systems. For simple set-ups, e.g., to measure the absolute rate of cosmic ray muons at sea level with the help of a scintillator telescope, it is sufficient to collect signals from the scintillators using appropriate thresholds. Muons may then be selected by a coincidence requirement on the set of scintillation counters. Other particles, like electrons or protons, can be rejected by requiring a certain penetration power for the particles. This can be realized in the form of an iron absorber placed between the scintillation counters.

In more complicated experiments a careful design of selective trigger electronics is mandatory. The purpose of this type of system is to prevent events without interest (i.e., background) from being recorded. Only events which fulfill certain requirements are kept and written to disk or tape for later analysis. In large experiments this is usually a multistep process.

Finally brief descriptions of detector monitoring, controlling and data acquisition (DAQ) and of slow control complete this chapter.

9.1   Readout methods

One of the first methods of seeing the reaction products from interactions of particles was the observation of short-duration light signals from zinc-sulfide screens if these were hit, e.g., by alpha-particles. This pioneering method, first introduced at the beginning of this century, was later used by Rutherford in his famous scattering experiments.

With more sophisticated detection techniques such as cloud, spark and bubble chambers, a long era of optical event recording and analyzing started around 1910 (see chapter 4). Even though the study of elementary particles and their interaction by photographic methods, as provided by the cloud, bubble and emulsion techniques, allows one to inspect the events visually, there is a major drawback associated with this approach in that the visual signals are not compatible with direct computer analysis. In contrast, a tedious and tiresome manual scanning of the event pictures is necessary to digitize these events and analyze them by appropriate pattern recognition methods.

A dramatic step forward in the development of modern readout techniques started in the late 1920s with the introduction of electronic amplifiers and counting circuits for radiation detectors like the proportional counter. Since this time tremendous progress has been made in electronics as well as in computing techniques.

9.1.1   Signal terminology

The signals provided by some detectors are entirely of an electrical nature. Other experimentally measurable quantities like light, temperature, pressure, magnetic field etc. must first be transformed into an electronic equivalent. Generally, the information is coded in the form of pulse signals, i.e., brief, temporal changes of current or voltage. In most cases the desired information is contained in the pulse shape, amplitude or in its relative timing with respect to other signals. Figure 9.1 shows a typical unipolar pulse and explains the commonly used terminology.

A typical time scale in the field of modern electronics covers the range from a few nanoseconds up to the fraction of a microsecond. The pulse height or amplitude is the maximal value of the pulse with respect to the instantaneous baseline. The baseline is the voltage or the current level in the absence of a signal. It can be arbitrarily adjusted by superposition of a constant voltage or current. The leading edge describes the rising signal from the baseline to the pulse maximum. The risetime is the time needed for the signal to rise from 10 % to 90 % of its full amplitude. The signal duration is usually characterized as the full width at half maximum.

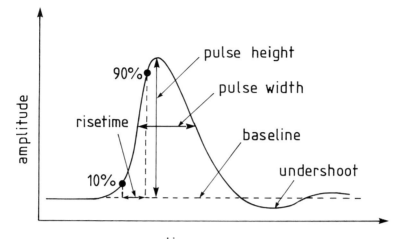

Fig. 9.1. Unipolar analog voltage pulse.

9.1.2 Amplifiers

Apart from some detector types like Geiger-Müller counters, flash tubes or streamer tubes, most detectors produce very small signals. The amount of charge liberated by a minimum ionizing particle in the active medium of a detector is of the order of a few femtocoulombs (corresponding to the production of typically 10 000 electrons, e.g., by an energy deposit of 300 keV in a gaseous detector). Given the usually low detector capacitance of a few picofarads, the generated signal amplitude is less than 1 mV. Such low voltage signals must be amplified before further processing, such as pulse height or shape analysis, can be carried out. The main function of a preamplifier is to receive the weak signal from a detector, amplify it and pass it on via a cable to the main part of the electronic processing system, which may consist of a main amplifier, pulse shaper, discriminator and counter.

There are several reasons for mounting the preamplifier as close as possible to the detector or minimizing the cable length. Any noise signals or spikes generated in the vicinity of the detector or in the cable connecting it to the input of the preamplifier will be amplified along with the genuine signal leading to an unfavorable signal-to-noise ratio. In addition, a long cable between detector and preamplifier will attenuate the signal and can cause serious impedance matching problems. In order to avoid pulse distortion also at the output of the preamplifier, a correct impedance match to the relatively small impedance of connecting cables must be guaranteed. If long cables must be used, the signal should be transmitted differentially. Noise signals generated on differential lines will cancel out to a large extent, so that the quality of the detector signal is preserved. The early grounding of one signal line has also the potential disadvantage of generating a ground loop.

The trend of modern readout technologies is not only to use preamplifiers at the detector but also to integrate main amplifiers, discriminators and even pulse height analysis into the front-end electronics. This is particularly true for the very large scale integration techniques (VLSI) used for the readout of, e.g., silicon-strip detectors. Here the front-end readout is fully integrated into the detector wafers and only digital signals, which are much less subject to electronic pick-up, are transmitted.

Depending on the detector and the field of application preamplifiers can be grouped basically into three different categories: voltage, charge and current sensitive preamplifiers. The first two types are most important for the applications discussed here. Voltage and charge sensitive preamplifiers differ in the type of feedback that is used. Since the open loop gain of an amplifier can vary significantly from one to another because of tolerances of the components, feedback is a necessary technique to obtain stable amplification. With the technique of negative feedback a fraction of the output voltage is fed to the input so that the effect of jitter or drift due to component variation is reduced. Using positive feedback would cause an unstable situation and lead to a ringing of the circuit. Voltage sensitive amplifiers use resistive feedback, while charge sensitive ones use capacitive feedback (see figure 9.2).

Voltage sensitive amplifiers are usually applied, if the detector has a large capacitance C_d and therefore is able to integrate the liberated charge and convert it into a voltage pulse ($U_{in} = Q/C_d$). For gain stability reasons it is important to keep the intrinsic capacitance of the input circuit constant during the operation.

Detectors such as semiconductor counters, with a small capacitance, which may even vary depending on the operation conditions, require charge sensitive preamplifiers. Such a type of preamplifier usually has a large and stable input capacitance which integrates the original charge signal and produces a voltage pulse.

Figure 9.2 shows a schematic diagram of the basic design of preamplifiers using feedback both for voltage and charge sensitive amplification.

The use of current sensitive preamplifiers is advisable for very low impedance devices and is therefore less useful with radiation detectors, which in general are of high impedance.

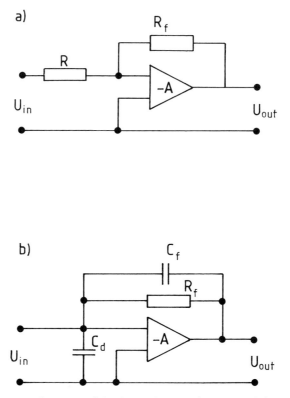

Fig. 9.2. Schematic diagram of the basic design of a preamplifier using feedback, a) voltage sensitive, b) charge sensitive. R_f, C_f are the feedback resistor and capacitor, respectively; C_d is the detector capacitance, $-A$ indicates an inverting amplifier.

9.1.3 Discriminators

One must distinguish between analog signals with variable pulse height and digital signals with well-defined amplitude. Apart from amplitude information, analog signals carry additional information based on their shape, i.e., time structure. Usually the signal amplitude is proportional to the energy loss of a particle in the detector and the pulse shape is related to the primary ionization statistics.

Digital or logic signals only have a discrete number of states, usually only two (binary logic). *Pulse present* is defined by the logic state 1 and *pulse absent* by the state 0. Due to the standardized shape of digital signals their processing is much easier compared to the handling of analog signals, which can vary in height and shape and are sensitive to pick-up, noise, cross-talk etc. Thus it is convenient to convert the analog information into a digital signal as soon as its additional information has been extracted. Digital pulses of a specific logic type have a well-defined amplitude and polarity and sometimes also a fixed duration (see table 9.1).

Table 9.1. *Signal levels for NIM (Nuclear Instrument Module), TTL (Transistor-Transistor Logic) and ECL (Emitter-Coupled Logic)*

Logic type	NIM (voltage on 50 Ω)	TTL	ECL
Logic 0	−1 to 1 mA (±0 V)	0 to 0.8 V	−0.90 V
Logic 1	−14 to −18 mA (−0.8 V)	2 to 5 V	−1.75 V

The main range of application of discriminators is rate counting and providing input for coincidence and timing circuits. Timing refers to the measurement of arrival times of signals with respect to each other or to a reference pulse. The correct timing of digital signals is especially important for logic circuits, like coincidences, anticoincidences or more complicated triggers (see section 9.2). Furthermore, any binary number can be transmitted serially as a string of logic signals or in parallel using an appropriate number of parallel lines.

In order to select which analog signals should be converted into digital form, discriminators are used. The most common version of such a circuit is the pulse height discriminator (see figure 9.3 [603]).

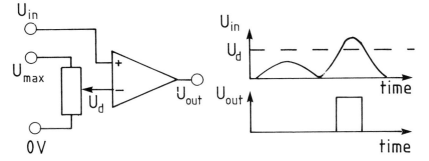

Fig. 9.3. The threshold discriminator, the threshold level U_d is adjustable between 0 V and U_{max} [603].

The pulse height discriminator accepts and transmits only signals whose amplitude exceeds a certain adjustable threshold voltage and rejects smaller pulses. Pulse height discrimination is normally done on the leading edge of a signal (leading-edge discriminator). This can lead to timing problems since the threshold level is reached at different times for signals of different amplitude. This problem can be overcome by requiring that a constant fraction of the signal must exceed a preselected value (constant-fraction discriminator). For timing purposes it may also be useful not to use the leading edge of a signal but rather the moment when the signal reaches its maximum value. This can easily be determined by differentiating the pulse and measuring the time when the differentiated signal crosses zero (zero-cross technique).

An application of a threshold discriminator is the single-channel analyzer or differential discriminator as illustrated in figure 9.4 [603]. Such a circuit selects the incoming analog signals according to their amplitudes. It contains two different threshold levels. Signals are only accepted if their amplitude exceeds the lower threshold and falls below the upper threshold. Such a single-channel analyzer can be used to measure amplitude spectra. Choosing a narrow, fixed window for signal acceptance and sweeping it systematically step by step across the full amplitude range allows one to count the number of signals accepted for each window setting, thus providing a histogram of the amplitude spectrum. This technique of pulse height analysis, however, is rather time consuming. It is much more convenient to use simultaneously a large number of different predefined amplitude windows (i.e., many single-channel analyzers). Such multichannel analyzers usually have between 1024 and 4096 channels.

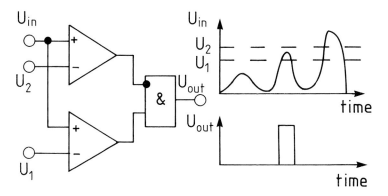

Fig. 9.4. The differential discriminator or single-channel analyzer; U_2 represents the upper, U_1 the lower threshold [603].

9.1.4 Converters

The fundamental interfaces between analog and digital electronics are circuits which convert the analog information into an equivalent digital form: analog-to-digital converters(ADC), flash ADCs (FADC) and time-to-digital converters (TDC). Digital information can also be transformed back into analog form by digital-to-analog converters (DAC).

Many different techniques are currently in use to perform the analog-to-digital conversion. One of the oldest and simplest methods is the ramp or Wilkinson conversion. The method is applicable only for low pulse repetition rates ($<$ 10 kHz). The analog input signal is used to charge a capacitor. After the loading process the capacitor is discharged at a constant rate. The time needed to completely discharge the capacitor is measured by a scaler counting the pulses from a gated oscillator. The number of pulses counted in the scaler is then proportional to the charge stored in the capacitor and hence also proportional to the amplitude of the input signal. The working principle of this method is illustrated in figure 9.5. The advantage of such a Wilkinson-ADC is its good linearity, high resolution (adjustable by the oscillator frequency) and simple, robust technical realization. The long conversion time (typically 100 μs) is certainly a disadvantage.

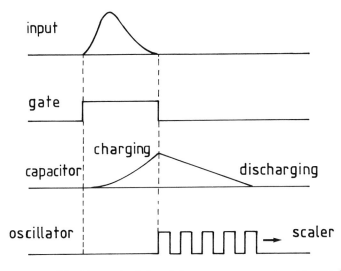

Fig. 9.5. Working principle of the ramp conversion (ADC) [7].

Another ADC-type also well suited for precise conversion of slow pulses is based on the dual slope method. This technique also employs, similar to the Wilkinson-ADC, the charging of a capacitor and its controlled discharging using a reference voltage for comparison [604]. The difference compared to the single-ramp method is that not only the discharging of the capacitor is digitized but, in addition, also its charging by the input signal. This dual slope technique provides good stability against low frequency noise. The result of the digitization is also largely independent of the oscillator frequency and the time constant used for the charging and discharging processes.

The conversion time of both methods for a resolution of n bits is proportional to $\tau \cdot 2^n$, where τ is the clock pulse period. For the currently used clock frequencies of 50 MHz to 200 MHz, required for an acceptable resolution, the digitization process takes several tens of microseconds. This is too slow for many applications.

In a much faster and widely used conversion technique the input signal is successively approximated by a set of reference voltages. This method represents to some extent a compromise between speed and resolution. The working principle of this ADC-technique is illustrated in figure 9.6 [603]. The incoming pulse is compared to a sequence of reference threshold voltages which are generated by a digital-to-analog converter (DAC). This device creates an analog voltage signal proportional to the set digital pattern. The setting of the bit register will be controlled by a logic module and follows a simple approximation rule. In the first step an analog signal corresponding to half of the full voltage range $(\frac{1}{2}U_0)$ is used for comparison with the input signal. The related bit pattern with the most significant bit set reads — for an eight bit register as an example — 10 000 000. If the input signal is larger than $\frac{1}{2}U_0$, this bit is kept and the next significant bit is set, i.e, 11 000 000, corresponding to $\frac{1}{2}U_0 + \frac{1}{2}(\frac{1}{2}U_0) = \frac{3}{4}U_0$, and the comparison is repeated. In case the signal falls below $\frac{1}{2}U_0$, the most significant bit is set to zero and next significant bit is set, i.e., 01 000 000, or $\frac{1}{4}U_0$ is used for comparison. This procedure is repeated until the least significant bit has been tested, in our example eight times, or, in general, n times for an n-bit conversion. The resulting digital number is stored in a buffer and can be read out serially.

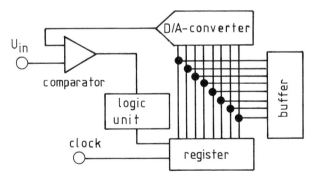

Fig. 9.6. An ADC with iterative approximation [603].

As has been shown, the conversion time is mainly governed by the number of steps n needed to compare the input signal with reference voltages. In order to complete the n-bit conversion in just one step, a bank of 2^n comparators working in parallel is required. Equally, a set of reference voltages subdividing the full amplitude range into 2^n, not necessarily equidistant steps, must be provided.

This basic concept of a flash ADC is shown in figure 9.7 [603]. Of course, the number of required components grows with the number of n like 2^n. However, with the advent of the modern integrated circuit technology, it is possible to integrate the huge number of components into a single chip.

The actual operation of the FADC is simple. A chain of resistors subdivides the voltage between the positive and negative reference level and defines a sequence of thresholds. One of the input channels of every comparator is set to a corresponding threshold and the input signal is fed to the other input channels of all comparators simultaneously. All comparators with input voltage larger than their reference voltage generate an output trigger signal and the leading comparator determines the measured value. The comparator outputs are connected to a gate unit which converts the received level pattern into binary digits.

In general, one is not only interested in a single value for the amplitude of the incoming signal. The main field of application for FADCs is to act as a waveform digitizer for fast detector signals and to record the shape of the signal as a function of time. In this case, the incoming signal is subdivided into time slices by a clock running with frequency of up to 1 GHz, corresponding to time intervals of 1 ns. For each time slice of the signal the comparison with reference voltages is done until the complete waveform is digitized. The information for each cycle is written into a fast memory. Quite obviously FADCs produce an enormous amount of data which must be handled by an associated data acquisition system.

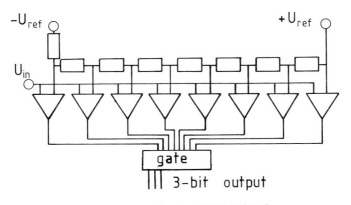

Fig. 9.7. A 3-bit FADC [603].

A time-to-digital converter is a circuit that can convert a time interval τ between two pulses (usually digital signals) directly into a digital number proportional to τ. The simplest method employs a high frequency oscillator (up to the GHz range) connected to a counter, which is started and stopped by the pulses. The time resolution of this method is limited by the oscillator frequency. For fast timing with scintillators or planar spark counters typical electronic resolutions in the range of several picoseconds are required.

In order to measure shorter time intervals, the use of time-to-amplitude converters (TAC) is more convenient. In the first step the time information is converted into a voltage amplitude. A stabilized reference current source charges a capacitor during the time between start and stop signal. The output voltage is then proportional to the time interval. In the second step this voltage can be digitized with an ADC in the usual fashion.

In addition to the analog-to-digital converters described above, there are also converters which adapt a digital signal of one logic type to another. These level adapters convert, e.g., NIM-signals to TTL-norm or vice versa. Their electronic realization is very simple. Such level adapters must be used, if units with different norm signals are to be combined in a common trigger logic.

9.2 Triggers

The purpose of a trigger is to make a particle detector sensitive for the recording of an event and to select events in the presence of background. In cosmic ray experiments the time at which interesting events may occur is not known. If a number of predefined signals from such a detector system occurs at the same time (in coincidence), a trigger is activated and the event is recorded. Since the rate of cosmic ray events is normally quite low, the trigger is not required to be very selective. For example, in an experiment looking for extensive air showers in an array of about 100 scintillation counters the trigger requirement could be simply a majority coincidence asking for a given number of scintillators out of the total to have fired.

For experiments at accelerators the arrival time of particles is, however, known. The event recording can be synchronized with the moment of interaction of the incident particles at a target. The triggering, indeed, can be entirely different for fixed target or collider experiments. The readout in fixed target situations will be particularly difficult because everything happens in the same short spill (the time span where the beam is extracted from the accelerator and hits the target). In colliders the particles are normally stored in bunches. Therefore, in this case, the events can only come in time with the bunch crossing guaranteeing a minimum time interval between two collisions. The use of fixed target techniques with beams extracted from an accelerator has definite advantages. Almost any external particle beam can be produced consisting, for example, of protons, pions, muons, photons, neutrinos, K^0's etc., depending on the physics that one would like to investigate. The disadvantage is that only a fraction of the particle's energy can be used for the production of new particles. For example, for protons of energy E_{lab} incident on a stationary hydrogen target the center of mass energy $\sqrt{s}$ available for particle production is only $\sqrt{s} \approx \sqrt{2mE_{lab}}$, where m is the target mass and $E_{lab} \gg mc^2$ is assumed.

In storage rings, on the other hand, one takes full advantage of the energy E of the colliding particles ($\sqrt{s} = 2 \cdot E$ for head-on collisions). In storage rings one particle beam is the target for the other beam and vice versa. The disadvantage with storage rings is that they usually can be operated with stable charged particles only (ep, e^+e^-, pp, $p\bar{p}$), although $\mu^+\mu^-$-colliders are also conceivable. The problem with colliders is to squeeze the beams to small transverse sizes so that the luminosity, and hence the interaction rate, is sufficiently large.

The time interval between two bunch collisions in a collider experiment is typically around $10\,\mu s$ for e^+e^- storage rings. Such an interval is considered quite long and provides ample time for a decision making process, i.e., a trigger. Triggering in e^+e^- colliders is usually done in a multistep process. Fast signals from, e.g., scintillation counters allow one to investigate whether some event of interest might have occured. Second level triggers incorporate subdetector information, which can only be obtained at a somewhat later time. In a third decision process, information from slow detectors, such as drift chambers, which only become available after several microseconds, are used for track finding. Based on the outcome of the highest level decision the event is accepted or not.

For ep or pp colliders the event rates are much higher and consequently the trigger must be more selective. From the readout point of view the problem is that the interaction processes with large cross sections have been normally well-studied in previous experiments. So they are considered as background which must be suppressed by the trigger so that the investigation of new processes, usually connected with extremely low cross sections, becomes feasible. A too selective trigger, however, can be strongly biased towards a searched-for signature and could prevent the discovery of new unexpected processes, whose detection may have been excluded in the design of the trigger. In the following, the elaborated decision making processes for the most demanding high collision rates will be described for the examples of ep and pp interactions with the HERA and the LHC-collider, respectively.

The need for sophisticated triggering will be painfully obvious, since the expected cross section for *new physics* reaches the level of at most 10^{-11}

of the total cross section. In order to make these rare processes accessible the highest achievable luminosity must be delivered to the experiments. The luminosity $\mathcal{L}$ is calculated from the number of particles N_i in the colliding bunches $(i = 1, 2)$, the number of bunches B and the revolution frequency f_0 to be

$$\mathcal{L} = \frac{B \cdot N_1 \cdot N_2 \cdot f_0}{A_{\text{eff}}}, \tag{9.1}$$

where A_{eff} is the effective bunch cross section, which depends on the bunch shape and bunch crossing angle. For $e^+ e^-$ colliders luminosities in excess of 10^{31} cm^{-2}s^{-1} have been achieved. This allows one to investigate cross sections of the order of picobarns (pb) for integrated luminosities of 10^{38} cm^{-2}, which can be accumulated in about one year of running.

The HERA *ep*-storage ring, for example, consists of two independent accelerators for 30 GeV electrons and 820 GeV protons. Both the electrons and protons are stored in 210 bunches each, where each bunch contains about 10^{11} particles. Given the circumference of 6.3 km the time interval between two bunch crossings is only about 100 ns. To avoid beam distortion and the excitation of beam oscillations (betatron oscillations) due to electron-proton beam coupling, the electrons and protons must collide with zero crossing angle. This, unfortunately, necessitates the use of electron bending magnets and focussing quadrupoles close to the interaction point and causes a high flux of synchrotron radiation photons. In addition, there are two other main sources of background events: strong interactions of protons with the residual gas in the vacuum pipe (beam-gas events) and the scattering of beam halo protons on the apertures of the collider (beam-wall events). All these sources give rise to a background rate of the order of 10 to 100 kHz. In comparison to this background the physics rate dominated by photoproduction (exchange of a photon or a Z^0 between the incoming electron and one of the partons inside the proton) produces a visible event rate of about 200 Hz. Due to this unfavorable signal to background ratio and because of the short bunch crossing interval, the HERA-experiments were forced to develop novel concepts of electronic readout and triggering. This is also of importance for experiments at the next collider generation, which will need to cope with even higher event and background rates.

Future colliders are designed to produce high energy *pp* collisions at luminosities from 10^{33} to $4 \cdot 10^{34}$ cm^{-2}s^{-1}. For the Large Hadron Collider (LHC), now being prepared at CERN, the expected inelastic cross section for high transverse momentum physics reaches 60 mb ($\sigma_{\text{tot}} = \sigma_{\text{inelastic}} + \sigma_{\text{elastic}} = (110 \pm 20)$ mb) at a center of mass energy of $\sqrt{s} = 16$ TeV. For the highest luminosity the interaction rates will be as high as 2 GHz. With a bunch spacing of 15 ns between 1 and 40 overlapping events are expected in every bunch crossing. In order to cope with such huge event rates and data volumes, an effective and highly selective trigger is required. Such a trigger can only be built with a multilevel decision architecture including extensive data pipelining.

The triggering proceeds in such a way that in a succession of steps the amount of data is continuously reduced to such a degree that can be handled by the subsequent higher trigger level. For example, the H1-experiment at HERA comprises more than 250 000 analog channels which are capable of generating nearly 3 Mbytes raw digitized information per event. Figure 9.8 [606] shows the main detector components of this *ep*-experiment in a view along the beam axis.

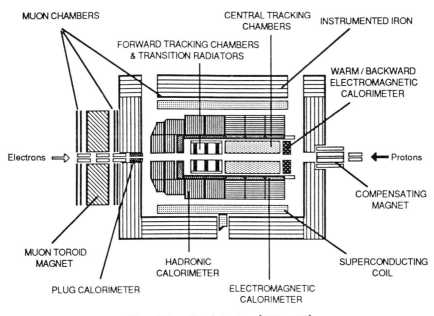

Fig. 9.8. H1 detector [605, 606].

This experiment provides a good example of how trigger architectures must be built to reduce the tape-writing rate to a manageable level. Therefore this special trigger system, which is typical for experiments at high rate colliders, will be described in somewhat more detail.

Before reaching a reasonable recording rate in the range of several hertz and a final event size of around 100 kbytes, various levels of hardware triggering and software filtering must be employed to reduce the raw data. Table 9.2 [606] summarizes information about the number of electronic readout channels and the amount of raw and formatted (zero suppressed)

Table 9.2. *H1 main detector components and anticipated event sizes [606]*

H1 Subdetector	Number of channels	Data raw (kbytes)	Formatted (kbytes)
Drift chambers	9 648	≈ 2 470	≈ 30
MWPC	3 936	≈ 2	≈ 2
Liquid Ar calorimeter	45 000	≈ 262	≈ 88
Calorimeter trigger	50 000	≈ 50	≈ 2
Backward calorimeter	1 500	≈ 8	≈ 2
Plug calorimeter	800	≈ 4	≈ 1
Muon detectors	160 000	≈ 20	≈ 1
Luminosity	256	≈ 66	≈ 0.2
Total	≈ 270 000	≈ 2 880	≈ 125

data for the various subdetectors.

The trigger decision is taken in four steps. Table 9.3 shows the triggering and filtering levels and the data reduction for this particular experiment. A very similar structure on a simplified scale can also be found in smaller experiments and those dealing with lower event rates as is the case in e^+e^- storage rings. In e^+e^- experiments a three-level trigger decision is usually adequate. For experiments with even lower rates, like in cosmic rays or experiments searching for proton decay, a single trigger level might even be sufficient.

Table 9.3. *The four levels of the H1 trigger (μp stands for microprocessor)*

Trigger level	L1	L2	L3	L4
Time	2 μs	20 μs	800 μs	≈ 800 ms
Dead time	0	20 μs	< 800 μs	≈ 0
Max. rate	1 kHz	200 Hz	50 Hz	5 Hz
Trigger	Subdetector hardware	Front-end hardware	Front-end software	Filter farm RISC-μp
Main function	Stops pipeline	Starts readout	Starts event building	Data logging

For an *ep* experiment with a four-level trigger of high complexity the first two levels are hardware triggers which are based on subdetector information available shortly after the interaction. Level three is a software trigger which makes use of trigger specific data. Finally, level four consists of a RISC-processor filter farm. A compromise made at the first level between crude information on the one hand and rapid decision on the other will be refined at the higher levels using detailed information from different subdetectors and combining them with the help of elaborate topological algorithms. This structuring of trigger levels is common to most particle physics experiments. In the following the different trigger levels are going to be discussed in somewhat more detail.

The short period of about 100 ns between two consecutive bunch crossings is much shorter than the time required to transfer the information via cable to some central place where a fast logical decision can be made (20 m standard cable correspond to 100 ns delay). So, not to lose information from events, the raw data is first sent into a pipeline. The complete data from subdetectors with longer response time, like drift chambers and calorimeters are accessible only 10 bunch crossings after the original *ep* interaction. The final level one trigger decision which selects the interesting *ep*-events is available only after 24 bunch crossings ($\approx 2.4\,\mu$s). This decision stops the recording phase of the various subdetector pipelines. Up to this point information from all channels of the whole detector must be stored.

The idea of pipelined processing is essential if information is not to be lost. It has been successfully employed in high performance computing and is adopted for event recording and storage. There are several different types of pipelines generally used for storing signals from various subdetectors. Some approaches make use of analog and other of digital pipelines (see figure 9.9).

In our example the drift time information coming from the tracking detectors is digitized by fast analog-to-digital converters (FADC) in steps of 10 ns, corresponding to 100 MHz. The digitized information is written into a fast random access memory. These memories operate as circular buffers. The increment of the 8 bit memory address is synchronized to the colliding frequency of about 10 MHz. So, after $2^8 \cdot 10\,\text{ns} = 2.56\,\mu$s, corresponding to about 25 bunch crossing periods, the memory will be overwritten unless a positive level one decision stops the pipeline and freezes the history of the last $2.5\,\mu$s in the memory. A similar FADC electronic readout concept is also used for the fast energy trigger from the calorimeter, only in this case, the digitization frequency is 10 MHz.

The numerous pads and wires from the multiwire proportional chambers generate single bit threshold information which is written into synchronized digital shift registers.

The pulse height information from different calorimeters is stored in analog delay lines and the digitization takes place at a later stage during the readout dead time.

The total pipeline length varies for the different subdetectors between 25 to 37 bunch crossing periods. This concept of a centrally clocked fully

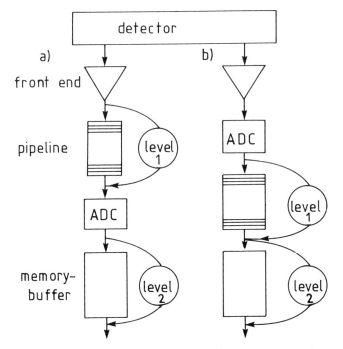

Fig. 9.9.   The analog a) or digital b) first level pipeline.

pipelined front-end system guarantees a first level trigger operation free of any dead time.

The aim of the first level trigger is to distinguish real *ep* interactions from background events on the basis of prompt signals from most of the subdetectors. Different trigger elements each based on the information from a certain subdetector are combined to form the first level trigger. Typical subdetectors used for pretrigger information are fast tracking chambers, electromagnetic and hadronic calorimeters and veto counters. The specific pieces of trigger information are, for example, the presence of charged tracks in the tracking system, their distance from the nominal vertex, a certain amount of electromagnetic or hadronic energy deposited in the calorimeters when summed over specific regions, hits in the muon chamber system and other characteristic, more or less complicated event properties. The individual pieces of information supplied by the different subdetectors are called trigger elements. Each subdetector usually provides up to 16 such trigger elements which are, for example, characterized by different energy thresholds (low, medium, high) in calorimeters or certain combinations of fired track chamber segments. These signals are fed into a central trigger logic where a programmable coincidence circuit generates so-called subtrigger signals. For the H1-experiment at HERA up to 128 different subtriggers can be formed by applying coincidence and threshold requirements. The subtriggers are predefined in such a way that certain event signatures, indicating interesting events, are searched for. Therefore, subtriggers which have fired, can be classified according to their physics origin. There are three event classes related to given subtriggers: physics events, monitor events (which are required for testing and monitoring the detector performance) and cosmic ray events which are useful for calibration purposes. If necessary, the rate of each subtrigger can be downscaled independently and the logical *OR* combination of all subtrigger signals provides the final level one trigger. This working scheme, synchronized with the bunch crossing signal, provides the possibility to deliver the trigger decision at every bunch crossing.

The higher levels two and three are only activated if a possibly interesting event is signaled by the level one trigger. We have seen that the level one trigger works completely free of dead time and its positive decision stops the data pipelining. The following level two trigger produces a dead time of about $20\,\mu$s. This trigger level is foreseen for analyzing complex events. It is based on topological information from individual subdetectors which are combined in this step. Using the advantages and the calculation power of microprocessors and making massive use of parallel decision algorithms eases establishment of a rapid trigger. These tools are ideally suited for fast trigger applications. Also a neural network approach which takes advantage of the multidimensional correlation between the trigger elements from different subdetectors is frequently employed for fast decisions.

In case the event under investigation survives the second level decision, the time consuming readout of detector elements starts. This readout includes, for example, the digitization of analog energy deposits in calorimeters and hit search algorithms based on digitized drift chamber information. Modern hit search algorithms select and store only signals from wires or pads which have fired (zero suppression). In this way the event size is reduced and only useful information is kept.

The largest processing time of about $1\,$ms results from the enormous granularity of calorimeters used for such detectors. The granularity is required because calorimeters should not only provide information on the total energy deposit in such a subdetector but also associate the measured energies to small calorimetric towers, so that the spatial dependence of the energy distribution is known. About $50\,000$ calorimetric readout channels can easily be obtained. A farm of digital signal processors working in parallel performs the zero suppression, pedestal subtraction, gain correction and calibration and sums the calorimeter signals. During this readout phase the unavoidable dead time is used for further analysis (level three trigger). The decision of this level three computation is typically available after a few hundred microseconds. If the level three rejects the event, the data pipelining will be immediately reactivated. Otherwise the event is taken over by the central data acquisition system.

The calculation both for trigger level two and three is based on the same information provided by the level one trigger. The increasing trigger complexity and time consumption from level one to level three reflects the increasing decision power of these systems.

The final trigger level four in our example experiment is built by a parallel array of several R3000 RISC processor boards — the so-called filter farm. This level is integrated into the central data acquisition system (see also section 9.4). It allows full access to event raw data from the memory buffers. The RISC-processors run in such a way that always one event is sent to a free processor board where it will be processed completely until a keep or reject decision is reached. The downloaded modular software is composed of either fast algorithms designed specifically for the filter farm or is part of the standard off-line reconstruction program. Level four performs the on-line reconstruction of charged tracks and the energy determination of the event using the full intrinsic detector resolution. In the next step elaborated filter criteria are applied to these quantities which enable the identification and removal of background events. The rejection criteria are based primarily on very detailed knowledge of the event quality, like the position of the interaction vertex along the beam line. Finally, the accepted event is sent back to the memory buffers where it is delivered via an optical fiber link to the main computer center.

9.3   On-line monitoring

At this stage the reconstructed data from all subdetectors is joined together for the first time so that the output from the filter farm is well suited for on-line monitoring and calibration purposes as well as for on-line event display. The particularly interesting interactions are recognized and flagged using simply a mask of level one trigger bits combined with more precise information from the filter farm. Such events can be passed on request to the on-line event display task.

Because of the detector complexity and the large number of electronic channels, the proper functioning of which must be guaranteed during data taking, there must be a simple procedure allowing for quick access to the desired monitoring information. Also a list of items which identifies pending problems of parts of the detectors and readout electronics must be easily available. For this purpose many monitor histograms are filled with on-line results from the reconstruction modules, which can be inspected on-line and compared to histograms from previous runs. A large number of processes automatically searches for anomalous behaviours and creates warning messages, if necessary.

Another output from the reconstruction procedure is the calibration data. The set of calibration constants are collected in one common database, which is regularly updated, to ensure that only the most recent calibration constants are used for the reconstruction process.

9.4 Data acquisition

Of course there is no sharp boundary between higher trigger level decisions and data acquisition tasks, and such an attempt made here appears somewhat artificial. Nevertheless one can say that triggers are to a large extent hardware oriented while data acquisition describes a field which is predominantly software related.

Data acquisition (DAQ) is known as the process of gathering and calibrating raw data from different parts of the detector and storing it on permanent storage media from where it can be centrally processed off-line. The worked out and well established method of recording the data is to assemble it into logical parts — so called events — which correspond to a single primary particle interaction. Subsequently the raw data must be transformed into physics observables like momentum, mass, energy, time etc. This is a primary task of the reconstruction procedure and will be dealt with in the next chapter.

The communication standard used for many DAQ-systems containing hundreds of processing elements is based mainly on the IEEE VMEbus industrial standard. This gives a high degree of flexibility for possible future upgrades. The basic logical structure of a typical DAQ-system, in this case for the H1-experiment at HERA, is shown in figure 9.10 [606].

The front-end digitizing electronics makes use of different bus types, e.g., VMEbus, Fastbus and other purpose built solutions; however, the VMEbus structure is applied nearly entirely throughout the system. The data from the front-end electronics of various subdetector systems are read out in parallel and delivered into individual, computer controlled subsystem crates. Each of these crates contains a memory buffer residing in the VMEbus and a fiber optical link to coordinating event builders. After digitization, data compression and formatting, full event records are built and transmitted over the fiber optical link. At this stage the data is ready for final filtering by level four trigger, as described above.

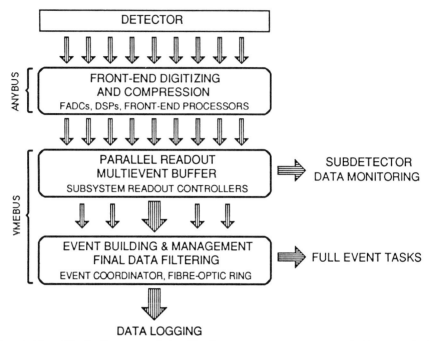

Fig. 9.10. Block diagram of the H1 data acquisition system (DSP stands for Digital Signal Processor) [606].

9.5 Slow control

The complexity and size of the present-day detectors, as they are used in particle physics experiments, requires that all detector operations and adjustments can be performed under remote software control. The concept of slow control includes the recording, monitoring and controlling of all parameters of the experiment which should stay at fixed predefined levels during data taking. These quantities are not directly associated with event data but have nevertheless significant influence of the final data quality. Within a large experiment a few main areas of slow control can be defined, such as:

- high voltage and current monitoring for all subdetectors

- the gas system for drift, multiwire proportional and muon chambers (gas flow, temperature, pressure, purity)

- the low voltage for all detector components (preamplifiers, shapers, line drivers and all other front-end electronics)

- the temperature of the front-end electronics including the detector cooling

- the monitoring and control of magnets

- the control of the cryogenic system for superconducting magnets or liquid noble gas detectors

- the interlock system.

Typically there are a few hundred quantities to be monitored by the control network and to be handled by the hardware alarm system.

During the initialization of each of the slow control subsystems, the nominal value and the allowed range of variation for each quantity must be fixed according to the settings stored in a central database. The slow control monitoring task searches for significant deviations of the currently measured values from predefined settings, and issues warnings or alarms according to preset levels. If significant deviations are detected, warning or alarm messages are transmitted to the data taker. These messages are also included into the event data to allow one to judge the quality of the data during the on-line or off-line analysis. Due to the sophisticated and potentially dangerous nature of certain quantities to be controlled, an automatic hardware alarm is generated to handle critical situations immediately, for example, to ramp down high voltage or magnetic fields without human intervention in case of emergency.

10
Data analysis

The purpose of physics is to get a number.

Anonymous

The field known as data analysis extends from the raw data to the publication of a physics result [607]. These raw data, a collection of selected signals, which are filtered and transmitted by the readout electronics, serve as input to the analysis. From this information physics quantities are derived. In present day high energy physics experiments, the outputs of several subdetectors are usually combined to give full information on the physics process under study. The information is then compared to a physics hypothesis or used to look for the unexpected and new.

In this chapter the various steps of a data analysis task will be demonstrated using mainly the ALEPH experiment, operating at the e^+e^- storage ring LEP (see section 11.12) as an example. The analysis of data from this experiment can be taken as a guideline for other experiments of comparable complexity. Data from experimental set-ups with fewer subdetectors require less sophistication. The general scheme of the data analysis described in the following sections can be easily translated to experiments working in completely different fields, like nuclear astrophysics, cosmic rays or radiation protection. Still, this chapter does not intend to cover all aspects of data analysis in a comprehensive fashion.

10.1 Input from detectors for data analysis

All physics analysis starts from the information supplied from the data acquisition system, stored on disks or tapes as raw data. For a collider experiment they comprise digitizings from all detector parts: for example, particles traversing a multiwire proportional chamber (see section 4.6) ionize the gas (chapter 1). The electrons initiate avalanches that create a

316

signal, when the charge is collected at the electrodes. Drift time, the wire and pad positions as well as the arrival time of the pulse and the charge at both ends of the wire may be written as raw data. Combining this input with calibration constants, such as drift velocity and the moment of the intersection t_0, the location of the electron initiating the avalanche both in the plane perpendicular to the wire and along the wire may be obtained. Similarly, in a TPC the tracks' position orthogonal to the magnetic field is determined by the position of the pads of the drift chambers closing the TPC at both ends. The drift time measurement completes the three-dimensional space point.

In addition to up to ten thousand of these digitizings, the shapes of signals induced by charged particles are also stored. They serve to aid the subsequent particle identification using the mass and momentum dependence of the ionization of charged particles.

In many high energy physics experiments, energy measurements are made using a combination of electromagnetic and hadronic calorimeters. Usually sampling calorimeters are used, where the absorbers are interleaved with chambers or scintillators providing analog information proportional to the energy deposited. The accuracy of the location where a particle has passed through the detector is limited by the granularity of the calorimeter. The granularity of electromagnetic calorimeters is superior to that of hadron calorimeters. The energy resolution for hadrons is inferior to that of electrons and photons because of the different nature of strong and electromagnetic interactions and because of the mixed composition of a hadronic shower (electromagnetic and hadronic particles in the shower). Thus it only makes sense to use fine granularity for compensating hadron calorimeters (see section 7.3). The number of readout channels that can be handled is another limiting factor for the number of calorimeter cells. Each cell that has fired and exceeds a certain level presents input to the data analysis.

If scintillators are used as sampling elements in calorimeters, they provide in addition timing information for particles traversing the detector. This is also true for other sampling elements like resistive plate chambers, planar spark counters, or spark chambers.

Additional information from, e.g., Cherenkov counters, where the position and diameter of Cherenkov rings is recorded, or transition radiation detectors, where the yield of X-ray photons is measured, eases particle identification and is added to the raw data.

This huge amount of information adds up to about 100 kbyte per event. For an interaction rate of the order of 10 to 100 kHz an elaborate trigger system must be used to match the data flow to the writing speed to tape, which is only several hertz. The alternatives of writing to disk or optical devices are a question of price and commercial availability. As a compromise one may write events from the data acquisition system via a buffer disk to tape.

Stored data must be written in a computer independent format. Collaborations of typically 400 physicists with tens of institutes need unlimited access to the data. In addition, the data should be readable with FORTRAN, the programming language widely used in physics. However, the fixed dimensioning of arrays and the incompatibility of different data types (real, integer, string) in FORTRAN is not well suited for the handling of detector data. For this purpose programs for dynamic management of data are implemented using FORTRAN. BOS [608], ZEBRA [609] and HYDRA [610] are the programs most commonly used. Information read from a certain detector, for example muon chambers, are stored in banks. They may contain the name of the bank, the number of the wire hit, the drift time, and the link to the next muon hit bank. To find all muon hits one must go through all muon hit banks using the link until there are no more banks. This allows a dynamical storage handling, in contrast to using an array dimensioned large enough to store the maximum possible number of hits expected in an event.

If faster data taking is required, the procedure described is split in two separate steps. In a first step raw data from the data acquisition system are written to disk or tape. This may include zero suppression. In a second step spatial coordinates from tracking chambers and energies from calorimeters are recorded: this process is called preprocessing.

10.2   Pattern recognition and track reconstruction

10.2.1   Track finding

Charged particles passing through a tracking device are reconstructed from their measured spatial coordinates. The complexity common to all tracking chambers is shown by an event in the central drift chamber of the UA1 experiment [611] in figure 10.1. The cylindrical chamber with length of 6 m and diameter of 2.2 m is shown cut along the beam line. The wires are orthogonal to the beam line and parallel to the dipole magnetic field of 0.7 T. Figure 10.1a shows the recorded hits in the chamber, while figure 10.1b shows the reconstructed tracks.

Two extreme attitudes can be taken to find tracks. The straightforward method of taking all possible combinations of hits is too time consuming. The number of combinations for thousands of hits is immense and all possible track candidates must be validated so as not to use hits twice, i.e., by several tracks. The other extreme point of view is the global method where a classification of all tracks is done simultaneously. For points close in space, function values are entered in an n-dimensional histogram.

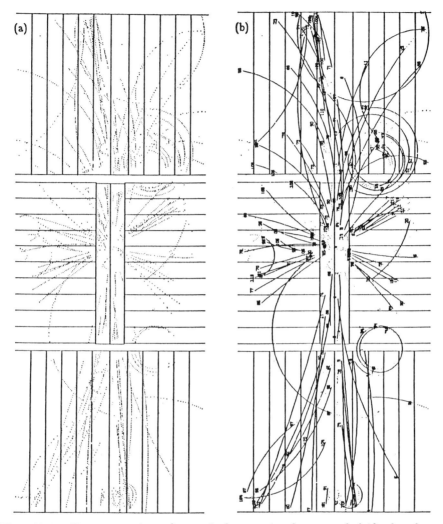

Fig. 10.1. Reconstruction of a typical event in the central drift chamber of UA1. The display shows a cut along the vertical beam line. The wire planes divide the chamber into 46 drift volumes. The magnetic field is perpendicular to the projection shown [611]. a) Displays the measured points and b) shows the reconstructed tracks.

Hits belonging to the same track should be close in parameter space. A simple example would be the reconstruction of tracks coming from the interaction point without a magnetic field. The ratio $\Delta y/\Delta x$ plotted in a histogram would show peaks at the values of the slopes expected for straight tracks.

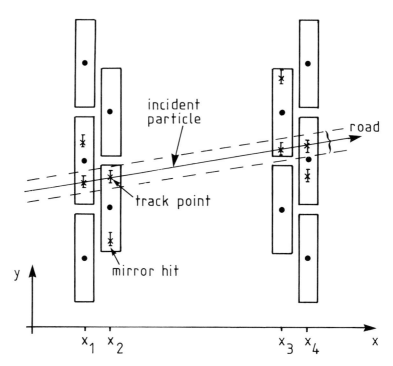

Fig. 10.2.   Track finding with the road method and straight line fit. Due to the left-right ambiguity of drift chambers two coordinates per hit are reconstructed, one being the true track point, the other a mirror hit.

In practice a method lying between these two approaches is chosen. Its implementation depends heavily on the chamber layout and physics involved.

One method that is commonly used is the road method. It is explained most easily for the example of the muon chambers consisting of two double layers of staggered drift tubes (see figure 10.2). Reconstructed spatial coordinates for a charged particle outside a magnetic field lie on a straight line. Possible tracks are found from the permutation list of four points lying on a road of width which corresponds roughly to the spatial resolution (mm or cm).

To all four points on a road with coordinates $x_1, \ldots, x_4$ one has measurements y_i with errors σ_i (here Gaussian errors are assumed). In a straight line fit [103, 612] the expected positions η_i with respect to the measured y_i are linear functions of the x_i:

$$\eta_i = y_i - \epsilon_i = x_i \cdot a_1 + 1 \cdot a_2 \, , \tag{10.1}$$

or

$$\vec{\eta} = \vec{y} - \vec{\epsilon} = \mathcal{X} \cdot \vec{a} , \tag{10.2}$$

where a_1 is the slope and a_2 is the axis intercept. For independent measurements the covariance matrix $\mathcal{C}_y$ is diagonal:

$$\mathcal{C}_y = \begin{pmatrix} \sigma_1^2 & 0 & 0 & 0 \\ 0 & \sigma_2^2 & 0 & 0 \\ 0 & 0 & \sigma_3^2 & 0 \\ 0 & 0 & 0 & \sigma_4^2 \end{pmatrix} =: \mathcal{G}_y^{-1} . \tag{10.3}$$

One obtains the values of $\vec{a}$ by the least-squares method, minimizing

$$\chi^2 = \vec{\epsilon}^T \mathcal{G}_y \vec{\epsilon} \tag{10.4}$$

which follows a χ^2-distribution with $4 - 2 = 2$ degrees of freedom:

$$\vec{a} = (\mathcal{X}^T \mathcal{G}_y \mathcal{X})^{-1} \mathcal{X}^T \mathcal{G}_y \vec{y}, \tag{10.5}$$

and the covariance matrix for $\vec{a}$ is given by

$$\mathcal{C}_a = (\mathcal{X}^T \mathcal{G}_y \mathcal{X})^{-1} =: \mathcal{G}_a^{-1} . \tag{10.6}$$

As shown in figure 10.2, several track candidates may be fitted to the data points, because of hit ambiguities. To resolve these, the χ^2 can be translated into a confidence limit for the hypothesis of a straight line to be true and one can keep tracks with a confidence level, for example, of more than 99 %. The more commonly used choice is to accept the candidate with the smallest χ^2. By this method mirror hits are excluded and ambiguities are resolved.

Points which have been used are marked so that they are not considered for the next track. When all four-point tracks have been found three-point tracks are searched for to allow for inefficiencies of the drift tube and to account for dead zones between them.

For larger chambers with many tracks, usually in a magnetic field, the following track finding strategy is adopted. The procedure starts in those places of the drift chamber, where the hit density is lowest, i.e., farthest away from the interaction point. In a first step three consecutive wires with hits are searched for. The expected trajectory of a charged particle in a magnetic field is a helix. As an approximation to a helix a parabola is fitted to the three hits. This is then extrapolated to the next wire layer or chamber segment. If a hit matching within the errors is found, a new parabola fit is performed. Five to ten consecutive points form a track segment or a chain. Consecutive implies at most two neighbouring wires do not have a hit. Chain finding is ended when no further points are found or do not pass certain quality criteria. When the track segment finding is complete the segments are linked by the track following method.

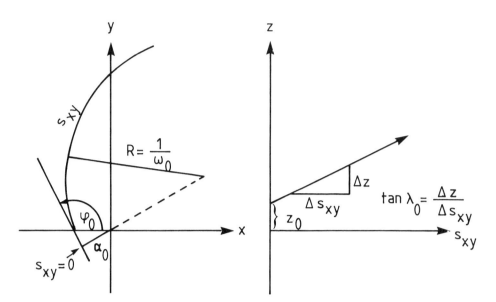

Fig. 10.3. Definition of helix parameters. On the left hand side the projection of the helix into the xy plane orthogonal to the magnetic field and the beam is given. The figure to the right shows the z-coordinate versus s_{xy}.

Chains on an arc are joined together and a helix is fitted. Points with large residuals, i.e., points that deviate too much in χ^2, are rejected and the helix fit is redone. The track is extrapolated to the closest approach of the interaction point. In the final fit variations in the magnetic field are included, and a more sophisticated track model is used. In the ALEPH experiment, for example, the closest approach to the beam line in $r\varphi$ is denoted by d_0 with the z-coordinate at that point z_0 (the z-coordinate is measured parallel to the magnetic field along the beam, see figure 10.3). The angle φ_0 of the track in the $r\varphi$ plane with respect to the x-axis at closest approach, the dip angle λ_0 at that point and the curvature ω_0 complete the helix parameters: $\vec{H} = (d_0, z_0, \varphi_0, \lambda_0, \omega_0)$. For some applications the set $(d_0, z_0, p_x, p_y, p_z)$ is used, with p_x, p_y, p_z being the components of the track's momentum at closest approach. This procedure also provides the covariance matrix $\mathcal{C}$ for the helix.

Another strategy for track finding is to use neural networks [613, 614]. Looking at figure 10.1 one sees that the human eye together with the image processing part of the brain can easily recognize the hits that form tracks. The eye ignores deficiencies of the chamber such as inefficiencies and noise hits. Consequently, it has been suggested that current bottlenecks in computing might be solved by methods that are modelled after the human brain. These programming methods are called neural networks, in which large numbers of gates — or neurons — are interconnected with each other with the synaptic strength T_{ij} ($T_{ii} = 0$). These strengths can have both positive (excitory) and negative (inhibitory) values. The connections of spatial coordinates in the tracking chamber S_i (the neurons) form the input to the network. The dynamics of the network is given by the updating rule:

$$S_i = \Theta(\sum_j T_{ij}S_j) \qquad (10.7)$$

with $S_i = 1$ for connections, which belong to a track, $S_i = 0$ otherwise. An energy function

$$E = -\frac{1}{2}\sum_{ij} T_{ij}S_iS_j \qquad (10.8)$$

is constructed and minimized; smooth tracks belong to global minima. To avoid being trapped in a local minimum a 'temperature' is introduced and equation (10.8) is solved iteratively. The idea of a temperature is to smooth out the Θ-function like behaviour of a discriminator, which accepts or rejects input information. It mimics the effect that a finite temperature has on the step-like Fermi-Dirac distribution at $T = 0$. Usually a 'constraint' term is added to the energy function, also called the 'cost' part, in the equation (10.8). A more detailed description for track finding with neural networks can be found in reference [615].

The knowledge of the position of the interaction vertex is of particular importance, if one is interested in determining the particles' lifetimes. For colliders the position of the incoming beams is known to ≈ 200 μm or better, while the length of the bunches may range from a few millimeters to half a meter. The vertex is fitted using all tracks with closest approach to the beam line of less than typically 200 μm. This restriction excludes particles not coming from the primary vertex such as $K_s^0, \Lambda^0, \bar{\Lambda}^0$, called V^0 and photon conversions, which produce a pair of oppositely charged tracks.

10.2.2 Energy determination

Once the momentum is known from the curvature in the magnetic field, the charged particle's energy is known as well, if one assumes that the particles are pions, which is correct in most cases. For neutral particles the measurement relies on calorimetric information. This is extensively discussed in chapter 7 and section 11.12. In this context it is important to differentiate between neutral calorimeter objects, like π^0 or photons, on the one hand, and electrons and calorimeter clusters from minimum-ionizing particles on the other hand. For the latter two, calorimeter objects can be associated with charged particles recorded in the tracking system.

10.2.3 Particle identification

Another important input for the analysis is the identification of particles. Various methods were described in chapter 6, such as energy-loss measurements dE/dx, use of Cherenkov counters and transition radiation detectors. The different longitudinal and lateral structure of energy deposition in calorimeters is used to separate electrons from hadrons. The simplest method is to introduce cuts in the corresponding shape parameters. More sophisticated procedures compare the lateral and longitudinal shower shape with a reference using a χ^2-test or neural networks. In this case (and in the physics analysis, see below), in contrast to track finding, multilayer feed-forward networks are used. (For pattern recognition feedback networks are applied.) The input neurons — each neuron represents an energy deposit in a calorimeter cell — are connected with weights to all neurons in a next layer and so forth until one obtains in the last layer one or a few output neurons. The result, which can vary between zero and one, indicates whether the input originated from a pion or an electron. The weights from the neuron connections are adjustable and are obtained by minimizing a cost function. This is done by an iterative learning algorithm called back-propagation [613, 614].

A comparison of these procedures to separate electrons from pions in a calorimeter can be found in reference [616].

The methods described so far have concentrated on stable particles. They are identified by their characteristic behaviour while traversing the detector or while being totally absorbed.

A more indirect method is used for particles with a non-zero decay length, i.e., typically a few hundred micrometers up to a meter. Their decay point, measured in a tracking chamber, is well separated from the primary interaction point. Their decay products are recorded with high precision and allow the reconstruction of the particle's properties. Typical candidates are weakly decaying particles such as B, D and V^0 (K_s^0, Λ^0) mesons and baryons (see section 4.11, figure 4.66). Converting photons produce a similar pattern: a photon may convert to an e^+e^- pair in the wall of a tracking chamber, beam pipe etc. The conversion probability in typical detectors is of the order of a few percent. Neglecting the masses of the positron and electron and the recoil of the nucleus, the e^+e^- tracks are parallel. This can be seen from the reconstructed photon mass squared: $m_\gamma^2 = 2p_{e^+}p_{e^-}(1 - \cos\theta)$, where θ is the opening angle between electron and positron. Figure 10.4 shows a sketch of a photon conversion in comparison to a Λ^0 decay. The two reconstructed tracks from a photon

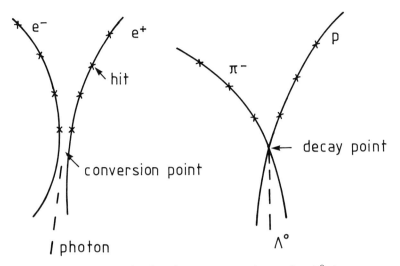

Fig. 10.4. Sketch of a photon conversion and a Λ^0 decay.

conversion can either intersect or may not have a common vertex because of measurement errors. The conversion point is found as the point where the two tracks are parallel ($m_\gamma^2 = 0$) in the plane orthogonal to the magnetic field. The photon momentum is the vector sum of the e^+ and e^- track at or closest to the conversion point.

For massive particles (e.g., Λ^0 with $m = 1.116\,\text{GeV}/c^2$) the opening angle is finite and the trajectories of proton and pion intersect. The closest approach of the two trajectories in space is a good approximation for the decay point. A more precise procedure, however, is to perform a geometrical fit using the parameters of the two tracks as obtained from the track fit ($\vec{H}_i$) and their error matrix (C_i) and to perform a χ^2 fit. With two tracks, including, for example, ten measurements $\mathcal{H} = (\vec{H}_1, \vec{H}_2)$, and nine parameters $\mathcal{Q} = (\vec{D}, \vec{p_1}, \vec{p_2})$ to be determined (decay point $\vec{D}$, and two momenta $\vec{p_1}, \vec{p_2}$), one has a fit with one degree of freedom. The calculation [103, 612, 617] is similar to the straight line fit discussed above. The covariance matrix, however, is non-diagonal as the five track variables are correlated. It is a 10×10 matrix consisting of two submatrices of dimension 5×5. A very important difference is that the expectation values of the nine parameters $\mathcal{Q}$ are not linear functions of the measurements $\mathcal{H}$. Therefore one must obtain the parameters by Taylor series expansion and approximate $\mathcal{X}$ (see equation (10.2)) from the first derivative $\delta\mathcal{H}/\delta\mathcal{Q}$. This matrix is evaluated at an assumed starting value of $\mathcal{Q}_0$, which is derived from an educated guess. Improved parameters $\mathcal{Q}_1$ are obtained using the least-squares method and the procedure is iterated.

With the Λ^0 mass known one can include the mass as a further constraint in the fit. In addition, the origin of the Λ^0 may be known; it is usually the primary vertex. Therefore a kinematical fit can use the fact that the direction of flight of the Λ^0 coincides with the momentum sum of the decay products $\vec{p_1} + \vec{p_2}$. This procedure allows one to obtain samples of V^0 with both high purity and high efficiency.

The identification of particles by their invariant mass alone is used for short-lived particles, where the decay point is not resolved from the collision point of the incident particles.

A particle that needs special treatment is the neutrino. It cannot be detected in the standard way. It has (almost) no interaction with matter and escapes detection. To find the energy of a possible neutrino one must detect all other particles in the detector. To each energy deposition in the calorimeter one assigns a vector with length proportional to the measured energy, and its direction given by a line connecting the interaction point with the fired calorimeter cell. A non-zero sum of these vectors indicates the presence and direction of missing energy. If this is the case, it may be attributed to a neutrino. It must be assumed that no particle escaped, for example through the beam pipe. Since this cannot be assured, especially for $p\bar{p}$ collider experiments, one usually restricts oneself to the analysis of the momentum transverse to the beam. In the hard scattering of proton and antiproton only one quark and antiquark collide. The other constituents fragment as jets close to the beam line and partially escape detection. Consequently, the event has a longitudinal imbalance and only the transverse momentum of the neutrino can be used. Certainly also other corrections have to be taken care of: muons deposit only a small fraction of their energy in the calorimeter. The missing energy must be corrected in this case using the difference between the muon momentum measured in the tracking chamber and its energy seen in the calorimeter.

One can conclude that particle identification proceeds hand in hand with the reconstruction of the whole event.

10.3   Event reconstruction

An event in a high energy collision comprises all information collected from the subdetectors. Momenta, energies and particle identification results are combined. As could be seen from the neutrino momentum determination additional information is obtained by combining that of the different subdetectors.

Muons and pions for instance can be distinguished by combining the information from two subdetectors. For this purpose a tracking chamber for the momentum determination, and a second chamber behind an absorber to filter the muons and stop the pions, can be used. In a jet where many particles are close in space and direction one must decide which track belongs to the muon chamber hits. This correlation of tracks from different subdetectors is called matching. All tracks pointing to the muon chambers are extrapolated and their position and direction, when traversing the muon chambers, is compared with the muon chamber hits. The one with the best average match $\chi^2 = (\chi_p^2 + \chi_a^2)/2$ is chosen, where

$$\chi_p^2 = \frac{(\Delta x_1)^2}{\sigma^2(\Delta x_1)} + \frac{(\Delta x_2)^2}{\sigma^2(\Delta x_2)}, \qquad (10.9)$$

$$\chi_a^2 = \frac{(\Delta \varphi)^2}{\sigma^2(\Delta \varphi)} + \frac{(\Delta \lambda)^2}{\sigma^2(\Delta \lambda)}, \qquad (10.10)$$

are the χ^2 for matching in position and angle between the muon-chamber track and the extrapolated track. Δx_i ($i = 1, 2$) are the differences of the coordinates between the extrapolated and the measured track in the muon chambers. $\Delta \varphi$ and $\Delta \lambda$ stand for the differences in the corresponding angles. The errors of the four variables, $\sigma(\Delta x_i)$, where $x_i = x_1, x_2, \varphi, \lambda$, consist of three components, which are added in quadrature $\sigma^2 = \sigma_e^2 + \sigma_\mu^2 + \sigma_{\text{sys}}^2$. Here σ_e denotes the error on the extrapolated track due to the uncertainty of the position measurement in the tracking chamber, the uncertainty from multiple scattering and the imperfect knowledge of the magnetic field. σ_μ is the error of the track in the muon chamber, while σ_{sys} is the uncertainty in the geometrical alignment of the subdetectors used. Alignment is the determination of the exact location of the subdetectors in space, their internal structure and their relative positions. The relative positions of the various detector parts are measured during assembly. The alignment precision is improved during data taking with cosmic rays or other low multiplicity events, $e^+e^- \to e^+e^-$, $e^+e^- \to \mu^+\mu^-$. To obtain a wire by wire correction in a tracking chamber $q\bar{q}$ events have to be used for statistical reasons.

Some subdetectors, in which detector parts overlap, are equipped with a built-in alignment. Further improvement is obtained with optical alignment and using lasers [618].

10.4 Event simulation

The measured data must be compared to a theoretical prediction to extract the physics. For a complicated detector or a combination of many subdetectors in multiparticle events, highly sophisticated selection criteria are applied in the data analysis. As a result it is difficult to compare subsets of data with an analytical theoretical prediction. In addition, the detector acceptance and response must be folded in.

This is why Monte Carlo simulation has become the most commonly used procedure to describe the data. The name Monte Carlo, the place with one of the most famous casinos in the world, indicates that all Monte Carlo simulations are based on random numbers.

Such a simulation is performed in two steps. The first part is common to all experiments at the same accelerator. It consists of an event generator describing the processes of particle production (e.g., $p\bar{p} \to W^{\pm}+$anything, $e^+e^- \to$ hadrons).

In the second part, the detector response to the incoming particles is simulated in great detail. The particles are propagated through the detector including their bending in the magnetic fields, their interaction in the detector material and their response in the active detector elements. This second step provides a result in a format identical to the real data taken with the detector. A comparison of the Monte Carlo output with the data will allow an interpretation of the physics involved, proving the assumptions put into the simulation right or wrong.

The purposes of the Monte Carlo, however, are varied.

- It helps to design and optimize a detector well suited for the physics one is going to study, regarding its rate, background, particle species etc.

- Before the first real data events are seen in the experiment, it allows the development and test of reconstruction and analysis programs.

- In the data analysis:

    - it is essential to develop strategies for the analysis and optimize cuts;

    - with acceptances and efficiencies determined from Monte Carlo the data from different experiments can be compared to each other;

    - it allows a comparison between theoretical predictions and experimental result: phenomena observed are either interpreted as known physics and free parameters of the theory or model can be determined, or used to indicate new physics.

10.4.1   Event generators

Event generators are supposed to describe all physics processes at a given accelerator in an experiment. For e^+e^- experiments at LEP generators that describe the formation of s-channel fermion-pair production, s and t-channel Bhabha scattering and $\gamma\gamma$ interactions are required. For background studies synchrotron radiation, cosmic ray and beam gas interaction Monte Carlos are available. Programs describing the signatures of

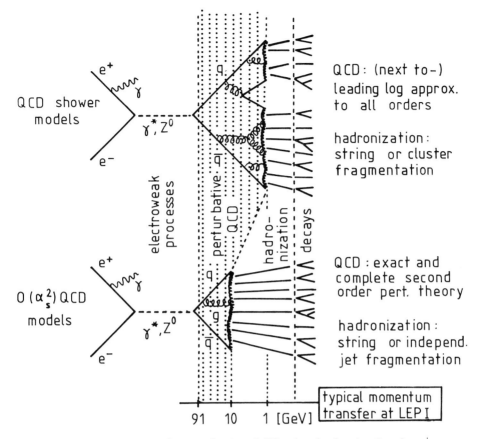

Fig. 10.5. Illustration of perturbative QCD plus hadronization in $e^+e^- \rightarrow Z^0 \rightarrow$ hadrons. The various phases of the process are indicated: electroweak interaction, perturbative QCD, hadronization and particle decay [620].

new phenomena predicted by theories such as supersymmetry or compositeness also exist [619].

To simulate a specific process, e.g., $e^+e^- \rightarrow Z^0 \rightarrow$ hadrons, one assumes that the process factorizes. The first part is the hard scattering of electrons and positrons to form partons, i.e., quarks and gluons. This process is based on perturbative quantum chromodynamics (QCD). The corresponding Feynman diagrams cannot be calculated to all orders, but are expanded in a series. Two complementary approaches are available (see figure (10.5)). In the matrix-element method diagrams are evaluated completely up to a fixed order in the strong coupling constant α_s. In the parton shower approach all leading contributions are summed. The created quarks and gluons are confined. Their transition to hadrons cannot be predicted by QCD and is described by phenomenological fragmentation models. Unstable final state hadrons are decayed according to their known lifetimes and branching ratios.

The simulation is started with a set of random numbers: a series of numbers that is unpredictable and irreproducible. True random numbers can be provided by a physical process such as radioactive decay. For practical reasons with all analysis done on computers a set of pseudo-random numbers is used. They are produced in a computer by numerical algorithms and are not strictly random but have a period of a billion or even more, which is sufficient for event generation. Such random numbers are made available as standard FORTRAN functions, e.g., RNDM(x) returning values equally distributed between zero and one. If the distribution of a certain physics quantity is needed, such as the angle of the created $q\bar{q}$ pair with respect to the beam axis, one can use the following technique. Suppose $f(x)$ ($f(x) \geq 0$) with $x_{\min} < x < x_{\max}$ is the anticipated distribution. For simplicity one can assume that the primitive function $F(x) = \int f(x)\mathrm{d}x$ is normalized $F(x_{\max}) - F(x_{\min}) = \int_{x_{\min}}^{x_{\max}} f(x)\mathrm{d}x = 1$ and $F(x_{\min}) = 0$. Then the inverse function $F^{-1}(x)$ will be distributed as $f(x)$ when x is thrown as RNDM(x), i.e., equally distributed between zero and one.

Unfortunately, only few functions $f(x)$ have such a pleasant behaviour that this transformation can be applied. Even the production of a $q\bar{q}$-pair has an angular dependence $(1 + \cos^2\theta)$ that cannot be obtained with this approach and therefore another technique must be used. With a function $f(x)$ with $0 < f(x) < f_{\max}$ one chooses first a random number x_1 in the range from $x_{\min}$ to $x_{\max}$:

$$x_1 = x_{\min} + \mathrm{RNDM}(x) \cdot (x_{\max} - x_{\min}) . \qquad (10.11)$$

If for a second random number $x_2 = \mathrm{RNDM}(x)$ the product $x_2 \cdot f_{\max}$ is less than $f(x_1)$ then x_1 is accepted as the desired random variable, if not it is rejected.

More ingenious procedures are required for more complicated functions $f(x)$, with spikes for instance. Various techniques are found in references [621, 622].

The simulation of cross sections always demands a positive function $f(x)$ to describe a probability. The reader, familiar with QCD, will realize that this may cause problems. The total cross section in first order α_s with two or three partons in the final state is finite. The three parton final state alone, however, is infrared and collinearly divergent, and a cutoff has to be introduced as a free parameter in the simulation to prevent the two quark cross section from becoming negative.

Altogether an event generator may have from ten to hundreds of free parameters. The process under consideration (here $e^+e^- \to Z^0 \to$ hadrons) is then simulated while all angular distributions, branchings, flavour compositions etc. are chosen using this random number technique. Certain parameters such as the energy of the incident beams must be given as input to the generator. Others, like the strong coupling constant α_s, have to be determined by tuning the Monte Carlo to the data. This will return the physical constants left free in the simulation.

A summary of event generators for LEP physics is to be found in reference [619]. For proton colliders ($p\bar{p}$ or e^-p) generators can be found in references [623, 624]. The latter require, in addition, structure functions [625] which describe the parton momentum distribution in the incident hadrons.

10.4.2 Detector simulation

The generated four vectors of particles serve as input to the detector simulation of the program package. Almost always the program GEANT [626, 627] is used as the core of this simulation. GEANT describes all important interactions of particles with matter. The electromagnetic interactions implemented are similar to those in EGS [628]. The EGS library was written especially for simulating electromagnetic interactions. The hadronic part in GEANT is based on GHEISHA [629], which has been appropriately modified to improve the simulation of particles with energy below 5 GeV [630] (HADRIN, NUCRIN, and FLUKA).

The input to the simulation is a detailed description of the detector including a possible magnetic field. The geometry is described and for each detector part the material's composition is given in terms of atomic number Z, mass A, and density.

The particles generated in the event generator are then tracked through the detector. For each step, e.g., advancement of the particle's trajectory by one centimeter, a decision is taken as to which interactions will occur. For the various possible processes a relative probability is defined, which is multiplied with a random number between zero and one. The process with the highest probability is simulated. For the process selected in this way further random variables are required to describe, e.g., the scattering angle, the ionization etc. The step size is a parameter which must be adjusted to the detector's granularity. A small step size would increase the computing time for the simulation, but it must in any case be small compared to the detector components. The tracking must, of course, include the response of the active material. This response, e.g., energy depositions and digitizings, is also simulated. In a last step this output is written into a format like that of digitization of the real data taken with the detector.

10.5   Physics analysis

The strategies for data analysis are very much dependent on the physics involved and the process under study. The selection of events and the extraction of a specific process requires accurate knowledge of the physics issues under consideration. Systematic effects and their impact on the desired precision call for a process specific treatment.

The basic strategies for physics analysis are very similar for many different applications. For example the discovery of neutrinos from the supernova explosion SN1987A required a careful selection of events in time with their arrival at earth [380, 631]. The signatures in the water Cherenkov detectors had to be consistent with ν_e and $\bar{\nu}_e$ processes in the expected 10 MeV range. A distinction had to be made to separate the supernova neutrinos from the interactions of atmospheric background neutrinos. This could be done since a large fraction of atmospheric neutrinos are muon neutrinos, which produce muons in their interaction. The subsequent decay of muons signaled unambiguously a background event.

The rate, energy distribution and arrival direction of the on-source neutrinos provided the possibility of determining the total number of neutrinos emitted from the supernova during the process of deleptonization and thermal emission. These results, along with data from observations in the optical range, allow one to draw conclusions on the type of supernova and details of the gravitational collapse of the known progenitor star.

In the following the main features of data analysis for an experiment in elementary particle physics will be explained* for the decay $\Sigma^0 \rightarrow \Lambda^0\gamma$, where the aim of the analysis is to determine the number of Σ^0 produced per hadronic event in the reaction $e^+e^- \rightarrow Z^0 \rightarrow$ hadrons at $E_{cm} = 91.2$ GeV. This analysis is based on data taken by the ALEPH experiment at LEP (see also section 11.12).

Hadronic events are characterized by two back-to-back groups of particles, called jets (see figures 4.47a and 4.60). Occasionally, when gluons are emitted a third or fourth jet may be seen. When the event is fully contained in the detector, i.e., events are selected in such a way that the $q\bar{q}$ axis, represented by the two-jet axis (thrust, sphericity [632]), has a large angle with respect to the beam, one observes a charged multiplicity of about 20 tracks. Most of the 91.2 GeV center of mass energy is visible in the calorimeter. Thus with cuts on the number of charged tracks (≥ 5), on the total energy ($E_{tot} \geq 20\% \cdot E_{cm}$) or the charged particles' energy coming from the primary vertex ($E_{tot} \geq 10\% \cdot E_{cm}$) hadronic events are selected with high efficiency. The background is suppressed below the

---

* In this analysis a specific charge state implies the charge conjugate state as well.

percent level. e^+e^- and $\mu^+\mu^-$ have a final state multiplicity of two, and $\tau^+\tau^-$ events are substantially suppressed as well; a τ decays with a probability of 86 % to one charged plus neutral particles ('1-prong') and has a branching ratio of about 14 % to '3-prong' final states. $\gamma\gamma$ events, where photons radiated by the incoming electron and positron interact, have lower charged multiplicities by a factor of three and less visible energy. The acceptance for cosmic ray and beam-gas events is negligible.

The Σ^0 decay exhibits a clear signature: a Λ^0, which can easily be detected by its long decay length and a low energy photon. The mass difference between Σ^0 and Λ^0 ($\Delta m = 1192.55\,\mathrm{MeV}/c^2$ - $1115.63\,\mathrm{MeV}/c^2$ $= 76.92\,\mathrm{MeV}/c^2$) is too low to allow a strong decay. The Σ^0 appears as a narrow resonance[†]. To exploit this fact, photons converting to an e^+e^- pair in the detector are used. The excellent momentum resolution of the time projection chamber results in a better energy assignment than directly measuring the photon energy in the calorimeter (section 11.12). The detection efficiency, however, is rather low.

To study specific physics questions, the detector may have already been tailored for these reactions in the design phase. For large general-purpose detectors at LEP the experiments are constructed in such a way as to study many different reactions simultaneously. Therefore, a special analysis like $\Sigma^0 \to \Lambda^0\gamma$ is not an isolated topic but rather an integral part of the set of many different analyses. This allows one to benefit from other similar analyses at the same time. The production of the Λ^0 was intensively studied at LEP [617, 633]. This is why a study on a new signal should work on lesser known aspects of the analyses. Σ^0 detection suffers from a low efficiency. Photon conversions, often only rejected as a background to V^0 studies, require a more detailed investigation and the largest error will come from this source.

The Σ^0 selection starts with the extraction of Λ^0 candidates. It is of advantage to use Λ^0 from a kinematical fit (see figure 10.6). This is known to select a clean sample with high efficiency. Also a requirement on the minimum decay length is made to avoid combinatorial background from tracks originating at the primary vertex. In addition, a cut around the nominal Λ^0 mass (e.g., $\pm 3\sigma$) and a track quality cut on the reconstruction of the Λ^0 decay ($\Lambda^0 \to p\pi^-$) will purify the sample even more.

As mentioned before the production of the Λ^0 is well studied and a χ^2-test of data to Monte Carlo could check, for example, whether the Λ^0 rate is correctly simulated. With the invariant mass distributions $m(\pi, p)$ of data and Monte Carlo a χ^2-test comparing data and Monte Carlo distributions is performed. In the Monte Carlo simulation the contribution

---

[†] The mean life of $7 \cdot 10^{-20}$ s is still too short to be useful for detection.

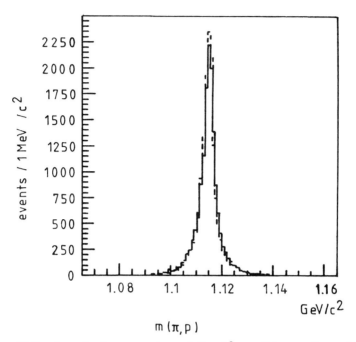

Fig. 10.6. Invariant mass $m(p, \pi^-)$ for Λ^0 candidates after a kinematical fit. The solid line indicates the distribution for data, while the dashed line represents the Monte Carlo simulation.

from Λ^0 and background is known. With the normalization of the Λ^0 and background as free parameters one can determine the multiplicity of the Λ^0.

The next task is to identify low energy photons from the Σ^0 decay. The absolute rate of photons in hadronic final states is not very well known. Understanding the photon conversions rate relies on the correct simulation of the material in the various subdetectors. This may not be perfect because, e.g., the position of cables is difficult to simulate. Therefore, to determine the absolute rate of conversions, a detector part must be selected as the conversion target, which is well known, uniform and homogeneous. The gas of the TPC is an excellent candidate for such a normalization, and it will allow a correction of a possibly imperfect Monte Carlo simulation.

Figure 10.7 shows the invariant mass distribution $m(\Sigma^0)$-$m(\Lambda^0)$ as a histogram [634, 635]. The mass difference has a better resolution, because of the low Q-value of the decay. The expectation for the Σ^0 signal from the Monte Carlo simulation is shown as dashed line.

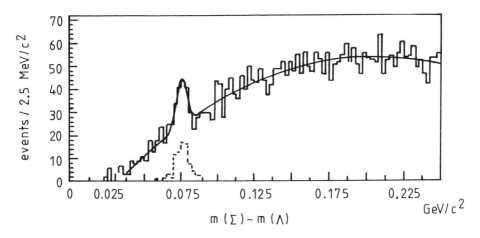

Fig. 10.7. Mass difference $\Delta m = m(\Sigma^0) - m(\Lambda^0)$. The dashed line indicates the expected Σ^0 signal from the Monte Carlo simulation.

The number of Σ^0 in the data is extracted by fitting an analytic function to the invariant mass spectrum (χ^2 or likelihood fit) [636]. For this purpose a sum of a third order polynomial and a Gaussian distribution is used. The polynomial describes the shape of the background while the Gaussian distribution represents the shape of the Σ^0 resonance. Here a Gaussian distribution for the Σ^0 is appropriate, because the detector resolution dominates the width of the peak. The fit of a function rather than a direct comparison is an advantage, because taking the mass of the peak and the width as free parameters implicitly takes the position and the width (i.e., resolution) from the data into account. By this procedure one finds in a sample of 977 240 hadronic events in total $90 \pm 17(stat)$ Σ^0.

To convert this number into the true number of Σ^0 per event one has to determine the efficiency of the selection and fit procedure. The number of Σ^0 produced in the simulation is known. Performing the same analysis as for data one obtains the efficiency as the ratio of reconstructed Σ^0 to Σ^0 initially generated: $\epsilon = n^{\text{rec}}(\text{MC})/n^{\text{gen}}(\text{MC})$. This leads to

$$n^{\text{true}}(\text{data}) = n^{\text{rec}}(\text{data})/\epsilon \tag{10.12}$$

yielding a number of Σ^0 per event of

$$n(\Sigma^0 \text{ per event}) = n^{\text{true}}(\text{data})/977\,240 . \tag{10.13}$$

The statistical error on the 90 observed Σ^0 was 17. In addition to the statistical error one must determine the systematic uncertainty of the analysis. In many analyses the systematic errors are only obtained by an educated guess. In the following a more thorough consideration will be given to the problem of systematic uncertainties. The Σ^0 analysis started with selection criteria on kinematical variables provided by intuition and experience. To obtain an idea of the systematic uncertainty, these criteria must be varied within reasonable intervals. The result should be invariant. The cuts must nonetheless restrict the selected sample to a set, where the simulation is well understood. In addition, the criteria are tuned in such a way to minimize the error. In principle, varying all cuts at the same time could optimize the selection. In practice, however, it has been shown that varying the cuts one by one is more practicable. This allows one to learn which cuts contribute to the systematic error most. Optimized cuts are usually found by iteration.

Further studies on systematic errors may be illustrated for the functional fit to the invariant mass of the Λ^0-γ spectrum. The signal could be described using the fixed, well-known mass and width of the Σ^0 including the detector resolution. Alternatively one could take the mass, width and shape from the Monte Carlo simulation or to leave it as a free parameter in the fit.

For rare events it is important to understand the background correctly. A way to simulate the background, for example, is to select Λ^0 candidates with a short decay length and combine them with conversions. Such a Λ^0 selection is dominated by wrong candidates from the combinatorial set. This is similar to the method known as 'wrong sign combination' used for many decay analyses. Mixing of Λ^0 candidates and photons from different events also provides a means to study the background. In addition, effects of a possible Ξ^0 reflection ($\Xi^0 \rightarrow \Lambda^0\pi^0, \pi^0 \rightarrow \gamma\gamma$), where one decay photon escapes detection, must be studied by using Monte Carlo simulations.

For an analysis, where the signal is extracted with low statistics, the statistical error of the simulation may have a substantial contribution on the result. If the production rate of the particle searched for is low, one may start a special Monte Carlo production. Either one generates a certain type of particle or selects interesting events, and passes this subset through the full detector simulation. The detector simulation is more time consuming (by one or two orders of magnitude) than the event generation and limits the Monte Carlo production on computers. When the efficiency of the selection is low, as it is the case for the presented Σ^0 analysis, a different approach is possible. The low selection efficiency for Σ^0 is due to the low conversion probability of photons into an e^+e^- pair. A more effective estimate is obtained by selecting Σ^0 in the simulation as before, but accepting all events no matter whether the photon has converted or not. One then weights each event with the conversion probability for a photon. This probability is determined from all photons, which predominantly come from π^0 decay, and does not suffer from low statistics. To obtain small systematic errors from this procedure, one has to parameterize the conversion probability as a function of momentum and angle to the beam line. The latter takes the angular dependence of the material thickness and composition into account.

Using different event generators is also important when studying systematics, especially in the case of the Σ^0 production, since the shape of the Σ^0 momentum spectrum is not well known. The difference in the results using two different models for fragmentation shows the influence of a reasonable change in the spectrum.

The visual inspection of reconstructed events, scanning, is also an important tool in understanding the background. Before the advent of detectors with complete electronic readout this was the standard procedure of analyzing events, e.g., from bubble chambers or spark chambers. Originally necessary for event classification and interpretation, it now guides pattern recognition and selection strategies on modern computers. Results, especially for selected rare decays, are verified by visual inspection. The discovery of $W^{\pm}$ in the UA1 experiment was guided by scanning. Scanning is also required for the control of the detector performance. During data taking it is part of the monitoring task (see chapter 9), and off-line it serves to check alignment, matching procedures and event reconstruction.

All studies of the systematics discussed above can be used as an analysis strategy on their own, and many more methods are conceivable. Instead of using Monte Carlo simulations one could use a reference signal with a known cross section to obtain the absolute production rate of a particle. Neural networks are well suited for the optimization of the selection. One can see that only imagination limits the ways of finding the best method to do physics analysis.

11
Applications of detector systems

There are a large number of applications for radiation detectors. They cover the field from medicine to space experiments, high energy physics and archaeology [1, 474, 637, 638].

In medicine and in particular in nuclear medicine, imaging devices are usually employed where the size and function of the inner organs can be determined, e.g., by registering γ-rays from radioactive tracers introduced into the body.

In geophysics it is possible to search for minerals by means of natural and induced γ-radioactivity.

In space experiments one is frequently concerned with measuring solar and galactic particles and γ-rays. In particular, the scanning of the radiation belts of the earth (van Allen belts) is of great importance for manned space missions. Many open questions of astrophysical interest can only be answered by experiments in space.

In the field of nuclear physics, methods of α, β, and γ-ray spectroscopy with semiconductor detectors and scintillation counters are dominant [15]. High energy and cosmic ray physics are the main fields of application of particle detectors [9, 10, 33, 639, 640, 641]. On the one hand, one explores elementary particles down to dimensions of 10^{-17} cm, and on the other, one tries by the measurement of PeV γ-rays (10^{15} eV) to obtain information on the sources of cosmic rays.

In archaeology absorption measurements of muons allow one to investigate otherwise inaccessible structures, like hollow spaces such as chambers in pyramids. In civil and underground engineering, muon absorption measurements allow one to determine the masses of buildings.

In the following, examples of experiments are presented which make use of the described detectors and measurement principles.

11.1 Radiation camera

The imaging of inner organs or bones of the human body by means of X-rays or γ-radiation is based on the radiation's specific absorption in various organs. If X-rays are used, the image obtained is essentially a shadow recorded by an X-ray film. X-rays are perfectly suited for the imaging of bones; the images of organs, however, suffer from a lack of contrast. This is related to the nearly identical absorption characteristics of tissue and organs.

If organ functions are to be investigated, radioactive tracers can be administered to the patient. These radionuclides will be deposited specifically in certain organs, thereby supplying an image of the organ and its possible malfunctions. Possible tracers for the skeleton are ^{90}Sr, for the thyroid gland ^{131}I or ^{99}Tc, for the kidney again ^{99}Tc, and ^{198}Au for the liver. In general, it is advisable to use γ-emitting tracers with short half lives to keep the radiation load on the patient as low as possible. The γ-radiation emitted from the organ under investigation has to be recorded by a special camera so that its image can be reconstructed.

A single small γ-ray detector, e.g., a scintillation counter, has fundamental disadvantages because it can only measure the activity of one picture element (pixel) at the time. In this method, much information remains unused, the time required for a complete picture of the organ is impractically long and the radiation load for the patient is large if many pixels have to be measured — and this is normally necessary for an excellent spatial resolution.

Therefore, a 'gamma camera' was developed which allows one to measure the total field of view with a single large area detector. Such a system, however, also requires the possibility to detect and reconstruct the point of origin of the γ-rays. One can use for this purpose a large NaI(Tl) inorganic scintillator, which is viewed by a matrix of photomultipliers (figure 11.1, [637, 642]). Gamma radiation coming from the human body is collimated by a multichannel collimator to maintain the information about the direction of incidence. The amount of light recorded by a certain photomultiplier is linearly related to the γ-activity of the organ part positioned beneath it. The light information of the photomultipliers provides a projected image of the organ based on its specific absorption for the γ-radiating tracer. Organ malfunctions are recognized by a characteristic modification of the γ-activity.

Positron emission tomography (PET) provides a means to reconstruct three-dimensional images of an organ. This method uses positron emitters for imaging. The positrons emitted from the radionuclides will stop within a very short range ($\approx$ mm) and annihilate with an electron from the tissue

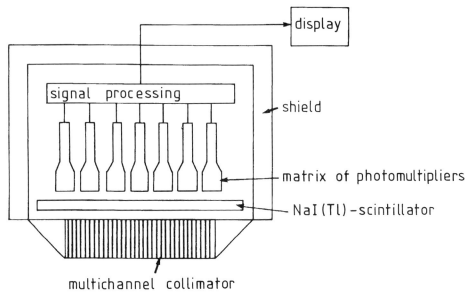

Fig. 11.1.   Sketch of a large area gamma camera [637, 642].

into two monoenergetic γ-rays

$$e^+ + e^- \rightarrow \gamma + \gamma \, . \tag{11.1}$$

Both γ-rays have $511\,\mathrm{keV}$ energy each, since the electron and positron masses are completely converted into radiation energy. Because of momentum conservation the γ-rays are emitted back-to-back. If both γ-rays are recorded in a segmented scintillation counter, which completely surrounds the organ to be investigated, the γ-rays must have been emitted from a line connecting the two fired modules. By measuring a large number of γ-pairs, the three-dimensional structure of the organ and its possible malfunctions can be recognized (see figure 11.2).

PET technology is also an excellent tool to probe, e.g., the structure of the brain, far more powerful than is possible by an electroencephalogram (EEG). In a PET scan, where blood or glucose is given a positron emitter tag and injected into the bloodstream of the patient, the brain functions can be thoroughly investigated. If the patient is observed performing various functions such as seeing, listening to music, speaking or thinking, the particular region of the brain primarily responsible for that activity will be preferentially supplied with the tagged blood or glucose to provide the energy needed for the mental process. The annihilation γ-rays emitted from these regions of mental activity allow one to reconstruct detailed pictures of regional brain glucose uptake, highlighting the brain areas associated with various mental tasks [643, 644].

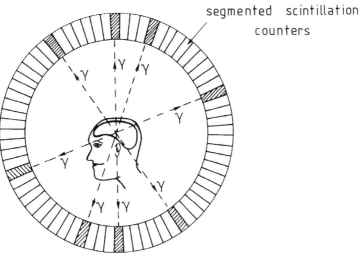

Fig. 11.2. Sketch of a positron emission tomograph.

11.2 Surface investigations with slow protons

A large number of non-destructive methods exist to determine the chemical composition of surfaces, one possibility being proton induced X-ray emission (PIXE). If slow charged particles traverse matter the probability for nuclear interactions is rather low. In most cases the protons lose their kinetic energy by ionizing collisions with atoms. In these ionization processes electrons from the K, L, and M-shells are liberated. If these shells are filled by electron transitions from higher shells, the excitation energy of the atom can be emitted in form of characteristic X-rays. This X-ray fluorescence represents a fingerprint of the target atom. The yield of K-shell X-rays increases with the atomic number. It varies between 15% at $Z = 20$ and reaches nearly 100% for $Z \geq 80$. The measurement of proton-induced characteristic X-rays is — quite in contrast to the application of electrons — characterized by a low background of continuous bremsstrahlung X-rays. Because of their high mass the probability for proton-induced bremsstrahlung is very low, so that the characteristic X-rays can be studied in a simple, clear and nearly background-free environment.

The X-rays can be recorded in lithium-drifted silicon semiconductor counters, which are characterized by a high energy resolution. An experimental set-up of a typical PIXE-system is sketched in figure 11.3 [130].

A proton beam of several microamperes with typical energies of several megaelectronvolts traverses a thin aluminum scattering foil, which widens the proton beam without a sizeable energy loss. The beam is then colli-

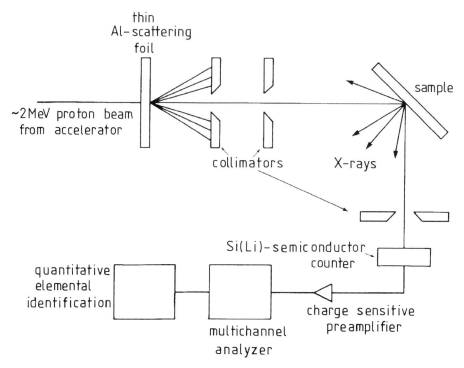

Fig. 11.3. Set-up of a PIXE-detector for the investigation of surface structure with slow protons [130].

mated and impinges on a selected area of the material to be investigated. A step-motor provides a means to move the sample in a well-defined way. This is required to investigate the homogeneity of an alloy over large areas.

The energy of characteristic X-rays increases with the atomic number Z according to

$$E_{\mathrm{K}} \propto (Z - 1)^2 \qquad (11.2)$$

(Moseley's law). The energy resolution of a scintillation counter and certainly that of a silicon lithium-drifted semiconductor counter is sufficient to separate characteristic X-rays of elements with Z differing by only one unit. Elements from phosphorus ($Z = 15$) up to lead ($Z = 82$) can be identified by this method down to concentrations of less than $1\,\mathrm{ppm}$ ($= 10^{-6}$).

The PIXE-technique is increasingly applied in the fields of biology, materials science, and archaeology and in all cases where a quick, sensitive, non-destructive method of surface investigation is required.

11.3   Tumor therapy with heavy particles

Classical methods for the treatment of tumors with γ-rays have the disadvantage that γ-radiation is exponentially attenuated in tissue, so that healthy surface tissue is more strongly affected than deep lying tissue. Although this effect can be reduced by rotating the patient or the source during the irradiation ('pendulum radiotherapy'), the largest fraction of the energy is nevertheless deposited in healthy tissue.

Charged particles have the great advantage that their energy loss increases towards the end of their range ('Bragg curve'). They therefore represent an ideal 'scalpel' for a radiotherapist. The range corresponding to the position where the maximum energy is deposited and where at the same time the largest tissue destruction is obtained can be adjusted by the energy of the charged particles.

The disadvantage of this method is that one needs an accelerator to produce the beam of heavy particles. In the following three different methods for the radiotherapy with heavy particles will be presented.

In the first method charged pions are produced by accelerated protons which collide with a light target. To obtain a reasonable yield of pions proton beams of around 500 MeV are used. A momentum spectrometer selects negative pions from the secondary particles produced at the target. These are then collimated and used to define a monoenergetic π^--beam used for irradiation (see figure 11.4).

Pions lose their energy in matter by ionization. Up to the end of their range their energy loss is relatively small. At the end of their range their energy loss increases considerably because of the $1/\beta^2$-term in the Bethe-Bloch relation (compare equation (1.12)). In addition, negative pions are captured by atoms, forming pionic atoms. By cascade transitions the pions reach orbitals very close to the nucleus and, finally, they are captured by the nucleus.

This process is much faster than the decay of free pions ($\pi^- \rightarrow \mu^- + \bar{\nu}_\mu$; $\tau_{\pi^-} = 26\,\text{ns}$). A large number of light fragments, like p, n, ^3He, $T(=^3\text{H})$ and α-particles can result from the pion capture (star formation). These will deposit their energy locally at the end of the pion's range. In addition, the relative biological efficiency of the fragments is rather high (see units of radiation measurements, chapter 3). Because of this effect, the Bragg peak of ionization is considerably amplified. The depth profile of the energy deposition of negative pions, showing the contributions of the various mechanisms, can be seen in figure 11.5. Also shown are the energy loss of muons and electrons [637, 645]. The star formation is localized somewhat deeper than the Bragg maximum.

The relative biological efficiency of negative pions was measured *in vivo*

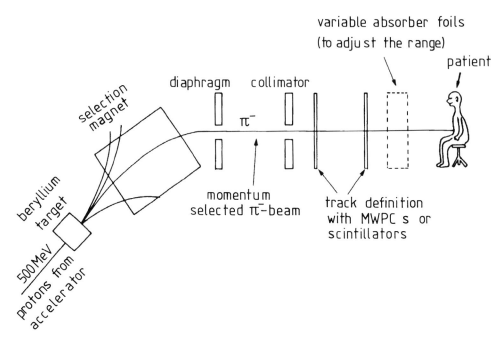

Fig. 11.4. Production of a particle beam for the treatment of tumors (MWPC = Multiwire Proportional Chamber).

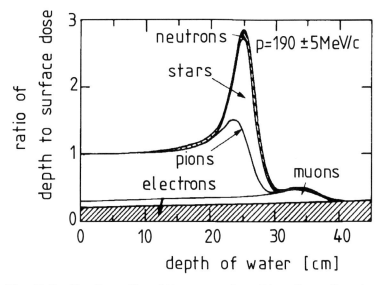

Fig. 11.5. Depth profiles of the energy deposition of negative pions, muons and electrons [637, 645]. A large number of light fragments can result from the pion capture ('star', see also figure 4.89).

and determined to be 2.4 for the energy loss of pions and 3.6 for the star formation. In addition to the much more favorable depth profile compared to γ-rays, one gains about a factor of three in destructive power for sick tissue [637, 646].

In addition to the radiotherapy with charged particles, fast neutrons are used for the treatment of brain tumors. It would be impossible to dissolve the tumors by a normal irradiation treatment without also destroying large parts of the healthy tissue in the neighbourhood of the tumor.

The neutron treatment works along the following lines. The tumor is sensitized with a boron compound before the neutron treatment is started. Neutrons have a large cross section for the reaction

$$n + {}^{10}\text{B} \rightarrow {}^{7}\text{Li} + \alpha + \gamma \,. \tag{11.3}$$

In this interaction, short-range α-particles with a high biological efficiency are produced. Reaction (11.3) produces 2 MeV α-particles with a range of several micrometers. This ensures that the destructive action of the α-particles is limited to the sick tissue. Clinical tests have shown that best results are obtained with epithermal neutrons (≈ 1 keV). Such neutron beams can be produced by interactions of 5 MeV protons on light target materials (e.g., lithium, beryllium) [647].

Cancer therapy can also be done with heavy ions. The treatment of cancer in this case is also based on the increased energy loss of charged particles at the end of their range. If a larger region of tissue inside the human body has to be destroyed, this can be achieved by varying the energy of the incident ions. The sick tissue can be covered in the transverse direction by scanning the cancer with a magnetic deflection system.

Such a magnetic deflection scanning device using heavy ion beams thus allows one to treat different depth layers by a stepwise increase or decrease of the particle energy [648].

11.4 Identification of isotopes in radioactive fallout

The γ-ray spectrum of a mixture of isotopes can be used to determine quantitatively the radionuclides it contains. Detectors well suited for this application are high resolution germanium semiconductor counters, into which lithium ions have been drifted, or high purity germanium crystals. The atomic number of germanium is sufficiently large so that the γ-rays emitted from the sample are absorbed with high probability via photoelectric effect, thereby producing distinct γ-ray lines. The well defined photopeaks are used for the identification of the radioisotope. Figure 11.6 shows part of the γ-ray spectrum of an air filter shortly after the reactor accident in Chernobyl [649]. Apart from the γ-ray lines originating from the natural radioactivity, some 'Chernobyl isotopes' like ^{137}Cs, ^{134}Cs, ^{131}I, ^{132}Te and ^{103}Ru are clearly recognizable by their characteristic γ-energies.

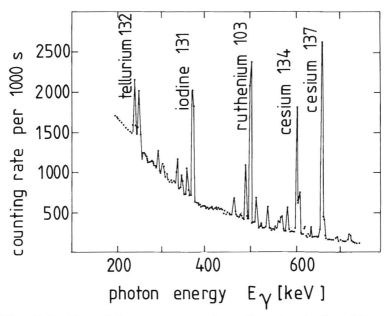

Fig. 11.6.　Part of the γ-spectrum of a radioactive air filter (the γ-energies of some 'Chernobyl isotopes' are indicated) [649].

The identification of pure β-ray emitters, which cannot be covered with this method, is possible with the use of silicon lithium-drifted semiconductor counters. Because of their relatively low atomic number ($Z = 14$) these detectors are relatively insensitive to γ-rays. Beta ray emitting isotopes can be quantitatively determined by a successive subtraction of calibration spectra. The identification of the isotopes is based on the characteristic maximum energies of the continuous β-ray spectra. The maximum energies can best be determined from the linearized electron spectra (Fermi–Kurie plot) [650].

11.5　Search for hidden chambers in pyramids

In the large Cheops pyramid in Egypt several chambers were found: the King's, Queen's, underground chamber and the so-called 'Grand Gallery' (figure 11.7). In the neighbouring Chephren pyramid, however, only one chamber, the Belzoni chamber (figure 11.8) could be discovered. Archae-

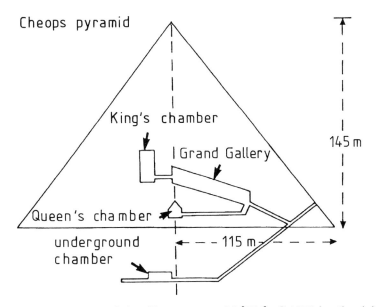

Fig. 11.7. Inner structure of the Cheops pyramid [651], ©1970 by the AAAS.

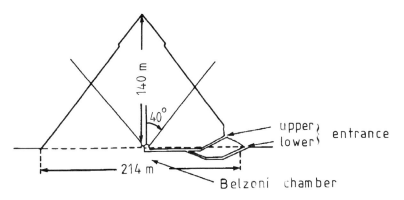

Fig. 11.8. Structure of the Chephren pyramid [651], ©1970 by the AAAS.

ologists suspected that there might exist further, undetected chambers in the Chephren pyramid.

It was suggested to 'X-ray' the pyramids using muons from cosmic radiation [651]. Cosmic ray muons can easily penetrate the material of the pyramid. Of course in this process their intensity is slightly reduced. The intensity reduction is related to the amount of material between the outer wall of the pyramid and the position of the detector. An enhanced relative intensity in a certain direction would indicate the presence of some hollow space which might represent an undiscovered chamber.

The intensity of muons as a function of depth $I(h)$ can be approximated by

$$I(h) = k \cdot h^{-\alpha} \qquad \text{with} \quad \alpha \approx 2 \,. \tag{11.4}$$

Differentiating equation (11.4) yields

$$\frac{\Delta I}{I} = -\alpha \frac{\Delta h}{h} \,. \tag{11.5}$$

In the case of the Chephren pyramid muons traversed typically about 100 m material before reaching the Belzoni chamber. Consequently, for an anticipated chamber height of $\Delta h = 5\,\text{m}$, a relative intensity enhancement compared to neighbouring directions of

$$\frac{\Delta I}{I} = -2 \frac{(-5\,\text{m})}{100} = 10\,\% \tag{11.6}$$

would be expected for a muon detector installed in the Belzoni chamber.

The detector used for this type of measurement (figure 11.9) consisted of a telescope $(2 \times 2\,\text{m}^2)$ of three large-area scintillation counters and four wire spark chambers [651, 652].

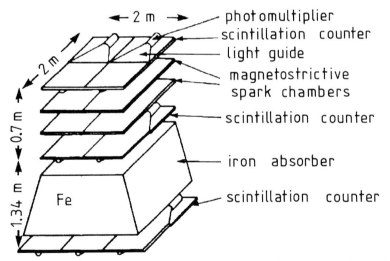

Fig. 11.9. Set-up of a muon absorption detector for the search for hidden chambers in the Chephren pyramid [651], ©1970 by the AAAS.

The spark chamber telescope was triggered by a three-fold coincidence of scintillation counters. The iron absorber prevented low energy muons from triggering the detector. Because of their large multiple scattering angles, low energy muons would only produce a fuzzy image of possible chambers. The spark chambers with magnetostrictive readout were used for the track reconstruction of the recorded muons.

The detector was installed approximately at the center of the base of the Chephren pyramid inside the Belzoni chamber (see figure 11.8). It had been suspected that just above the Belzoni chamber there might be additional cavities. Therefore the range of acceptance of the muon telescope was restricted to zenith angles of about 40° with complete azimuthal coverage. The measured azimuthal variation of the intensity for a fixed zenith angle clearly shows the corners of the pyramid, thus proving the working principle of the method. The section of the pyramid scanned by the detector was subdivided into cells of $3° \times 3°$. In total, several million muons were recorded. The azimuthal and zenith angle variation of the muon flux was compared to a simulated intensity distribution, which took into account the known details of the pyramid structure and the properties of the detector. This allowed one to determine deviations from the expected muon rates. Since the angular distributions of cosmic ray muons agreed with the simulation within the statistics of measurement, no further chambers in the pyramid could be revealed. The first measurement only covered a fraction of the pyramid volume, but later the total volume was subjected to a 'muon X-ray photography'. This measurement also showed that within the resolution of the telescope no further chambers existed in the Chephren pyramid.

11.6   Experimental proof of $\nu_e \neq \nu_\mu$

Neutrinos are produced in weak interactions, e.g., in the beta decay of the free neutron

$$n \rightarrow p + e^- + \bar{\nu} \tag{11.7}$$

and in the decay of charged pions

$$\pi^+ \rightarrow \mu^+ + \nu \tag{11.8}$$
$$\pi^- \rightarrow \mu^- + \bar{\nu} \,.$$

(For reasons of lepton number conservation one has to distinguish between neutrinos (ν) and antineutrinos ($\bar{\nu}$).) The question now arises whether the antineutrinos produced in the beta decay and π^- decay are identical particles or whether there is a difference between the electron and muon like neutrinos?

A pioneering experiment at the AGS accelerator (Alternating Gradient Synchrotron) in Brookhaven with optical spark chambers showed that electron and muon neutrinos are in fact distinct particles. The Brookhaven experiment used neutrinos from the decay of pions. The

15 GeV proton beam of the accelerator collided with a beryllium target, producing — among other particles — positive and negative pions (figure 11.10, [653]).

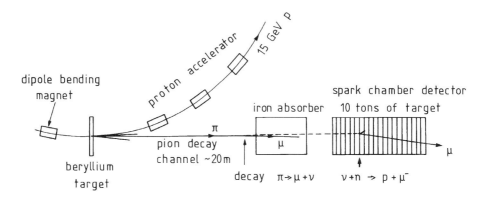

Fig. 11.10. Production of a neutrino beam at the 15 GeV AGS-proton synchrotron [653].

Charged pions decay with a lifetime of $\tau_0 = 26\,\mathrm{ns}$ ($c\tau_0 = 7.8\,\mathrm{m}$) into muons and neutrinos. In a decay channel of $\approx 20\,\mathrm{m}$ length practically all pions have decayed. The muons produced in this decay were stopped in an iron absorber so that only neutrinos could emerge from the iron block.

Let us assume for the moment that there is no difference between electron and muon neutrinos. Under this assumption, neutrinos would be expected to be able to initiate the following reactions:

$$
\begin{aligned}
\nu + n &\rightarrow p + e^- \\
\bar{\nu} + p &\rightarrow n + e^+ \\
\nu + n &\rightarrow p + \mu^- \\
\bar{\nu} + p &\rightarrow n + \mu^+ \, .
\end{aligned}
\tag{11.9}
$$

If, however, electron and muon neutrinos were distinct particles, neutrinos from the pion decay could only produce muons.

The cross sections for neutrino-nucleon interactions in the GeV-range are only of the order of magnitude $10^{-38}\,\mathrm{cm}^2$. Therefore, to cause the neutrinos to interact at all in the spark chamber detector, it had to be quite large and very massive. Ten one-ton modules of optical spark chambers with aluminum absorbers were used for the detection of the neutrinos. To reduce the background of cosmic rays, anti-coincidence counters were installed. The spark chamber detector can clearly identify muons and electrons. Muons are characterized by a straight track almost without interaction in the detector, while electrons initiate electromagnetic cascades with multiparticle production (see figure 7.53). The experiment showed that neutrinos from the pion decay only produced muons, thereby proving that electron and muon neutrinos are distinct elementary particles.

Figure 11.11 shows the 'historical' record of a neutrino interaction in the spark chamber detector [653]. A long-range muon produced in the neutrino interaction is clearly visible. At the primary vertex a small amount of hadronic activity is seen, which means that the interaction of the neutrino was inelastic, possibly

$$\nu_\mu + n \rightarrow \mu^- + p + \pi^0 \,, \tag{11.10}$$

with subsequent local shower development by the π^0 decay into two photons.

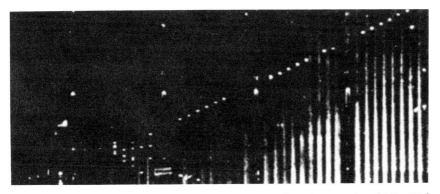

Fig. 11.11. Muon production in a neutrino-nucleon interaction [653, 654].

Later, the experimental result was confirmed in an experiment at the European Organisation for Nuclear Research (CERN). Figure 11.12 shows a neutrino interaction (ν_μ) in the CERN experiment, in which a high energy muon is generated via the reaction

$$\nu_\mu + n \rightarrow p + \mu^- \,, \tag{11.11}$$

which produces a straight track in the spark chamber system. The recoil proton can also be clearly identified from its short straight track [104, 655].

11.7 Spark chamber telescope for high energy γ-rays

In the field of γ-ray astronomy the detection of point sources that emit photons in the MeV-range and at even higher energies is an interesting topic. The determination of the γ-ray spectra emitted from the source may also provide a clue about the acceleration mechanism for charged particles and the production of energetic γ-rays [656, 657]. For energies in excess of several megaelectronvolts the electron-positron pair production

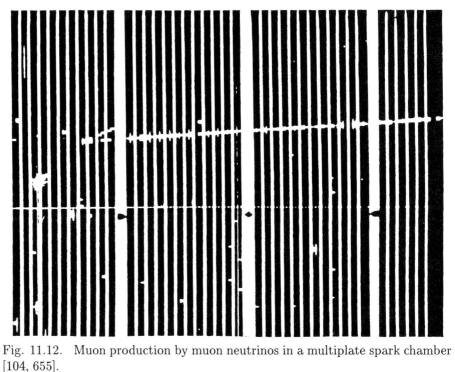

Fig. 11.12. Muon production by muon neutrinos in a multiplate spark chamber [104, 655].

is the dominating photon interaction process. The schematic set-up of a detector for γ-ray astronomy is shown in figure 11.13 [658, 659].

The telescope is triggered by a coincidence of three scintillation counters B_1, B_2, and C with an anti-coincidence requirement of the counter A. This selects photons that converted in the spark chamber volume Sp. In the multiplate wire spark chamber with ferrite-core readout the produced e^+e^- pair is registered, and the incident direction of the γ-ray is reconstructed from the tracks of the electron and positron. The total absorption scintillator calorimeter E is made from a thick cesium-iodide crystal doped with thallium. Its task is to determine the energy of the γ-ray by summing up the energies of both electrons.

Such a spark chamber telescope (figure 11.14) was on board the COS-B satellite [658] launched in 1975. It has recorded γ-rays in the energy range between $30 \, \text{MeV} \leq E_\gamma \leq 1000 \, \text{MeV}$ from the Milky Way. COS-B had a highly eccentric orbit with an apogee of $95\,000 \, \text{km}$. At this distance the background originating from the earth's atmosphere is negligible.

The COS-B satellite could identify the galactic center as a strong γ-ray source. In addition, point sources like Cygnus X3, Vela X1, Geminga and the Crab Nebula could be detected [658].

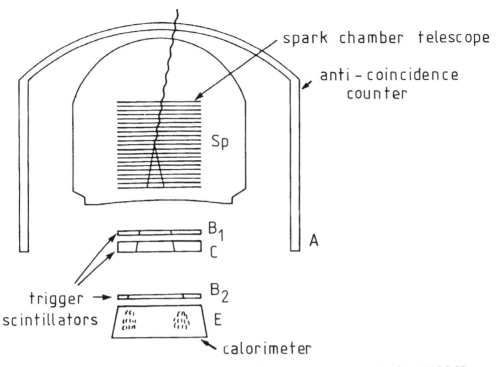

Fig. 11.13.   Sketch of a photon detector for γ-ray astronomy in the 100 MeV-range [658, 659].

Fig. 11.14.   Photograph of the COS-B-detector [660].

Figure 11.15 shows the intensity distribution of γ-rays with energies greater than 100 MeV as a function of the galactic length in a band of $\pm 10°$ galactic latitude. These data were recorded with the SAS-2 satellite [661]. The solid line is the result of a simulation, which assumes that the flux of cosmic γ-rays is proportional to the density of the interstellar gas. In this representation the Vela pulsar appears as the brightest γ-ray source in the energy range greater than 100 MeV.

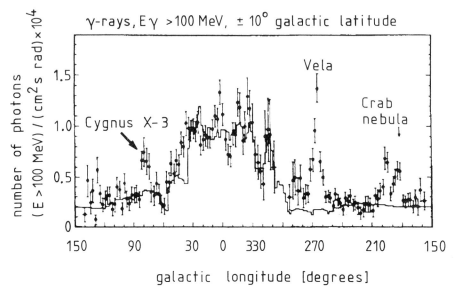

Fig. 11.15. Distribution of γ-rays with energies greater than 100 MeV as a function of galactic longitude [661].

11.8 Measurement of extensive air showers with the fly's eye detector

High energy charged particles and photons produce hadronic and electromagnetic cascades in the atmosphere. In a classical technique for registering these extensive air showers (EAS) the shower particles are sampled by a large number of scintillation counters normally installed at sea level [567]. The scintillation counters typically cover 1 % of the lateral shower distribution and give information on the number of shower particles at a depth far beyond the shower maximum. Clearly the energy of the primary particle initiating the cascade can only be inferred with a large measurement error. It would be much better to detect the complete longitudinal development of the shower in the atmosphere. Such a measurement can be done for energies in excess of 10^{17} eV, if the scintillation light produced by the shower particles in the atmosphere is registered (figure 11.16). This can be achieved with the 'fly's eye' experiment, which consists of 67 mirrors of 1.6 m diameter each [478, 662, 663, 664]. Each mirror has in its focal plane 12 to 14 photomultipliers. The individual mirrors have slightly overlapping fields of view. An extensive air shower passing through the atmosphere in the vicinity of the fly's eye experiment is only seen by some of the photomultipliers. From the fired phototubes, the longitudinal profile of the air shower can be reconstructed. The total recorded light yield is proportional to the shower energy [665].

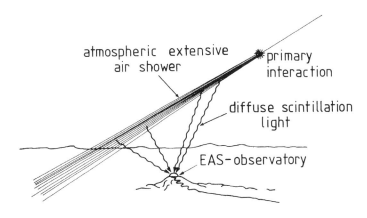

Fig. 11.16. Measurement principle for extensive air showers (EAS) via the scintillation light produced in the atmosphere.

Such a fly's eye experiment is installed in Utah, USA, for the measurement of high energy primary cosmic rays (figure 11.17). The disadvantage connected with this measurement technique is that the detection of the weak scintillation light can only be done on clear, moonless nights.

The individual mirrors of the fly's eye can also be separately operated as Cherenkov telescopes. With such telescopes the Cherenkov radiation of highly relativistic shower particles in the atmosphere is measured. Cherenkov mirror telescopes provide a means to detect γ-ray point sources, which emit in the energy range in excess of 1 TeV. A high angular resolution of these telescopes allows one to suppress the large background of hadron-induced showers, which is isotropically distributed over the sky, and to identify γ-ray induced cascades from point sources unambiguously. In this particular case, one takes advantage of the fact that γ-rays travel along straight lines in the galaxy, while charged primary cosmic rays do not carry any directional information on their origin because they become randomized by irregular galactic magnetic fields.

Fig. 11.17. Photograph of the 'fly's eye' experiment [478, 666].

11.9   Search for proton decay with water Cherenkov counters

In theories that attempt to unify the electroweak and strong interactions, the proton is no longer stable. In some models it can decay, violating baryon and lepton conservation number, according to

$$p \to e^+ + \pi^0 \ . \tag{11.12}$$

The predicted proton lifetime of the order of 10^{30} years requires large volume detectors to be able to see such rare decays. One possibility for the construction of such a detector is provided by large volume water Cherenkov counters (several thousand tons of water). These Cherenkov detectors contain a sufficiently large number of protons to be able to see several proton decays in a measurement time of several years if the theoretical prediction is correct. The proton decay products are sufficiently fast to emit Cherenkov light.

Large volume water Cherenkov detectors require ultra-pure water of high transparency to be able to register the Cherenkov light using a large number of photomultipliers. The phototubes can either be installed in the volume or at the inner surfaces of the detector. Directional information and vertex reconstruction of the decay products is made possible by fast timing methods on the phototubes. Short range charged particles from nucleon decays produce a characteristic ring of Cherenkov light (see figure 11.18), where the outer radius r_a is used to determine the distance of the decay vertex from the detector wall and the inner radius r_i approximately reflects the range of the charged particle in water until it falls below the Cherenkov threshold. The measured light yield allows one to determine the energy of the particles.

Two such water Cherenkov detectors are installed in the Kamioka-zinc mine in Japan (KamiokaNDE = Kamioka Nucleon Decay Experiment) and the Morton-Thiokol salt mine in Ohio, USA (Irvine-Michigan-Brookhaven (IMB) Experiment) [380, 631, 381].

In spite of running these detectors over several years, no proton decay was detected. From this result, a new limit on the lifetime of the proton was determined to be $\tau \geq 10^{32}$ years.

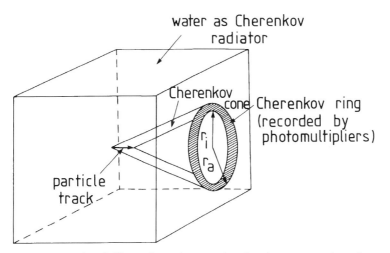

Fig. 11.18. Principle of Cherenkov ring production in an experiment searching for proton decay.

The large volume water Cherenkov counters have been spectacularly successful, however, in registering neutrinos emitted by the supernova 1987A. The KamiokaNDE experiment was even able to detect solar neutrinos because of its low detection threshold for electron energies [381].

11.10 Radio-carbon dating

The dating of archaeological objects of biological origin can be performed with the radio-carbon dating method [667, 668]. The earth's atmosphere contains in its carbon dioxide the continuously produced radioisotope ^{14}C. This radioisotope is produced by secondary neutrons in cosmic radiation via the reaction

$$n + {}^{14}_{7}\text{N} \rightarrow {}^{14}_{6}\text{C} + p \;. \tag{11.13}$$

^{14}C is a β^--emitter with a half life of 5730 years. It decays back into nitrogen according to

$$^{14}_{6}\text{C} \rightarrow {}^{14}_{7}\text{N} + e^- + \bar{\nu}_e \;. \tag{11.14}$$

In this way a concentration ratio of

$$r = \frac{N({}^{14}_{6}\text{C})}{N({}^{12}_{6}\text{C})} = 1.2 \cdot 10^{-12} \tag{11.15}$$

is formed. All plants and, as a consequence of eating vegetable matter also animals and humans, have ^{14}C. Therefore, the isotopic ratio produced in the atmosphere is also formed in the entire biosphere. With the death of a living organism the radio-carbon incorporation comes to an end. The radioactive decay of ^{14}C now reduces the ^{14}C/^{12}C-ratio. By comparing the ^{14}C-activity of an archaeological object and a biological object of the present time, the age of the object can be determined.

An experimental problem arises because of the low beta activity of archaeological objects. The maximum energy of the electrons emitted in the ^{14}C-decay is only 155 keV. Therefore, a very sensitive detector is required for their detection. If the radioisotope ^{14}C is part of a gas (^{14}CO$_2$), a methane flow counter can be used (a so called low-level counter). This detector has to be shielded passively by lead and actively by anti-coincidence counters against background radiation. The methane flow counter is constructed in such a way that the sample to be investigated — which does not necessarily have to be in the gaseous state — is introduced into the detector volume. This is to prevent energy losses of electrons when entering the counter. A steady methane flow through the detector guarantees a stable gas amplification.

Due to systematic and statistical errors, radio-carbon dating is possible for archaeological objects with ages between 1000 and 75 000 years. In recent times, however, it has to be considered that the concentration ratio r is altered by burning ^{14}C-poor fossil fuels and also by nuclear bomb tests in the atmosphere. As a consequence r is no longer constant in time. Therefore, a time calibration must first be performed. This can be done by measuring the radio-carbon content of a sample of known age [667].

11.11   Average dosimetry

Occasionally the problem arises of determining a radiation dose after radiation accidents if no dosimeter information was available. It is possible to estimate the body dose received after the accident has happened by the hair activation method [65]. Hair contains sulphur with a concentration of 48 mg S per gram hair. By neutron irradiation (e.g., after reactor accidents) the sulphur can be converted to phosphorus according to

$$n + {}^{32}\text{S} \to {}^{32}\text{P} + p \,. \qquad (11.16)$$

In this reaction the radioisotope ^{32}P is produced, which has a half life of 14.3 days. In addition to this particular reaction, the radioactive isotope ^{31}Si is formed in the following manner,

$$n + {}^{32}\text{S} \to {}^{31}\text{Si} + \alpha \,. \qquad (11.17)$$

The ^{31}Si isotope renders the determination of the phosphorus activity difficult. The half life of ^{31}Si, however, is only 2.6 hours. Therefore one waits for a certain amount of time until this activity has decayed before attempting to measure the ^{32}P activity. In case of surface contaminations, careful cleaning of the hair has to precede the activity measurement.

^{32}P is a pure β-ray emitter. The maximum energy of the electrons in this decay is 1.71 MeV. Because of the normally low expected event rates, a detector with high efficiency and low background is required. An actively and passively shielded end-window counter is a suitable candidate for this kind of measurement. Knowing the activation cross section for the reaction (11.16), the measured ^{32}P activity can be used to infer the radiation dose received.

11.12   The electron-positron storage ring experiment ALEPH

The ALEPH experiment represents a multipurpose detector for the investigation of electron-positron interactions at the presently largest electron-positron storage ring LEP (= Large Electron-Positron collider) [270]. The ALEPH detector surrounds the interaction point almost completely (99.8 % solid angle coverage). In the first year of operation, the principal aim of the experiment was to determine the number of neutrino generations, the parameters of the electroweak theory, to search for the Higgs boson and possibly to discover further elementary particles, partly predicted by theory. Figure 11.19 shows the ALEPH detector. Electrons and positrons collide in the evacuated beam pipe at the center of the detector. In the central region the beam pipe is made of beryllium, in the adjacent parts of aluminum. A superconducting solenoidal coil of 7 m length and 5.30 m diameter (including the cryostat) produces a longitudinal magnetic field of 1.5 Tesla at a current of 5000 Ampères.

The tracking detectors for charged particles are situated inside the coil. To minimize effects of multiple scattering, which would limit the momentum resolution, the tracking detectors are extremely thin. The track

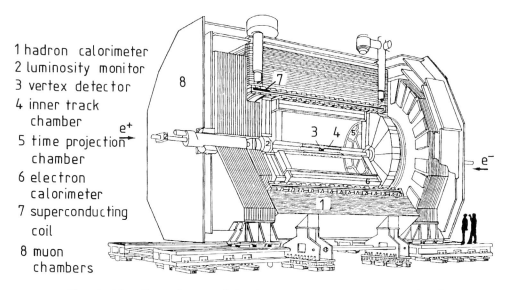

1 hadron calorimeter
2 luminosity monitor
3 vertex detector
4 inner track chamber
5 time projection chamber
6 electron calorimeter
7 superconducting coil
8 muon chambers

e^+

e^-

Fig. 11.19. General view of the ALEPH experiment at LEP [270].

measurement starts very close to the interaction point by a vertex detector, consisting of two layers of silicon strip counters. The silicon detectors are arranged on two concentric rings with radii of 9.6 and 11.3 cm. The inner ring contains 12, the outer 15, silicon modules. Each detector is read out on both faces by orthogonally segmented electrodes ('strips'). In total, the vertex detector contains 82 944 readout channels. The tracks are measured with an accuracy of 12 μm in azimuthal direction and with 10 μm along the beam axis for normal incident particles. The main purpose of the silicon vertex detector is the determination of the lifetimes of short-lived hadrons by reconstruction of secondary vertices.

The silicon strip detector is followed by a cylindrical multiwire drift chamber, 2 m long and covering the radial range from 16 to 26 cm. It has 960 signal wires altogether, arranged in 8 concentric layers. A spatial resolution of 100 μm is reached in the $r\varphi$ plane. The coordinate along the axial wires is determined by measuring the differences of the signal propagation time at both ends of the wires. This allows an accuracy of about 3 cm. This inner tracking chamber supplies precise information about the tracks near the interaction point. In addition, it also provides fast information about tracks originating from the vertex, which is used for the trigger system.

Possibly the most important component of the ALEPH experiment is the time projection chamber TPC (figure 11.20). The TPC is the central tracking detector with an inner radius of 31 cm, an outer radius of 180 cm, and a length of 4.7 m. The electric field is formed by potential strips on the inner and outer cylinder of the chamber ('field cage'). The charge carriers produced by the ionizing particles drift in the longitudinal electric field to the endplates which are instrumented with multiwire proportional chambers. Each endplate consists of two rings of wire chamber sectors, the inner ring has 6, the outer one 12 sectors. The wires run essentially in the azimuthal direction. Behind the anode wires, cathode pads are arranged which allow a determination of the azimuthal coordinate along the wire. The radial coordinate is given by the number of the fired wire or pad row. The coordinate parallel to the beam is obtained from the electron drift time to the endplates.

In radial direction r there are 21 cathode pads and 352 signal wires. The electrodes supply not only tracking but also energy loss information. The energy loss measurements from the wires provide the possibility of particle identification.

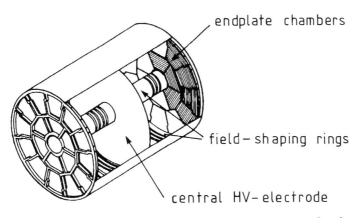

endplate chambers

field-shaping rings

central HV-electrode

Fig. 11.20.   The ALEPH time projection chamber [270].

There are in total 41 004 cathode pads and 6336 signal wires.

For transverse tracks, position resolutions of 160 μm in the $r\varphi$ plane and 1 mm along z are obtained (φ is the azimuthal angle and z is measured along the cylinder axis). Transverse momenta are determined with a resolution of

$$\sigma(p)/p = 1.2 \cdot 10^{-3} \cdot p \, [\text{GeV}/c] \,. \tag{11.18}$$

If, in addition, the information from the inner tracking chamber and vertex detector are used for track reconstruction, one obtains

$$\sigma(p)/p = 6 \cdot 10^{-4} \cdot p \, [\text{GeV}/c] \,. \tag{11.19}$$

The time projection chamber is filled with a counting gas consisting of 91 % argon and 9 % methane. The chamber is calibrated by a laser system. This system produces straight ionization tracks at five different polar and five different azimuthal angles. On the basis of reconstructed laser tracks, the drift velocity and field inhomogeneities can be monitored [669].

The electromagnetic calorimeter is situated inside the magnet coil in order to obtain the best possible energy resolution for electrons and photons or for particles which decay into photons (e.g., $\pi^0 \rightarrow \gamma\gamma$). This calorimeter is a sampling device made up from alternating lead absorbers and multiwire proportional chambers with a total thickness of 22 radiation lengths. The anode wires of the multiwire proportional chamber layers, as well as the cathode pads mounted behind the wires, are read out. The cathode pads are arranged in a projective geometry pointing to the interaction point. The pads are connected to form calorimetric towers, subtending a solid angle of $1° \times 1°$. In the central part, the barrel consists of 12 large calorimeter modules and covers the largest fraction of the solid angle. Together with the endcaps, which consist of 12 pie-shaped modules, the electromagnetic calorimeter encloses the time projection chamber almost completely.

The multiwire proportional chambers in the electromagnetic calorimeter are operated with a gas mixture of xenon (80 %) and CO_2 (20 %).

The calorimeter attains an energy resolution of

$$\frac{\sigma(E)}{E} = \frac{17\,\%}{\sqrt{E}} \oplus 1.6\,\% \; ; \quad (E \text{ in GeV}) \tag{11.20}$$

($\oplus$ means that the energy dependent sampling error and the constant term must be added in quadrature). By segmenting the readout planes into $3 \times 3\,\text{cm}^2$ cathode pads, the calorimeter provides at the same time tracking information, yielding an azimuthal spatial resolution of

$$\sigma = 6.8\,\text{mm}/\sqrt{E}\,, \quad (E \text{ in GeV})\,. \tag{11.21}$$

By using particle identification estimators as described in section 7.4, the electromagnetic calorimeter provides a pion suppression of 10^3 for an electron acceptance of 95 % for 10 GeV particles.

The hadron calorimeter is mounted outside the superconducting coil. It is constructed from iron plates of 5 cm thickness which serve a fourfold purpose. The iron represents the flux return yoke for the magnetic field. It provides the hadron energy measurements and is a filter for muons. Finally it forms the mechanical structure for all other subdetectors. Just like the electromagnetic calorimeter it comes in two parts: the endcaps and the barrel. The development of hadron cascades is measured in layers of streamer tubes. The anode wires of the streamer tubes and also the induced signals on segmented cathodes are read out. The streamer tubes are filled with a gas mixture of argon (13 %), carbon dioxide (57 %) and isobutane (30 %). The hadron calorimeter has an energy resolution of typically

$$\frac{\sigma(E)}{E} = 84\,\%/\sqrt{E}\,, \quad (E \text{ in GeV}) \tag{11.22}$$

and an azimuthal spatial resolution of 3.5 mm.

Outside the iron of the hadron calorimeter are two double layers of muon chambers. These chambers are constructed in the same way as the detection layers of the hadron calorimeter. Electrons and hadrons are usually completely absorbed in the calorimeters. Only penetrating muons can reach the layers of the muon chambers with a significant probability. In addition, muons can also be identified by their characteristic energy depositions in the electromagnetic and hadronic calorimeters. This also provides a means to distinguish muons from electrons and hadrons. Figure 11.21 shows the charge distributions of 20 GeV muons and pions in the hadron calorimeter, which would already allow an excellent μ/π-separation. If, in addition, one requires signals in the muon chambers — which normally cannot be reached by pions — and if also the lateral structure of energy deposition in the hadron calorimeter is taken into account, a clear separation of muons and pions is possible.

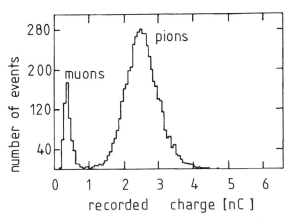

Fig. 11.21. Charge distribution of 20 GeV pions and muons in the ALEPH hadron calorimeter [270].

In the forward region the luminosity monitor covers that part of the solid angle which was left open by the electromagnetic calorimeter. The luminosity monitor consisted of a nine-layer tracking system mounted in front of an electromagnetic calorimeter. The purpose of the luminosity monitor is to determine the number of electrons and positrons which can collide at the interaction point. This is obtained by measuring the rate of small-angle e^+e^- scattering processes (Bhabha scattering), which is very well known theoretically. The luminosity determined in this way is used for the normalization of cross sections of other processes recorded, e.g., in the central detector. Apart from the luminosity determination the monitor also identifies and measures the energy of electrons and photons and determines their track coordinates. In 1992 after the installation of a smaller beam pipe in ALEPH the nine-layer tracking system was replaced by a silicon tungsten calorimeter, which allows to determine the luminosity with even higher precision.

The trigger system decides in a multistep process based on selected data whether an interesting event has occurred. In case of a positive decision all subdetectors of ALEPH are read out. The trigger rate in the measurements on the Z^0-resonance is about 1 Hz. Figure 11.22 shows various projections of a Z^0-production event and its subsequent decay into hadrons.

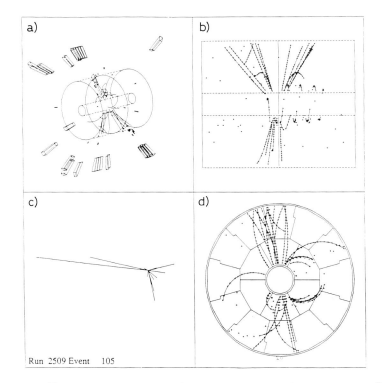

Fig. 11.22. Various projections and reconstructions of a hadronic Z^0-decay in the ALEPH experiment. (a) Three-dimensional view with tracks in the TPC (Time Projection Chamber) and energy deposits in the calorimeters. (b) rz-view in the TPC. (c) Momentum diagram; the length of the lines is proportional to the momentum of the particles. (d) $r\varphi$-view in the TPC [270].

The ALEPH detector in total is 11 m long and 9.3 m high. It has approximately a cylindrical shape and weighs 3000 tons.

Experiments of this size can only be built and financed by large collaborations. The ALEPH collaboration has about 400 members from 32 different institutions. The construction cost of the ALEPH detector amounted to approximately £50 million.

Résumé

The scope of detection techniques is very wide and diverse. Depending on the aim of the measurement, different physics effects are used. Basically each physics phenomenon can be the basis for a particle detector. If complex experimental problems are to be solved, it is desirable to develop a multipurpose detector which allows one to unify a large variety of different measurement techniques. This would include a high (possibly 100 %) efficiency, excellent time, spatial and energy resolution with particle identification. For certain energies these requirements can be fulfilled, e.g., with suitably instrumented calorimeters. Calorimetric detectors for the multi-GeV and for the eV-range, however, have to be basically different.

The discovery of new physics phenomena allows one to develop new detector concepts and to investigate difficult physics problems. For example, superconductivity provides a means to measure extremely small energy depositions with high resolution. The improvement of such measurement techniques, e.g., for the discovery and detection of cosmological neutrinos, would be of large astrophysical interest.

In addition to the measurement of low energy particles, the detection of extremely small changes of length may be of considerable importance. If one searches for gravitational waves, relative changes in length of $\Delta\ell/\ell \approx 10^{-21}$ have to be detected. If antennas with a typical size of one meter were chosen, this would correspond to a measurement accuracy of 10^{-21} m, or one millionth of the diameter of a typical atomic nucleus. This ambitious goal has not yet been reached but it is expected to be attained in the near future.

Since it would be bold to assume that the physical world is completely understood (in the past and also recently [670] this idea has been put forward several times), there will always be new effects and phenomena. Experts in the field of particle detection will pick up these effects and use them as a basis for the development of new particle detectors. For this reason a description of detection techniques can only be a snapshot. 'Old' detectors will 'die out' and new measurement devices will move to the forefront of research. Occasionally an old detector, already believed to be discarded, will experience a renaissance. The holographic readout of vertex-bubble chambers for three-dimensional event reconstruction is an excellent example of this. But also in this case it was a new effect, namely the holographic readout technique, that has triggered this development.

12
Glossary

The glossary summarizes the most important properties of detectors along with their main fields of application. An abridged description of the characteristic interactions of particles is also presented.

12.1 Interactions of charged particles and radiation with matter

Charged particles interact mainly with the electrons of matter. The atomic electrons are either excited to higher energy levels ('excitation') or liberated from the atomic shell ('ionization') by the charged particles. High energy ionization electrons which are able themselves to ionize are called δ-rays or 'knock-on electrons'. In addition to the ionization and excitation of atomic electrons bremsstrahlung plays a particular rôle, especially for primary electrons as charged particles.

Energy loss by **ionization and excitation** is described by the Bethe-Bloch formula. The basic features describing the mean energy loss per unit length $(\mathrm{d}E/\mathrm{d}x)$ for heavy particles are given by

$$-\left.\frac{\mathrm{d}E}{\mathrm{d}x}\right|_{\text{Ion}} \propto z^2 \cdot \frac{Z}{A} \cdot \frac{1}{\beta^2} \left\{ \ln a \cdot \gamma^2 \beta^2 - \beta^2 - \frac{\delta}{2} \right\} ,$$

where

z	—	charge of the incident particle
Z, A	—	atomic number and atomic weight of the target
β, γ	—	velocity and Lorentz factor of the incident particle
δ	—	parameter describing the density effect
a	—	parameter depending on the mass of the incident particle, the electron mass and the ionization energy of the absorber.

Typical values of the energy loss by ionization and excitation are around $2\,\text{MeV}/(\text{g}/\text{cm}^2)$. The energy loss in a given material layer is not distributed according to a Gaussian but is characterized, in particular for thin absorber layers, by a high asymmetry (Landau distribution).

Detectors only measure the energy deposited in the sensitive volume. This is not necessarily the same as the energy loss of the particle in the detector, since a fraction of the energy can escape from the detector volume as, e.g., δ-rays.

The energy loss of a charged particle in a detector leads to a certain number of free charge carriers n_T given by

$$n_T = \frac{\Delta E}{W} \,,$$

where ΔE is the energy deposited in the detector and W is a characteristic energy which is required for the production of a charge carrier pair ($W \approx 30\,\text{eV}$ in gases; $3.6\,\text{eV}$ in silicon; $2.8\,\text{eV}$ in germanium).

The second important interaction process of charged particles is particularly important for light particles. The **bremsstrahlung energy loss** can essentially be parametrized by

$$-\frac{\mathrm{d}E}{\mathrm{d}x}\bigg|_{\text{brems}} \propto z^2 \cdot \frac{Z^2}{A} \cdot \frac{1}{m_0^2} \cdot E \,,$$

where m_0 and E are the projectile mass and energy, respectively. For electrons ($z = 1$) one defines

$$-\frac{\mathrm{d}E}{\mathrm{d}x}\bigg|_{\text{brems}} = \frac{E}{X_0} \,,$$

where X_0 is the **radiation length** characteristic for the absorber material.

The **critical energy** E_c characteristic for the absorber material is defined as the energy at which the energy loss of electrons by ionization and excitation on the one hand and bremsstrahlung on the other hand are equal:

$$-\frac{\mathrm{d}E}{\mathrm{d}x}(E_c)\bigg|_{\text{ion}} = -\frac{\mathrm{d}E}{\mathrm{d}x}(E_c)\bigg|_{\text{brems}} = \frac{E_c}{X_0} \,.$$

Multiple Coulomb scattering of charged particles in matter leads to a deviation from a straight trajectory. It can be described by an r.m.s. planar scattering angle

$$\sigma_\theta = \sqrt{\langle \theta^2 \rangle} = \frac{13.6}{p\beta}\sqrt{\frac{x}{X_0}} \,,$$

where

p, β — momentum (in MeV$/c$) and velocity of the particle

x — material traversed in units of radiation lengths X_0.

In addition to the interaction processes mentioned so far, direct electron-pair production and photonuclear interactions come into play at high energies. Energy loss by Cherenkov radiation and transition radiation are of considerable interest for the construction of detectors but they play only a minor rôle as far the energy loss of charged particles is concerned.

Neutral particles like neutrons or neutrinos first have to produce charged particles in interactions before they can be detected via the interaction processes described above.

Photons of low energy (< 100 keV) are detected via the photoelectric effect. The cross section for the **photoelectric effect** can be approximated by

$$\sigma^{\text{photo}} \propto \frac{Z^5}{E_\gamma^{7/2}} \; ,$$

where at high γ-energies the dependence flattens to $\propto E_\gamma^{-1}$. In the photoelectric effect one electron (usually from the K-shell) is removed from the atom. As a consequence of the rearrangement in the atomic shell either characteristic X-rays or Auger electrons are emitted.

In the region of medium photon energies (100 keV $- 1$ MeV) the scattering on quasifree electrons dominates (**'Compton scattering'**). The cross section for the Compton effect can be approximated by

$$\sigma^{\text{Compton}} \propto Z \cdot \frac{\ln E_\gamma}{E_\gamma} \; .$$

At high energies ($\gg 1$ MeV) electron-pair production is the dominating interaction process of photons

$$\sigma^{\text{pair}} \propto Z^2 \cdot \ln E_\gamma \; .$$

The above photoprocesses lead to an absorption of X-ray or γ-radiation which can be described by an absorption law for the photon intensity according to

$$I = I_0 e^{-\mu x} \; .$$

μ is a characteristic absorption coefficient which is related to the cross sections for the photoelectric effect, Compton effect and pair production. Compton scattering plays a special rôle since the photon is not completely absorbed after the interaction like in the photoelectric effect or for pair production, but only shifted to a lower energy. This requires the introduction and distinction of attenuation and absorption coefficients.

Charged and also neutral particles can produce further particles in inelastic interaction processes. The strong interactions of hadrons can be described by characteristic absorption and nuclear interaction lengths.

The electrons produced by ionization — e.g., in gaseous detectors — are thermalized by collisions with the gas molecules. Subsequently they are normally guided by an electric field to the electrodes. The directed motion of electrons in the electric field is called drift. Drift velocities of electrons in typical gases for usual field strengths are of the order of $5\,\mathrm{cm}/\mu\mathrm{s}$. During the drift the charged particles (i.e., electrons and ions) are subject to transverse and longitudinal diffusion caused by collisions with gas molecules.

The presence of magnetic fields causes the electrons to deviate from a drift parallel to the electric field.

Low admixtures of electronegative gases can have a considerable influence on the properties of gas detectors.

12.2   Characteristic quantities of detectors

The quality of a detector can be expressed by its measurement resolution for time, track accuracy, energy and other characteristics. Spatial resolutions of $10-20\,\mu\mathrm{m}$ can be obtained in silicon strip counters and small drift chambers. Time resolutions in the subnanosecond range are achievable with planar spark counters. Energy resolutions in the eV-range can be reached with cryogenic calorimeters.

In addition to resolutions, the efficiency, uniformity and time stability of detectors are of great importance.

12.3   Units of radiation measurement

The radioactive decay of atomic nuclei (or particles) is described by the decay law

$$N = N_0 e^{-t/\tau}$$

with the lifetime $\tau = \frac{1}{\lambda}$ (where λ is the decay constant). The half-life $T_{1/2}$ is smaller than the lifetime ($T_{1/2} = \tau \cdot \ln 2$).

The activity $A(t)$ of an element is

$$A(t) = -\frac{\mathrm{d}N}{\mathrm{d}t} = \lambda \cdot N$$

with the unit Becquerel ($= 1$ decay per second).

The energy dose D is defined by the absorbed radiation energy dW per unit mass

$$D = \frac{dW}{\rho dV} = \frac{dW}{dm} \ .$$

D is measured in **Grays** (1 Gray = 1 J/kg). The old unit of the absorbed dose was rad (100 rad = 1 Gray).

The biological effect of an energy absorption can be different for different particle types. If the physical energy absorption is weighted by the relative biological efficiency (RBE) one obtains the equivalent dose H which is measured in **Sieverts** (Sv)

$$H \ [\text{Sv}] = RBE \cdot D \ [\text{Gy}]$$

The old unit of the equivalent dose was rem (1 Sv = 100 rem). The **equivalent radiation dose due to natural radioactivity** amounts to about 3 mSv per year. Persons working in **regions of controlled access** are typically limited to a maximum of 50 mSv per year. The lethal dose for humans (50 % probability of death within 30 days) is around 4000 mSv.

12.4   Detectors for track and ionization measurement

12.4.1   Ionization chambers

Planar ionization chambers

Application: Measurement of energy loss and — for total absorption — of the energy; e.g., α-spectroscopy; measurement of personal radiation dose.

Construction: Constant homogeneous electric field; constant drift velocity of electrons and ions; no gas amplification. The counting medium is usually gaseous but can also be liquid (argon, krypton, xenon) or solid. In liquids extremely high purity (ppb) is required.

Measurement principle, readout: The charge carriers collected in the electric field induce charge signals on the electrodes. The resulting current produces a voltage signal across a resistor.

Advantages: Simple construction and readout.

Disadvantage: Extremely low signals in gas ionization chambers require sensitive low noise amplifiers. The signal amplitude depends on the point of incidence (can be corrected by the introduction of a Frisch grid).

Cylindrical ionization chamber

Application: See planar ionization chamber.

Construction: Cylindrically symmetric set-up with inhomogeneous $1/r$-field (potential $U \propto \ln r$); variable, field-dependent (i.e., position dependent) drift velocity; no gas amplification.

Measurement principle, readout: Charge signals are induced on the electrodes. The main contribution to the signal originates from the movement of *electrons*.

Advantages, disadvantages: See planar ionization chamber.

12.4.2 *Proportional counter*

Application: X-ray spectroscopy, neutron detection (BF_3-counter).

Construction: Similar to the cylindrical ionization chamber, but higher field strength and lower anode wire diameter. Gas amplification 10^3 to 10^6 in the $1/r$-field. The avalanche formation occurs at the anode wire at the position of particle passage and *does not* spread out laterally.

Measurement principle, readout: The gas amplification (without consideration of photons) is described by the first Townsend coefficient α

$$A = e^{\alpha x} \, ,$$

where the path length x is measured from the starting point of the avalanche. If photons have an influence on the avalanche formation the gas amplification can be expressed by

$$A_\gamma = \frac{A}{1 - \gamma A} \, ,$$

where γ is the probability per electron that one photoelectron is produced in the avalanche (the second Townsend coefficient). The proportional counter is operated in the proportional region only if

$$\gamma A \ll 1 \, .$$

The avalanches develop on the last mean free paths in the vicinity of the anode wire (10 - 20 μm). The signal originates mainly from the *ions* moving in the direction of the cathode.

Advantages: Low primary ionization is sufficient for particle detection.

Disadvantages: The energy resolution is limited by fluctuations in the primary ionization and charge multiplication. In general, large time constants are involved (differentiation of the signal is usually required).

12.4.3　Geiger counters

Application: Counting passages of particles.

Construction: Like the proportional counter but operated at high field strengths; gas amplifications $\approx 10^8$.

Measurement principle, readout: Because of the high field strength and gas amplification, photons produced during the avalanche formation play a decisive rôle. The discharge propagates laterally along the anode wire because of the effect of photons. An ion channel along the total length of the anode wire is formed. The discharge can be stopped electronically or by the addition of quench gases. There is no longer a proportionality between the primary ionization and the output pulse height.

Advantages: Simple construction and operation.

Disadvantages: No multiparticle registering possible, except through interruption of the discharge along the anode wire with nylon fibers (limited Geiger mode); high dead time; low efficiency for photons.

12.4.4　Streamer tube (not to be confused with the Streamer *chamber*)

Application: Sampling element in electron and hadron calorimeters. Detector element in large nucleon decay experiments.

Construction: Similar to Geiger counters, usually quadratic cross section for tube. Gas amplification $\approx 10^{10}$; still the avalanche is only limited to the point of particle passage because the quench gas fraction is high ($> 50\%$); thick anode wires ($\phi = 100\,\mu$m); high anode voltage ($\approx 5\,$kV).

Measurement principle, readout: Discontinuous transition from the proportional to the streamer mode bypassing the Geiger mode.

Advantages: Extremely simple construction and operation; large signals (processing without preamplifiers possible); broad counting rate and efficiency plateau; stable working point; multiparticle detection possible. The cathodes are typically segmented into pads.

Disadvantages: High operation voltages required ($\approx 5\,\mathrm{kV}$); quenching gases in general flammable (isobutane).

12.4.5 Particle detection in liquids

Application: Mainly as sampling element in electron and hadron calorimeters, but also for β- and γ-ray spectroscopy in the MeV-range.

Construction: Large area, flat modules with LAr, LKr or LXe; also certain 'warm' liquids (e.g., TMS) allow electron drift.

Measurement principle, readout: Collection of the produced charge carriers; the small signals have to be amplified with low-noise charge sensitive amplifiers.

Advantages: Simple construction; the detector volumes can be sealed; high radiation hardness.

Disadvantages: Cryogenic equipment at liquid nitrogen temperatures for the operation of liquid noble gas chambers is required. Extremely high demands concerning the purity of the counting liquid.

12.4.6 Multiwire proportional chamber

Application: Track detector with the possibility of measuring the energy loss. Suitable for high rate experiments if small sense wire spacing is used or in the case of microstrip detectors.

Construction: Planar layers of proportional counters without partition walls.

Measurement principle, readout: Analogous to the proportional counter; with high-speed readout (FADC = Flash ADC) the spatial structure of the ionization can be resolved.

Advantages: Simple, robust construction; use of standard electronics.

Disadvantages: Electrostatic repulsion of anode wires; limited mechanical wire stability; sag for long anode wires for horizontal construction.

Variations: 1) Straw chambers (aluminized mylar straws with central anode wire); reduces risk for broken wires.
2) Segmentation of cathodes possible to obtain spatial coordinates.
3) Miniaturization as microstrip detectors possible with 'anode

wires' on plastics or ceramic substrates; electrode structures normally produced using industrial microlithographic methods. Possible problems with ion deposition on dielectrics which may distort the field.

12.4.7   Planar drift chamber

Application: Track detector with energy loss measurement.

Construction: For the improvement of the field quality compared to the multiwire proportional chamber, potential wires are introduced between the anode wires. In general, far fewer wires are used than in a multiwire proportional chamber.

Measurement principle, readout: In addition to the readout as in the multiwire proportional chambers, the drift time of the produced charge carriers is measured. This allows — even at larger wire spacings — a higher spatial resolution.

Advantages: Drastic reduction of the number of anode wires; high track resolution.

Disadvantages: Spatial dependence of the track resolution due to charge carrier diffusion and primary ionization statistics; left-right ambiguity of drift time measurement (curable by double layers or staggered anode wires).

Variations: 1) 'Electrodeless' chambers: field shaping by intentional ion deposition on insulating chamber walls.
2) Time expansion chambers: introduction of a grid to separate the drift space from the amplification space allowing adjustable drift velocities.
3) Induction drift chamber: use of anode and potential wires with very small spacing. Readout of induced signals on the potential wires to solve the left-right ambiguity; high rate capability.

12.4.8   Cylindrical wire chambers

Cylindrical proportional and drift chambers

Application: Central detectors in storage ring experiments with high track resolution; large solid angle coverage around the primary vertex.

Construction: Concentric layers of proportional chambers (or drift chambers). The drift cells are approximately trapezoidal or hexagonal. Electric and magnetic fields (for momentum measurement) are usually perpendicular to each other.

Measurement principle, readout: The same as in planar proportional or drift chambers. The coordinate along the wire can be determined by charge division, by measuring the signal propagation time on the wire, or by stereo wires. Compact multiwire drift modules with high-rate capability can be constructed.

Advantages and Disadvantages: See sections 4.6 and 4.7; $\vec{E} \times \vec{B}$-effect complicates track reconstruction.

Jet drift chambers

Application: Central detector in storage ring experiments with excellent particle identification properties via multiple measurements of the energy loss.

Construction: Azimuthal segmentation of a cylindrical volume into pie-shaped drift spaces; electric drift field and magnetic field for momentum measurement are orthogonal. Field shaping by potential strips; staggered anode wires to resolve the left-right ambiguity.

Measurement principle, readout: As in normal drift chambers; particle identification by multiple dE/dx measurement.

Advantages: See sections 4.6 and 4.7.

Disadvantages: See sections 4.6 and 4.7; $\vec{E} \times \vec{B}$-effect complicates track reconstruction.

Time projection chamber (TPC)

Application: Practically 'massless' central detector mostly used in storage ring experiments; accurate three-dimensional track reconstruction; electric drift field and magnetic field (for track bending) are parallel.

Construction, measurement principle, readout: There are neither anode nor potential wires in the sensitive volume of the detector. The produced charge carriers drift to the endcap detectors (in general multiwire proportional chambers) which supply two track coordinates; the third coordinate is derived from the drift time.

Advantages: Apart from the counting gas there is no material in the
sensitive volume (low multiple scattering, high momentum resolution; extremely low photon conversion probability). Availability of
three-dimensional coordinates, energy-loss sampling and high spatial
resolution.

Disadvantages: Positive ions drifting back into the sensitive volume will
distort the electric field (can be avoided by an additional grid ('gating')); because of the long drift times the TPC cannot be operated
in a high rate environment.

Variation: 1) The TPC can also be operated with liquid noble gases as a
detector medium and supplies digital three-dimensional 'pictures' of
bubble chamber quality (requires extremely low-noise readout since
in liquids no gas amplification occurs).
2) An optical picture of tracks can be obtained in gas-filled time
projection chambers by high field strengths in the endcap chambers.
For this purpose a chamber gas is chosen that produces a maximum
number of photons in the visible range during avalanche formation
('imaging chamber').

12.4.9   Imaging chamber

This detector is based on the same working principle as the time projection chamber. In this application the track information is transformed
into an optical image at the endcap detectors which are operated at high
gas gain with strong photon multiplication. The readout is done via image
intensifiers with a video camera.

12.4.10   Ageing effects in wire chambers

— Ageing in wire chambers is caused by the production of molecule
fragments in microplasma discharges during avalanche formation. Depositions of carbon, silicates or oxides on anode, potential and cathode wires
can be formed.

— Ageing effects can be suppressed by a suitable choice of gases and
gas admixtures (e.g., noble gases with additions containing oxygen). In
addition one must be careful to avoid substances which tend to form
polymers (e.g., carbon containing polymers, silicon compounds, halides
and sulphur containing compounds).

— Ageing effects can also be reduced by taking care in chamber set-up
and a careful selection of all components used for chamber construction
and the gas supply system.

12.4.11 Bubble chamber

Application: Precise optical tracking of charged particles; studies of rare complex events.

Construction: Liquid gas close to the boiling point; superheating of liquid by synchronization of the bubble chamber expansion with the moment of particle incidence into the chamber.

Measurement principle, readout: The bubbles formed along the particle track in the superheated liquid are photographed stereoscopically.

Advantages: High spatial resolution; measurement of rare and complex events; lifetime determination of short-lived particles possible.

Disadvantages: Extremely tedious analysis of photographically recorded events; cannot be triggered but only synchronized; insufficient mass for the absorption of high energy particles.

Variation: Holographic readout allows three-dimensional event reconstruction with excellent spatial resolution (several μm).

12.4.12 Cloud chamber

Application: Measurement of rare events in cosmic rays; demonstration experiment, historical importance.

Construction: Gas-vapor mixture close to the saturation vapor pressure. Additional detectors (e.g., scintillation counters) can trigger the expansion to reach the supersaturated state of the vapor.

Measurement principle, readout: The droplets formed along the ionization track in the supersaturated vapor are photographed stereoscopically.

Advantage: The cloud chamber can be triggered.

Disadvantage: Very long dead and cycle times; tiresome evaluation of cloud chamber photographs.

Variation: In non-triggerable diffusion cloud chambers a permanent zone of supersaturation can be produced.

12.4.13   Streamer chamber

Application: Investigation of complex events with bubble chamber quality in a detector which can be triggered.

Construction: Large-volume detector in a homogeneous strong electric field. A high voltage signal of very short duration induces streamer discharges along the ionization track of charged particles.

Measurement principle, readout: The luminous streamers are photographed stereoscopically.

Advantages: High-quality photographs of complex events. Diffusion suppression by addition of oxygen; targets can be mounted inside the sensitive volume of the detector.

Disadvantages: Demanding event analysis; the very short high voltage signals ($100\,\mathrm{kV}$ amplitude, $2\,\mathrm{ns}$ duration) may interfere with the performance of other detectors.

12.4.14   Neon flash chamber

Applications: Investigation of rare events in cosmic rays; studies of neutrino interactions; search for nucleon decay.

Construction: Neon or neon/helium filled, sealed cylindrical glass tubes or spheres ('Conversi spheres'), or polypropylene tubes with normal gas flow operation.

Measurement principle, readout: A high voltage pulse applied to the chamber causes those tubes which have been hit by charged particles to light up in full length. The discharge can be photographed or read out electronically.

Advantages: Extremely simple construction; large volumes can be instrumented at low cost.

Disadvantages: Long dead times; low spatial resolution; no three-dimensional space points but only projections.

12.4.15   Spark chamber

Applications: Somewhat older track detector for the investigation of cosmic ray events; spectacular demonstration experiment.

Construction: Planar, parallel electrodes mounted in a gas filled volume. The spark chamber is usually triggered by a coincidence of external detectors (e.g., scintillation counters).

Measurement principle, readout: The high gas amplification causes a plasma channel to develop along the particle track; spark formation occurs. The readout of chambers with continuous electrodes is done photographically. For wire spark chambers a magnetostrictive readout or readout via ferrite cores is possible.

Advantages: Simple construction.

Disadvantages: Low multitrack efficiency, can be improved by current limitation ('glass spark chamber'); tedious analysis of optically recorded events.

12.4.16   Nuclear emulsion

Application: Permanently sensitive detector; mostly used in cosmic rays or as vertex detector at accelerators with high spatial resolution.

Construction: Silver bromide or silver chloride microcrystals embedded in gelatine.

Measurement principle, readout: Detection of charged particles similar to light recording in photographic films; development and fixation of tracks. Analysis is done under the microscope or with a CCD camera with subsequent semi-automatic pattern recognition.

Advantages: 100 % efficient; permanently sensitive; simple in construction; high spatial resolution.

Disadvantages: Non-triggerable; tedious event analysis.

12.4.17   Silver halide crystals

Silver halide crystals are frequently used as an alternative or in addition to nuclear emulsions. These crystals allow larger detector volumes compared to nuclear emulsions. Within certain limits track information can be selectively stored by external triggers (irradiation with light). Uninteresting non-conserved tracks fade away.

12.4.18   X-ray films

X-ray films — similar to nuclear emulsions — are frequently used as sampling element in calorimetric detectors for cosmic ray experiments.

12.4.19   Thermoluminescence detectors

Thermoluminescence detectors are an alternative to X-ray films. They can also be used as sampling elements in calorimeters in cosmic ray experiments. In a different configuration they are also used in the field of radiation protection to measure the personal radiation dose.

12.4.20   Radiophotoluminescence detectors

Silver-activated phosphate glasses are typically used in the field of radiation protection. The absorption of radiation produces stable photoluminescence centers whose number is proportional to the absorbed dose. Reading the received dose does not erase the stored information.

12.4.21   Plastic detectors

Application: Heavy ion physics and cosmic rays; search for magnetic monopoles; radon concentration measurement.

Construction: Foils of cellulose nitrate usually in stacks.

Measurement principle, readout: The local damage of the plastic material caused by the ionizing particle is etched in sodium hydroxide. This makes the particle track visible. The readout is done as in nuclear emulsions.

Advantages: Extremely simple, robust detector; perfectly suited for satellite and balloon-borne experiments; permanently sensitive; adjustable threshold to suppress the detection of weakly ionizing particles.

Disadvantages: Non-triggerable; tedious event analysis.

12.5   Time measurement

12.5.1   Photomultiplier

Applications: Measurement of faint light signals; counting of single photons; time measurement.

Construction: Photocathode of metals with low work function; chain of dynodes with high secondary emission coefficient.

Measurement principle, readout: Detection of photons via photoelectric effect with subsequent amplification by secondary emission at dynodes.

Advantages: Large voltage signals for single photons; fast rise times with the possibility of time measurement in the subnanosecond range.

Disadvantages: Time resolution is limited by path length differences between photocathode and first dynode in large photomultipliers. Quantum efficiency only about 20 %. Operation even in weak magnetic fields only with strong shielding possible (can be overcome with transmission-type photomultipliers).

Variation: Microchannel electron multiplier ('channel plate'); less sensitive to magnetic fields because of compact construction; electron multiplication on a continuous dynode; used as image intensifier, e.g., in special cameras (night vision cameras).

12.5.2 Scintillators

Applications: γ-spectroscopy and electromagnetic calorimetry with inorganic scintillators; triggering and time measurement with plastic scintillators; sampling element in calorimeters.

Construction: One must distinguish between three different types of scintillators:
1) inorganic scintillators such as doped crystals (NaI(Tl), ...);
2) organic scintillators such as three-component mixtures consisting of a scintillator, wavelength shifter and base material (liquid or polymerized);
3) gas scintillation counters.

Measurement principle, readout: The energy of charged particles or photons is converted into photons in the visible range; in inorganic scintillators this is an effect of the crystal lattice; in organic and gas scintillators it is simply fluorescence. The readout of the light is done with photomultipliers or channel plates.

Advantages: The geometry of a scintillator can easily be adapted to the physics application. High time resolution for plastic scintillators (subnanosecond range); self-triggering for gas scintillation drift chambers (electroluminescence chambers).

Disadvantage: Long decay times for inorganic scintillators; sensitivity to radiation for high absorbed doses.

Variation: Scintillating fibers as sampling elements in calorimeters which can simultaneously be used for tracking.

12.5.3 Planar spark counters

Applications: Trigger counters; precise timing.

Construction: Pair of planar electrodes at narrow distance under high
constant voltage.

Measurement principle, readout: The particle passage produces a
plasma channel between the electrodes via ionization and gas amplification. The resulting current is measured as voltage signal across
a resistor.

Advantages: Extremely high time resolution (30 ps); high multitrack
efficiency for electrodes made of semiconductor material; tracking
information for segmented electrodes.

Variation: Resistive plate chambers (electrodes of resistive material);
parallel plate avalanche chambers (lower gas gain than in spark counters).

12.6   Particle identification

The aim of particle identification detectors is to determine the mass m_0
and charge z of particles. Usually this is achieved by combining information from different detectors. The main inputs to this kind of measurement are

   a) the momentum p determined in magnetic fields $p = \gamma m_0 \beta c$;
      (β = velocity; γ = Lorentz factor of the particle)

   b) particle's time of flight τ $\tau = s/(\beta \cdot c)$;
      (s — flight path)

   c) mean energy loss per unit length $-\dfrac{\mathrm{d}E}{\mathrm{d}x} \propto \dfrac{z^2}{\beta^2} \ln \gamma$;

   d) kinetic energy in calorimeters $E_{\mathrm{kin}} = (\gamma - 1)m_0 c^2$;

   e) Cherenkov light yield $\propto z^2 \sin^2 \theta_{\mathrm{c}}$;
      ($\theta_{\mathrm{c}} = \arccos 1/n\beta$; n — index of refraction)

   f) yield of transition radiation photons ($\propto \gamma$).

The measurement and identification of neutral particles (neutrons, photons, neutrinos, etc.) is done via conversion into the charged particles on
suitable targets or inside the detector volume.

12.6.1 Neutron detection

Applications: Detection of neutrons in various energy ranges for radiation protection, at nuclear reactors or in elementary particle physics.

Construction: Borontrifluoride counters; coated cellulose nitrate foils, or LiI(Eu)-doped scintillators.

Measurement principle: Neutrons — as electrically neutral particles — are induced to produce charged particles in interactions, which are then registered with standard detection techniques.

Disadvantages: Neutron detectors typically have a low detection efficiency.

12.6.2 Neutrino detectors

Similar to neutron counters neutrinos first have to produce charged particles in interactions, which then are detected with standard methods. Extremely low efficiency because of very small neutrino cross sections.

12.6.3 Time of flight measurement

Applications: Identification of particles of different mass with known momenta.

Construction, measurement principle, readout: Two scintillation or planar spark counters for start-stop measurements; readout with time-to-amplitude converters.

Advantages: Simple construction.

Disadvantages: Only usable for 'low' velocities ($\beta < 0.99$; $\gamma < 10$).

12.6.4 Cherenkov counters

Applications: Mass determination (threshold Cherenkov counters) in momentum selected beams; velocity determination (differential Cherenkov counter).

Construction: Solid, liquid or gaseous transparent radiators; phase mixtures (aerogels) to cover indices of refraction not available in usual materials.

Measurement principle, readout: Cherenkov light emission for particles with $v > c/n$ (n — index of refraction) due to asymmetric polarization of the radiator material. Readout with photomultipliers or photosensitive multiwire proportional chambers.

Advantage: Simple method of mass determination; variable and adjustable threshold for gas Cherenkov counters via gas pressure; Cherenkov light emission can also be used for calorimetric detectors; also imaging systems possible (ring-imaging Cherenkov counter (RICH)).

Disadvantages: Low photon yield (compared to scintillation); Cherenkov counters only measure the velocity β (apart from z); this limits the application to not too high energies.

12.6.5 Transition radiation detectors

Application: Measurement of the Lorentz factor γ for particle identification.

Construction: Arrangement of foils or porous dielectrics with the number of transition layers as large as possible (discontinuity in the dielectric constant).

Measurement principle, readout: Emission of electromagnetic radiation at boundaries of materials with different dielectric constants. Readout by multiwire proportional chambers filled with xenon or krypton for effective photon detection.

Advantages: The number, or more precisely, the total energy radiated as transition radiation photons is proportional to the *energy* of the charged particle. The emitted photons are in the X-ray range and therefore are easy to detect.

Disadvantages: Separation of the energy loss from transition radiation and from ionization is difficult.

12.6.6 Multiple dE/dx-measurement

Application: Particle identification.

Construction: Multilayer detector for individual dE/dx measurements.

Measurement principle, readout: The Landau distributions of the energy loss are interpreted as probability distributions. For a fixed momentum different particles are characterized by different energy loss distributions. The reconstruction of these distributions with as large a number of measurements as possible allows for particle identification. In a simplified method the truncated mean of the energy loss distribution can be used for particle identification.

Advantages: The dE/dx measurements can be obtained as a by-product in multiwire proportional, jet, or time projection chambers. The measurement principle is simple.

Disadvantages: In certain kinematical ranges the mean energy losses for different charged particles overlap appreciably. The density effect of the energy loss leads to the same dE/dx distribution for all singly charged particles at high energies ($\beta\gamma \approx$ several hundred) even in gases.

12.7 Energy measurement

12.7.1 Solid state counters

Applications: α, β and γ-spectroscopy; energy-loss measurement and particle identification; high resolution track detector (silicon strip counter).

Construction: Silicon or germanium crystals doped with electron donor or electron acceptor impurities (lithium, with only *one* electron in the outer shell, is a donor).

Measurement principle, readout: Solid state detectors work like solid state ionization chambers; charged particles (or neutrals via interactions) produce electron-hole pairs, which are collected in the electric field. The readout requires low-noise, charge-sensitive amplifiers.

Advantages: High density; low energy required for the production of an electron-hole pair ($3.6\,\text{eV}$ for Si and $2.8\,\text{eV}$ for Ge) leading to an excellent energy resolution; precise track measurement with silicon strip detectors.

Disadvantages: Cryogenic equipment (for germanium counters) at liquid nitrogen temperatures required; radiation sensitive.

Variations: 1) Compared to electron-hole pair production, even lower W-values can be achieved if the break-up of Cooper pairs is used for the measurement of the energy loss leading to improved energy resolution; however, cryogenics at extremely low temperatures required.

2) Purely cryogenic calorimeters for the measurement of extremely low energy absorptions via temperature rise of the absorber; the

readout of such systems is difficult because of the low signals involved.

3) Pixel detectors: readout of the charge depositions in silicon detectors segmented by two-dimensional electrode structures.

12.7.2   Electron calorimeters

Application: Measurement of electron and photon energies in the range above several hundred MeV.

Construction: Total absorption detectors in which the energy of electrons and photons is deposited via bremsstrahlung and pair production. In sampling calorimeters the energy deposition is usually only sampled in constant longitudinal depths.

Measurement principle, readout: Depending on the type of sampling detector used, the deposited energy is recorded as charge signal (e.g., liquid argon chambers) or as light signal (scintillators) and correspondingly processed. For the complete absorption of 10 GeV electrons or photons about 20 radiation lengths are required.

Advantages: Compact construction; the relative energy resolution *improves* with increasing energy ($\sigma/E \propto 1/\sqrt{E}$).

Disadvantages: Sampling fluctuations, Landau fluctuations as well as longitudinal and lateral leakage deteriorates or limits the energy resolution.

Variation: By using a segmented readout, calorimeters can also provide excellent spatial resolution. In particular, 'spaghetti calorimeters' should be mentioned in this respect.

12.7.3   Hadron calorimeters

Applications: Measurement of hadron energies above 1 GeV; muon identification.

Construction: Total absorption detector or sampling calorimeter; all materials with short nuclear interaction lengths can be considered as sampling absorbers (e.g., uranium, tungsten; also iron and copper).

Measurement principle, readout: Hadrons with energies > 1 GeV deposit their energy via inelastic nuclear processes in hadronic cascades. This energy is, just as in electron calorimeters, measured via the produced charge or light signals in the active detector volume.

Advantages: Improvement of the relative energy resolution with increasing energy.

Disadvantages: Substantial sampling fluctuations; large fractions of the energy remain 'invisible' due to the break-up of nuclear bonds and due to neutral long-lived particles or muons escaping from the detector volume. Therefore, the energy resolution of hadron calorimeters does not reach that of electron-photon calorimeters.

Variation: By compensation methods the signal amplitudes of electron or photon and hadron induced cascades for fixed energy can be equalized. This is obtained by partially regaining the invisible energy. This compensation is of importance for the correct energy measurement in jets with unknown particle composition.

12.7.4 Particle identification with calorimeters

Particle identification with calorimeters is based on the different longitudinal and lateral development of electromagnetic and hadronic cascades.

Muons can be distinguished from electrons, pions, kaons, and protons by their high penetration power.

12.7.5 Calibration and monitoring of calorimeters

Calorimeters have to be calibrated. This is normally done at test beams with particles of known identity and momentum. In the low energy range β and γ-rays from radioisotopes can also be used for calibration purposes. To guarantee time stability, the calibration parameters have to be permanently monitored during data taking. This requires special on-line calibration procedures.

12.7.6 Cryogenic calorimeters

Applications: Detection of low energy particles or measurement of extremely low energy losses.

Construction: Detectors that experience a detectable change of state even for extremely low energy absorptions.

Measurement principle: Break-up of Cooper pairs by energy depositions; transitions from the superconducting to the normal conducting state in superheated superconducting granules; detection of phonons in solids.

Readout: With extremely low-noise electronic circuits, e.g., SQUIDs (Superconducting Quantum Interference Devices).

Advantages: Exploitation in cosmology for the detection of 'dark matter' candidates. Also usable for non-ionizing particles.

Disadvantages: Extreme cooling required (milli-Kelvin range).

12.8   Momentum measurement

Applications: Momentum spectrometer for fixed target experiments at accelerators, for investigations in cosmic rays and at storage rings.

Construction: A magnet volume is either instrumented with track detectors or the trajectories of incoming and outgoing charged particles are measured with position sensitive detectors.

Measurement principle, readout: Detectors determine the track of charged particles in a magnetic field; the track bending together with the strength of the magnetic field allows one to calculate the momentum.

Advantages: For momenta in the GeV/c-range high momentum resolutions are obtained. The momentum determination is essential for particle identification.

Disadvantages: The momentum resolution is limited by multiple scattering in the magnet and in the detectors, as well as by the limited spatial resolution of the detectors. The momentum resolution *deteriorates* with momentum ($\sigma/p \propto p$). For high momenta the required detector length becomes increasingly large.

12.9   Electronics

The readout of particle detectors can be considered as an integral part of the detection system. There is a clear tendency to integrate even sophisticated electronics into the front-end part of a detector. The front-end electronics usually consists of preamplifiers, but discriminators can also be integrated. The information contained in analog signals is normally extracted by analog-to-digital converters (ADCs). With fast flash ADCs even the time structure of signals can be resolved with high accuracy. Logic decisions are normally made in places which are also accessible during data taking. These logic devices usually have to handle large numbers of input signals and are consequently configured in different levels. These trigger levels — which can be just coincidences in the most simple case — allow a stepwise decision on whether to accept an event or not. Modern trigger systems also make massive use of microprocessors for the handling of complex event signatures. Events which pass the trigger decision are handed over to the data acquisition system.

For good data quality, on-line monitoring and slow control are mandatory.

For simpler detection techniques the amount of electronics can be substantially reduced. The operation of visual detectors uses only very few electronic circuits and some detectors, like nuclear emulsions or plastic detectors, require no electronics at all.

12.10   Data analysis

The raw data provided by the detectors consist of a collection of analog and digital signals and preprocessed results from the on-line data acquisition. The task of the data analysis is to translate this raw information off-line into physics quantities.

The detector data are first used to determine, e.g., the energy, momentum, arrival direction and identity of particles which have been recorded. This then allows one to reconstruct complete events. These can be compared with some expectation which is obtained by combining physics events generators based on a theory with detector simulation. A comparison between recorded and simulated data can be used to fix parameters which are not given by the theory. A possible disagreement requires the modification of the model under test, or it may hint at the discovery of new physics.

In many applications the experimental technique only collects information that is used to construct a model, which can then be tested by comparing predictions of the model with further observations.

Appendix A
Table of fundamental physical constants

[From Particle Data Group; *Phys. Lett.* B **239** (1990) 1; *Phys. Rev.* D **45** (1992) 1; B.N. Taylor & E.R. Cohen, *Journ. Research Nat. Inst. Standards and Technology* **95** (1990) 497; *Handbook of Chemistry and Physics*, R.C. Weast & M.J. Astle (eds.), CRC Press, (1973).]

speed of light*	c	$299\,792\,458\,\text{m/s}$
Planck's constant	h	$6.626\,075\,5 \cdot 10^{-34}\,\text{J}\cdot\text{s}$
		$\pm 0.000\,004\,0 \cdot 10^{-34}\,\text{J}\cdot\text{s}$
Planck's constant, reduced	$\hbar = \dfrac{h}{2\pi}$	$1.054\,572\,66 \cdot 10^{-34}\,\text{J}\cdot\text{s}$
		$\pm 0.000\,000\,63 \cdot 10^{-34}\,\text{J}\cdot\text{s}$
electron charge†	e	$1.602\,177\,33 \cdot 10^{-19}\,\text{C}$
		$\pm 0.000\,000\,49 \cdot 10^{-19}\,\text{C}$
		$= 4.803\,206\,8 \cdot 10^{-10}\,\text{esu}$
		$\pm 0.000\,001\,5 \cdot 10^{-10}\,\text{esu}$
gravitational constant	G	$6.672\,59 \cdot 10^{-11}\text{m}^3\,/\left(\text{kg s}^2\right)$
		$\pm 0.000\,85 \cdot 10^{-11}\text{m}^3\,/\left(\text{kg s}^2\right)$
Avogadro number	N_A	$6.022\,136\,7 \cdot 10^{23}\ \text{per mol}$
		$\pm 0.000\,003\,6 \cdot 10^{23}\ \text{per mol}$

---

* The value of the velocity of light forms the basis for the definition of the length unit, the meter. 1 meter is now defined to be the distance travelled by light in $1/299\,792\,458$ second. The quoted value for the speed of light is therefore exact and without error.

† esu = electrostatic charge unit.

Boltzmann constant	k	$1.380\,650\,1 \cdot 10^{-23}$ J/K $\pm 0.000\,002\,3 \cdot 10^{-23}$ J/K
molar gas constant	$R(=kN_A)$	$8.314\,472$ J/(K·mol) $\pm 0.000\,014$ J/(K·mol)
molar volume, ideal gas at STP[‡]	V_{mol}	$22.413\,992 \cdot 10^{-3}$ m³/mol $\pm 0.000\,038 \cdot 10^{-3}$ m³/mol
permittivity of free space[§]	ε_0	$8.854\,187\,817 \cdot 10^{-12}$ F/m
permeability of free space	μ_0	$12.566\,370\,614 \cdot 10^{-7}$ N/A²
Stefan-Boltzmann constant	$\sigma = \dfrac{\pi^2 k^4}{60 \hbar^3 c^2}$	$5.670\,397 \cdot 10^{-8}$ W/(m² K⁴) $\pm 0.000\,039 \cdot 10^{-8}$ W/(m² K⁴)
electron mass	m_e	$0.510\,999\,06$ MeV/c^2 $\pm 0.000\,000\,15$ MeV/c^2 $= 9.109\,389\,7 \cdot 10^{-31}$ kg $\pm 0.000\,005\,4 \cdot 10^{-31}$ kg
proton mass	m_p	$938.272\,31$ MeV/c^2 $\pm 0.000\,28$ MeV/c^2 $= 1.672\,623\,1 \cdot 10^{-27}$ kg $\pm 0.000\,001\,0 \cdot 10^{-27}$ kg
unified atomic mass unit (u)	$(1\mathrm{g}/N_A)$	$931.494\,32$ MeV/c^2 $\pm 0.000\,28$ MeV/c^2 $= 1.660\,540\,2 \cdot 10^{-27}$ kg $\pm 0.000\,001\,0 \cdot 10^{-27}$ kg

---

[‡] Standard temperature and pressure.

[§] Because of the fact that the velocity of light c is without error by definition, and because μ_0 is defined to be $\mu_0 = 4\pi \cdot 10^{-7}$ NA^{-2}, ε_0 is also exact.

charge to mass ratio of the electron	e/m_e	$1.758\ 819\ 62 \cdot 10^{11}\ \dfrac{\text{C}}{\text{kg}}$ $\pm 0.000\ 000\ 53 \cdot 10^{11}\ \dfrac{\text{C}}{\text{kg}}$
fine structure constant[¶] α	$\alpha^{-1} = \left(\dfrac{e^2}{4\pi\epsilon_0 \hbar c} \right)^{-1}$	$137.035\ 992\ 22$ $\pm 0.000\ 000\ 94$
classical electron radius	$r_e = \dfrac{e^2}{4\pi\epsilon_0 m_e c^2}$	$2.817\ 940\ 92 \cdot 10^{-15}$ m $\pm 0.000\ 000\ 38 \cdot 10^{-15}$ m
electron Compton wavelength	$\dfrac{\lambda_e}{2\pi} = \dfrac{\hbar}{m_e c} = \dfrac{r_e}{\alpha}$	$3.861\ 593\ 23 \cdot 10^{-13}$ m $\pm 0.000\ 000\ 35 \cdot 10^{-13}$ m
Bohr radius	$r_0 = \dfrac{4\pi\epsilon_0 \hbar^2}{m_e e^2} = \dfrac{r_e}{\alpha^2}$	$0.529\ 177\ 249 \cdot 10^{-10}$ m $\pm 0.000\ 000\ 024 \cdot 10^{-10}$ m
Rydberg energy	$E_{\text{Ry}} = m_e c^2 \alpha^2 / 2$	$13.605\ 698\ 1$ eV $\pm 0.000\ 004\ 0$ eV
Bohr magneton	$\mu_{\text{B}} = e\hbar/2m_e$	$5.788\ 382\ 63 \cdot 10^{-11}\ \dfrac{\text{MeV}}{\text{T}}$ $\pm 0.000\ 000\ 52 \cdot 10^{-11}\ \dfrac{\text{MeV}}{\text{T}}$
gravitational acceleration, sea level[‖]	g	$9.806\ 65$ m/s^2
mass of earth	M	$5.979 \cdot 10^{24}$ kg $\pm 0.004 \cdot 10^{24}$ kg
solar mass	$M_{\odot}$	$1.98892 \cdot 10^{30}$ kg $\pm 0.00025 \cdot 10^{30}$ kg

---

[¶] At a four-momentum transfer squared $q^2 = -m_e^2$. At $q^2 = -m_W^2$ the value is approximately $1/128$, where $m_W = 80.22\,\text{GeV}/c^2$ is the mass of the W-boson.

[‖] Exact by definition. Actually g varies for different locations on earth. At the equator $g \approx 9.75$ m/s^2, at the poles $g \approx 9.85$ m/s^2.

Appendix B
Definition and conversion
of physical units

Physical quantity	Name of unit and symbol
activity A	1 Becquerel (Bq) = 1 decay per second (s^{-1})
	1 Curie (Ci) = $3.7 \cdot 10^{10}$ Bq
work, energy W	1 Joule (J) = 1 W s = 1 N m
	1 erg = 10^{-7} J
	1 eV = $1.602\ 177 \cdot 10^{-19}$ J
	1 cal = 4.186 J
density ϱ	1 kg/m^3 = 10^{-3} g/cm^3
pressure p	1 Pascal (Pa) = 1 N/m^2
	1 bar = 10^5 Pa
	1 atm = $1.013\ 25 \cdot 10^5$ Pa
	1 Torr (mmHg) = $1.333\ 224 \cdot 10^2$ Pa
	1 kp/m^2 = 9.806 65 Pa
charge	1 Coulomb (C)
	1 C = $2.997\ 924\ 58 \cdot 10^9$ electrostatic charge units (esu)

unit of absorbed dose D 1 Gray (Gy) = 1 J/kg
1 rad = 0.01 Gy

unit of equivalent dose H 1 Sievert (Sv) = 1 J/kg
$(H[\text{Sv}] = RBE \cdot D[\text{Gy}]$;
RBE = relative biological efficiency)
1 rem = 0.01 Sv

unit of ion dose I 1 I = 1 C/kg
1 Röntgen (r) = $2.58 \cdot 10^{-4}$ C/kg
$= 8.77 \cdot 10^{-3}$ Gy
(for absorption in air)

entropy S 1 J/K

electric field strength E 1 V/m

magnetic field strength H 1 A/m
1 Oersted (Oe) = 79.59 A/m

magnetic induction B 1 Tesla (T) = 1 V s/m^2 = 1 Wb/m^2
1 Gauss (G) = 10^{-4} T

magnetic flux ϕ 1 Weber (Wb) = 1 V s

inductance L 1 Henry (H) = 1 V s/A = 1 Wb/A

capacitance C 1 Farad (F) = 1 C/V

force F 1 Newton (N) = 10^5 dyn

length l 1 m = 10^{10} Ångström (Å)
1 fermi (fm) = 10^{-15} m
(= 1 femtometer)
1 astronomical unit (AU)
$= 1.496 \cdot 10^{11}$ m
1 parsec (pc) = $3.085\,68 \cdot 10^{16}$ m
$= 3.26$ light years
$= 1\text{AU}/1\text{arcsec}$
1 light year (ly) = 0.3066 pc

power P 1 Watt (W) = 1 N m/s = 1 J/s

mass m 1 kg $= 10^3$ g

electric potential U 1 Volt (V)

electric current I 1 Ampère (A)

temperature T 1 Kelvin (K)
Celsius $(°C) = T[K] - 273.16$

electric resistance Ω 1 Ohm $(\Omega) = 1$ V/A

specific resistivity ϱ 1 $\Omega \cdot$ cm

time t 1 s

cross section σ 1 barn $= 10^{-24}$ cm^2

References

[1] K. Kleinknecht, *Detektoren für Teilchenstrahlung*, Teubner (1984, 1987, and 1992), and *Detectors for Particle Radiation*, Cambridge University Press (1986)

[2] O.C. Allkofer, *Teilchendetektoren*, Thiemig, München (1971)

[3] R. Fernow, *Introduction to Experimental Particle Physics*, Cambridge University Press (1986/89)

[4] P. Rice-Evans, *Spark, Streamer, Proportional and Drift Chambers*, Richelieu Press, London (1974)

[5] B. Sitar, G.I. Merson, V.A. Chechin & Yu.A. Budagov, *Ionization Measurements in High Energy Physics*, (in Russian), Energoatomizdat, Moskau (1988)

[6] B. Sitar, G.I. Merson, V.A. Chechin & Yu.A. Budagov, *Ionization Measurements in High Energy Physics*, Springer Tracts in Modern Physics Vol. 124 (1993)

[7] W.R. Leo, *Techniques for Nuclear and Particle Physics Experiments*, Springer, Berlin (1987)

[8] R.S. Gilmore, *Single Particle Detection and Measurement*, Taylor and Francis, London (1992)

[9] T. Ferbel (ed.), *Experimental Techniques in High Energy Nuclear and Particle Physics*, World Scientific, Singapore (1991)

[10] F. Sauli (ed.), *Instrumentation in High Energy Physics*, World Scientific, Singapore (1992)

[11] C.F.G. Delaney & E.C. Finch, *Radiation Detectors*, Oxford Science Publications, Clarendon Press, Oxford (1992)

[12] R.C. Fernow, *Fundamental Principles of Particle Detectors*, Summer School on Hadron Spectroscopy, University of Maryland, 1988; BNL-Preprint, BNL-42114 (1988)

[13] G.F. Knoll, *Radiation Detection and Measurement*, John Wiley & Sons Inc. (Wiley Interscience), New York (1979)

[14] D.M. Ritson, *Techniques of High Energy Physics*, Interscience Publishers Inc., New York (1961)

[15] K. Siegbahn (ed.), *Alpha, Beta and Gamma-Ray Spectroscopy*, Vol. 1 and Vol. 2, North Holland Publ. Comp., Amsterdam (1968)

[16] J.C. Anjos, D. Hartill, F. Sauli & M. Sheaff (eds.), *Instrumentation in Elementary Particle Physics*, World Scientific, Singapore (1992)

[17] G. Charpak & F. Sauli, High-Resolution Electronic Particle Detectors, *Ann. Rev. Nucl. Phys. Sci.* **34** (1984) 285

[18] W.J. Price, *Nuclear Radiation Detectors*, 2nd edition, McGraw-Hill, New York (1964)

[19] S.A. Korff, *Electron and Nuclear Counters*, 2nd edition, Van Nostrand, Princeton, New Jersey (1955)

[20] H. Neuert, *Kernphysikalische Meßverfahren zum Nachweis für Teilchen und Quanten*, G. Braun, Karlsruhe (1966)

[21] W. Stolz, *Messung ionisierender Strahlung: Grundlagen und Methoden*, Akademie-Verlag, Berlin (1985)

[22] E. Fenyves & O. Haimann, *The Physical Principles of Nuclear Radiation Measurements*, Akadémiai Kiadó, Budapest (1969)

[23] P.J. Ouseph, *Introduction to Nuclear Radiation Detectors*, Plenum Press, New York/London (1975)

[24] W.H. Tait, *Radiation Detection*, Butterworths, London (1980)

[25] C.W. Fabjan, *Detectors for Elementary Particle Physics*, CERN-PPE/94-61 (1994)

[26] G. Charpak, Electronic Imaging of Ionizing Radiation with Limited Avalanches in Gases, Nobel-Lecture 1992, CERN-PPE/93-25 (1993); and *Rev. Mod. Phys.* **65** (1993) 591

[27] B. Rossi, *High Energy Particles*, Prentice-Hall (1952)

[28] H.A. Bethe, Theorie des Durchgangs schneller Korpuskularstrahlen durch Materie, *Ann. d. Phys.* **5** (1930) 325

[29] H.A. Bethe, Bremsformel für Elektronen mit relativistischen Geschwindigkeiten, *Z. Phys.* **76** (1932) 293

[30] F. Bloch, Bremsvermögen von Atomen mit mehreren Elektronen, *Z. Phys.* **81** (1933) 363

[31] R.M. Sternheimer & R.F. Peierls, General Expression for the Density Effect for the Ionization Loss of Charged Particles, *Phys. Rev.* B **3** (1971) 3681

[32] E.A. Uehling, Penetration of Heavy Charged Particles in Matter, *Ann. Rev. Nucl. Part. Sci.* Vol. **4** (1954) 315

[33] S. Hayakawa, *Cosmic Ray Physics*, John Wiley & Sons Inc. (Wiley Interscience) (1969)

[34] Particle Data Group, Review of Particle Properties, *Phys. Lett.* **239** (1990) 1

[35] Particle Data Group, Review of Particle Properties, *Phys. Rev.* D **45** (1992) 1, and Particle Data Group, *Phys. Rev.* D **46** (1992) 5210 (Errata)

[36] C. Serre, *Evaluation de la Perte D'Energie et du Parcours de Particules Chargées Traversant un Absorbant Quelconque*, CERN 67-5 (1967)

[37] P. Marmier, *Kernphysik I*, Zürich (1977)

[38] L. Landau, On the Energy Loss of Fast Particles by Ionization, *J. Phys. USSR* **8** (1944) 201

[39] R.S. Kölbig, *Landau Distribution*, CERN Program Library G 110, CERN Program Library Section (1985)

[40] P.V. Vavilov, Ionization Losses of High Energy Heavy Particles, *Sov. Phys. JETP* **5** (1957) 749

[41] R. Werthenbach, *Elektromagnetische Wechselwirkungen von 200 GeV Myonen in einem Streamerrohr-Kalorimeter*, Diploma-Thesis, University of Siegen (1987)

[42] S. Behrends & A.C. Melissinos, *Properties of Argon-Ethane/Methane Mixtures for Use in Proportional Counters*, University of Rochester, Preprint UR-776 (1981)

[43] J.E. Moyal, *Theory of Ionization Fluctuations*, Ser. 7, Vol. 46, No. 374, March 1955

[44] R.K. Bock *et al.* (eds.), *Formulae and Methods in Experimental Data Evaluation*, General Glossary, Vol. 1 (1984) 110

[45] Y. Iga *et al.*, Energy Loss Measurements for Charged Particles and a New Approach Based on Experimental Results, *NIM* **213** (1983) 531

[46] S.I. Striganov, Ionization Straggling of High Energy Muons in Thick Absorbers, *NIM* A **322** (1992) 225

[47] C. Grupen, Electromagnetic Interactions of High Energy Cosmic Ray Muons, *Fortschr. d. Physik* **23** (1976) 127

[48] U. Fano, Penetration of Photons, Alpha Particles and Mesons, *Ann. Rev. Nucl. Sci.* **13** (1963) 1

[49] G. Musiol, J. Ranft, R. Reif & D. Seeliger, *Kern- und Elementarteilchenphysik*, VCH Verlagsgesellschaft, Weinheim (1988)

[50] W. Heitler, *The Quantum Theory of Radiation*, Oxford (1954)

[51] F. Sauli, *Principles of Operation of Multiwire Proportional and Drift Chambers*, CERN 77-09 (1977) and references therein

[52] N.I. Koschkin & M.G. Schirkewitsch, *Elementare Physik*, Hanser, München/Wien (1987)

[53] U. Fano, Ionization Yield of Radiations. II. The Fluctuation of the Number of Ions, *Phys. Rev.* **72** (1947) 26

[54] A.H. Walenta, *Review of the Physics and Technology of Charged Particle Detectors*, Preprint University of Siegen SI-83-23 (1983)

[55] H.A. Bethe, Molière's Theory of Multiple Scattering, *Phys. Rev.* **89** (1953) 1256

[56] H.A. Bethe & W. Heitler, Stopping of Fast Particles and Creation of Electron Pairs, *Proc. R. Soc. Lond.* A **146** (1934) 83

[57] E. Lohrmann, *Hochenergiephysik*, Teubner, Stuttgart (1978, 1981, 1986, 1992)

[58] U. Amaldi, Fluctuations in Calorimetric Measurements, *Phys. Scripta* **23** (1981) 409

[59] W. Lohmann, R. Kopp & R. Voss, *Energy Loss of Muons in the Energy Range 1 – 10.000 GeV*, CERN 85-03 (1985)

[60] M.J. Tannenbaum, *Simple Formulas for the Energy Loss of Ultrarelativistic Muons by Direct Pair Production*, Brookhaven National Laboratory, BNL-44554 (1990)

[61] W.K. Sakumoto *et al.*, Measurement of TeV Muon Energy Loss in Iron, University of Rochester UR-1209 (1991), and *Phys. Rev.* D **45** (1992) 3042

[62] K. Mitsui, Muon Energy Loss Distribution and its Applications to the Muon Energy Determination, *Phys. Rev.* D **45** (1992) 3051

[63] G. Hertz, *Lehrbuch der Kernphysik*, Bd. 1, Teubner, Leipzig (1966)

[64] J.S. Marshall, A.G. Ward, Absorption Curves and Ranges for Homogeneous β-Rays, *Canad. J. Res.* A **15** (1937) 39

[65] E. Sauter, *Grundlagen des Strahlenschutzes*, Thiemig, München (1982)

[66] G. Joos & E. Schopper, *Grundriß der Photographie und ihrer Anwendungen, besonders in der Kernphysik*, Frankfurt am Main (1958)

[67] A.G. Wright, A Study of Muons Underground and Their Energy Spectrum at Sea Level, Polytechnic of North London Preprint (1974); and *J. Phys.* A **7** (1974) 2085

[68] P. Marmier & E. Sheldon, *Physics of Nuclei and Particles* Vol. 1, Academic Press, New York (1969)

[69] O. Klein & Y. Nishina, Über die Streuung von Strahlung durch freie Elektronen nach der neuen relativistischen Quantenmechanik von Dirac, *Z. Phys.* **52** (1929) 853

[70] W.S.C. Williams, *Nuclear and Particle Physics*, Clarendon Press, Oxford (1991)

[71] C. Grupen & E. Hell, Lecture Notes, *Kernphysik*, University of Siegen (1983)

[72] H.A. Bethe & J. Ashkin, Passage of Radiation through Matter, in *Experimental Nucl. Phys.* (ed. E. Segré), John Wiley & Sons Inc. (Wiley Interscience), New York Vol. **1** (1953) 166ff

[73] R.D. Evans, *The Atomic Nucleus*, McGraw-Hill, New York (1955)

[74] G.W. Grodstein, *X-Ray Attenuation Coefficients from 10 keV to 100 MeV*, Circ. Natl. Bur. Stand. No. 583 (1957)

[75] G.R. White, *X-ray Attenuation Coefficients from 10 keV to 100 MeV*, Natl. Bur. Standards (U.S.) Rept. 1003 (1952)

[76] Particle Data Group, Review of Particle Properties, *Phys. Lett.* **111** B (1982)

[77] V. Palladino & B. Sadoulet, *Application of the Classical Theory of Electrons in Gases to Multiwire Proportional and Drift Chambers*, LBL-3013, UC-37, TID-4500-R62 (1974)

[78] W. Blum & L. Rolandi, *Particle Detection with Drift Chambers*, Springer Monograph XV, Berlin/New York (1993)

[79] A. Peisert & F. Sauli, *Drift and Diffusion in Gases: a Compilation*, CERN-Report 84-08 (1984)

[80] L.G. Huxley & R.W. Crompton, *The Diffusion and Drift of Electrons in Gases*, John Wiley & Sons Inc. (Wiley Interscience), New York (1974)

[81] E.W. McDaniel & E.A. Mason, *The Mobility and Diffusion of Ions in Gases*, John Wiley & Sons Inc. (Wiley Interscience), New York (1973)

[82] V. Palladino & B. Sadoulet, Application of the Classical Theory of Electrons in Gases to Multiwire Proportional and Drift Chambers, *NIM* **128** (1975) 323

[83] J. Townsend, *Electrons in Gases*, Hutchinson, London (1947)

[84] S.C. Brown, *Basic Data of Plasma Physics*, MIT-Press, Cambridge, Mass. (1959)

[85] C. Ramsauer & R. Kollath, Die Winkelverteilung bei der Streuung langsamer Elektronen an Gasmolekülen, *Ann. Phys.* **12** (1932) 35

[86] C. Ramsauer & R. Kollath, Über den Wirkungsquerschnitt der Edelgasmoleküle gegenüber Elektronen unterhalb 1 Volt, *Ann. Phys.* **3** (1929) 536

[87] C. Ramsauer, Über den Wirkungsquerschnitt der Gasmoleküle gegenüber langsamen Elektronen, *Ann. Phys.* **66** (1921) 546

[88] E. Brüche *et al.*, Über den Wirkungsquerschnitt der Edelgase Ar, Ne, He gegenüber langsamen Elektronen, *Ann. Phys.* **84** (1927) 279

[89] C.E. Normand, The Absorption Coefficient for Slow Electrons in Gases, *Phys. Rev.* **35** (1930) 1217

[90] L. Colli & U. Facchini, Drift Velocity of Electrons in Argon, *Rev. Sci. Instr.* **23** (1952) 39

[91] J.M. Kirshner & D.S. Toffolo, Drift Velocity of Electrons in Argon and Argon Mixtures, *J. Appl. Phys.* **23** (1952) 594

[92] H.W. Fulbright, Ionization Chambers in Nuclear Physics, in S. Flügge (ed.), *Handbuch der Physik*, Springer, Berlin Band **XLV** (1958) 1

[93] J. Fehlmann & G. Viertel, ETH-Zürich-Report, *Compilation of Data for Drift Chamber Operation* (1983)

[94] W.N. English & G.C. Hanna, Grid Ionization Chamber Measurement of Electron Drift Velocities in Gas Mixtures, *Canad. J. Phys.* **31** (1937) 768

[95] A. Breskin *et al.*, Recent Observations and Measurements with High-Accuracy Drift Chambers, *NIM* **124** (1975) 189

[96] A. Breskin *et al.*, Further Results on the Operation of High-Accuracy Drift Chambers, *NIM* **119** (1974) 9

[97] G. Charpak & F. Sauli, High-Accuracy Drift Chambers and their Use in Strong Magnetic Fields, *NIM* **108** (1973) 413

[98] J. Vávra *et al.*, Measurement of Electron Drift Parameters for Helium and CF_4-Based Gases, *NIM* A **324** (1993) 113

[99] T. Kunst *et al.*, Precision Measurements of Magnetic Deflection Angles and Drift Velocities in Crossed Electric and Magnetic Fields, *NIM* A **423** (1993) 127

[100] V.H. Regener, Statistical Significance of Small Samples of Cosmic Ray Counts, *Phys. Rev.* **84** (1951) 161

[101] G. Zech, Upper Limits in Experiments with Background or Measurement Errors, *NIM* A **277** (1989) 608

[102] O. Helene, Upper Limit of Peak Area, *NIM* **212** (1983) 319

[103] S. Brandt, *Datenanalyse*, Bibliographisches Institut, Mannheim/Leipzig (1992)

[104] O.C. Allkofer, W.D. Dau & C. Grupen, *Spark Chambers*, Thiemig, München (1969)

[105] W. Minder, *Dosimetrie der Strahlungen radioaktiver Stoffe*, Springer, Wien (1961)

[106] R.H. Thomas & V. Perez-Mendez, *Advances in Radiation Protection and Dosimetry in Medicine*, Plenum-Press, New York/London (1980)

[107] J.R. Greening, *Fundamentals of Radiation Dosimetry*, Adam Hilger Ltd., Bristol (1981)

[108] D.G. Miller, *Radioactivity and Radiation Detection*, Gordon and Breach Science Publ., New York/Paris/London (1972)

[109] W. Jacobi, *Strahlenschutzpraxis, Teil I - Grundlagen*, Thiemig, München (1968)

[110] M. Oberhofer, *Strahlenschutzpraxis, Teil II - Meßtechnik*, Thiemig, München (1968/1972)

[111] M. Oberhofer, *Strahlenschutzpraxis, Teil III - Umgang mit Strahlern*, Thiemig, München (1968)

[112] M. Frank & W. Stolz, *Festkörperdosimetrie ionisierender Strahlung*, Teubner, Leipzig (1969)

[113] P.F. Sharp, P.P. Dendy & W.I. Keyes, *Radionuclide Imaging Techniques*, Academic Press Inc., London (1985)

[114] G.E. Adams, D.K. Bewley & J.B. Boag, *Charged Particle Tracks in Solids and Liquids*, Proc. of the Second L.H. Gray Conference, Cambridge (1969), publ. 1970

[115] B. Rajewsky (ed.), *Wissenschaftliche Grundlagen des Strahlenschutzes*, G. Braun, Karlsruhe (1957)

[116] E. Sauter, *Grundlagen des Strahlenschutzes*, Siemens AG, Berlin/München (1971)

[117] H. Yagoda, *Radioactive Measurements with Nuclear Emulsions*, John Wiley & Sons Inc. (Wiley Interscience), New York, and Chapman and Hall Ltd. London (1949)

[118] R. Kiefer & R. Maushart, *Radiation Protection Measurement*, Pergamon Press, Oxford 1972

[119] F.H. Attix & W.C. Roesch, *Radiation Dosimetry, Vol. II, Instrumentation*, Academic Press, New York/London (1966)

[120] C.M. Lederer, *Table of Isotopes*, John Wiley & Sons Inc. (Wiley Interscience), New York (1967)

[121] H. Landolt & R. Börnstein, *Atomkerne und Elementarteilchen*, Band 5, Springer (1952)

[122] R.C. Weast & M.J. Astle (eds.), *Handbook of Chemistry and Physics*, CRC-Press (1979)

[123] O.C. Allkofer, *Introduction to Cosmic Radiation*, Thiemig, München (1975)

[124] D.M. Websdale & P.R. Hobson (eds.), Position-Sensitive Detectors, 2nd Conf. London UK Sept. 1990; *NIM* A **310** (1991) 1-575; and P.R. Hobson, A. Faruqi & G.W. Fraser (eds.), Position-Sensitive Detectors, 3rd Conf. London UK Sept. 1993; *NIM* A **348** (1994) 207-746

[125] W. Bartl, G. Neuhofer, M. Regler & A. Taurok (eds.), Proc. 6th Int. Conf. on Wire Chambers, Vienna 1992, *NIM* A **323** (1992) 1-552

[126] B. Rossi & H. Staub, *Ionization Chambers and Counters*, McGraw-Hill, New York (1949)

[127] D.M. Wilkinson, *Ionization Chambers and Counters*, Cambridge University Press (1950)

[128] D. McCormick, *Fast Ion Chambers for SLC*, SLAC-Pub-6296 (1993)

[129] M. Fishman & D. Reagan, The SLAC Long Ion Chamber for Machine Protection, *IEEE Trans. Nucl. Sci.* NS-**14** (1967) 1096

[130] S.E. Hunt, *Nuclear Physics for Engineers and Scientists*, John Wiley & Sons Inc. (Wiley Interscience), New York (1987)

[131] J.S. Townsend, *Electricity of Gases*, Clarendon Press, Oxford (1915)

[132] S.C. Brown, *Introduction to Electrical Discharges in Gases*, John Wiley & Sons Inc. (Wiley Interscience), New York (1966)

[133] H. Raether, *Electron Avalanches and Breakdown in Gases*, Butterworths, London (1964)

[134] F. Llewellyn Jones, Ionization Growth and Breakdown, in S. Flügge (ed.), *Handbuch der Physik*, Springer, Berlin, Band **XXII** (1956) 1-40

[135] L.B. Loeb, Electrical Breakdown of Gases with Steady or Direct Current Impulse Potentials, in S. Flügge (ed.), *Handbuch der Physik*, Springer, Berlin, Band **XXII** (1956) 445-528

[136] H. Raether, Die Elektronenlawine und ihre Entwicklung, *Ergeb. Exakt. Naturwiss.*, Springer, Berlin/Heidelberg Band **22** (1949) 73-120

[137] J. Berkowitz, *Photoabsorption, Photoionization and Photoelectron Spectroscopy*, Academic Press, New York (1979)

[138] G.F. Marr, *Photoionization Processes in Gases*, Academic Press, New York (1967)

[139] L.J. Kieffer & G.H. Dunn, Electron Impact Ionization Cross-Section Data for Atoms, Atomic Ions, and Diatomic Molecules: I. Experimental Data, *Rev. Mod. Phys.* **38** (1966) 1

[140] D. Rapp & P. Englander-Golden, Total Cross-Section for Ionization and Attachment in Gases by Electron Impact. I. Positive Ionization, *J. Chem. Phys.* **43** (1965) 1464

[141] G. Francis, *Ionization Phenomena in Gases*, Butterworths Science Publ., London (1960)

[142] A. Sharma & F. Sauli, A Measurement of the First Townsend Coefficient in Ar-based Mixtures at High Fields, *NIM* A **323** (1992) 280

[143] A. Sharma & F. Sauli, First Townsend Coefficient Measured in Argon Based Mixtures at High Fields, CERN-PPE/93-50 (1993), and *NIM* A **334** (1993) 420

[144] S.C. Brown, *Basic Data of Plasma Physics*, Wiley, New York (1959)

[145] A. von Engel, Ionization in Gases by Electrons in Electric Fields, in S. Flügge (ed.), *Handbuch der Physik*, Elektronen-Emission; Gasentladungen I, Bd. **XXI**, Springer, Berlin (1956), 530

[146] A. Arefev *et al.*, *A Measurement of the First Townsend Coefficient in CF_4, CO_2, and CF_4/CO_2-Mixtures at High, Uniform Electric Field*, RD5 Collaboration, CERN-PPE/93-082 (1993)

[147] E. Bagge & O.C. Allkofer, Das Ansprechvermögen von Parallel-Platten Funkenzählern für schwach ionisierende Teilchen, *Atomenergie* **2** (1957) 1

[148] T.Z. Kowalski, Generalized Parametrization of the Gas Gain in Proportional Counters, *NIM* A **243** (1986) 501; On the Generalized Gas Gain Formula for Proportional Counters, *NIM* A **244** (1986) 533;

Measurement and Parametrization of the Gas Gain in Proportional Counters, *NIM* A **234** (1985) 521

[149] T. Aoyama, Generalized Gas Gain Formula for Proportional Counters, *NIM* A **234** (1985) 125

[150] H.E. Rose & S.A. Korff, Investigation of Properties of Proportional Counters, *Phys. Rev.* Vol. **59** (1941) 850

[151] A. Williams & R.I. Sara, Parameters Effecting the Resolution of a Proportional Counter, *Int. J. Appl. Radiation Isotopes* **13** (1962) 229

[152] M.W. Charles, Gas Gain Measurements in Proportional Counters, *J. Phys.* E **5** (1972) 95

[153] A. Zastawny, Gas Amplification in a Proportional Counter with Carbon Dioxide, *J. Sci. Instr.* **43** (1966) 179

[154] L.G. Kristov, Measurement of the Gas Gain in Proportional Counters, *Doklady Bulg. Acad. Sci.* **10** (1947) 453

[155] L.B. Loeb, *Basis Processes of Gaseous Electronics*, University of California Press, Berkeley (1961)

[156] G.A. Schröder, Discharge in Plasma Physics, in *Summer School Univ. of New England* (ed. S.C. Haydon), The University of New England, Armidale (1964)

[157] M. Salehi, *Nuklididentifizierung durch Halbleiterspektrometer*, Diploma-Thesis, University of Siegen, (1990)

[158] G.C. Smith *et al.*, High Rate, High Resolution, Two-Dimensional Gas Proportional Detectors for X-Ray Synchrotron Radiation Experiments, *NIM* A **323** (1992) 78

[159] H. Geiger, Method of Counting α and β-Rays, *Verh. d. Deutsch. Phys. Ges.* **15** (1913) 534

[160] E. Rutherford & H. Geiger, α-Particles from Radio-active Substances, *Proc. R. Soc. Lond.* **81** (1908) 141

[161] E. Iarocci, Plastic Streamer Tubes and Their Applications in High Energy Physics, *NIM* **217** (1983) 30

[162] G. Battistoni *et al.*, Operation of Limited Streamer Tubes, *NIM* **164** (1979) 57

[163] G.D. Alekseev, Investigation of Self-Quenching Streamer Discharge in a Wire Chamber, *NIM* **177** (1980) 385

[164] R. Baumgart *et al.*, The Response of a Streamer Tube Sampling Calorimeter to Electrons, *NIM* A **239** (1985) 513; Performance Characteristics of an Electromagnetic Streamer Tube Calorimeter, *NIM* A **256** (1987) 254; Interactions of 200 GeV Muons in an Electromagnetic Streamer Tube Calorimeter, *NIM* A **258** (1987) 51; Test of an Iron/Streamer Tube Calorimeter with Electrons and Pions of Energy between 1 and 100 GeV, *NIM* A **268** (1988) 105

[165] R. Baumgart *et al.*, Properties of Streamers in Streamer Tubes, *NIM* **222** (1984) 448

[166] CERN-Courier, *Dubna: Self Quenching Streamers Revisited*, Vol. **21** No. 8 (1981) 358

[167] D. Achterberg *et al.*, The Helix Tube Chamber, DESY 78/15 (1978) and *NIM* **156** (1978) 287

[168] T. Doke (ed.), Liquid Radiation Detectors, *NIM* A **327** (1993) 1

[169] T. Doke, A Historical View on the R&D for Liquid Rare Gas Detectors, *NIM* A **327** (1993) 113

[170] T.S. Virdee, Calorimeters Using Room Temperature and Noble Liquids, *NIM* A **323** (1992) 22

[171] J. Engler, H. Keim & B. Wild, Performance Test of a TMS Calorimeter, *NIM* A **252** (1986) 29

[172] M.G. Albrow *et al.*, Performance of a Uranium/Tetramethylpentane Electromagnetic Calorimeter, *NIM* A **265** (1988) 303

[173] K. Ankowiak *et al.*, Construction and Performance of a Position Detector for the UA1 Uranium-TMP Calorimeter, *NIM* A **279** (1989) 83

[174] M. Pripstein, *Developments in Warm Liquid Calorimetry*, Lawrence-Berkeley Laboratory, LBL-Report, LBL-30282 (1991); and B. Aubert *et al.*, *Warm Liquid Calorimetry*, Proc. 25th Int. Conf. on High Energy Physics, Singapore, Vol. 2 (1991) 1368; and B. Aubert *et al.*, A Search for Materials Compatible with Warm Liquids, *NIM* A **316** (1992) 165

[175] G. Bressi *et al.*, Electron Multiplication in Liquid Argon on a Tip Array, *NIM* A **310** (1991) 613

[176] R.A. Muller *et al.*, Liquid Filled Proportional Counter, *Phys. Rev. Lett.* **27** (1971) 532

[177] S.E. Derenzo *et al.*, *Liquid Xenon-Filled Wire Chambers for Medical Imaging Applications*, Lawrence Berkeley Lab. LBL-2092 (1973)

[178] E. Aprile, K.L. Giboni & C. Rubbia, *A Study of Ionization Electrons Drifting Large Distances in Liquid and Solid Argon*, Harvard University Preprint, May 1985

[179] G. Charpak *et al.*, The Use of Multiwire Proportional Chambers to Select and Localize Charged Particles, *NIM* **62** (1968) 202

[180] W. Bartl, G. Neuhofer, M. Regler & A. Taurok (eds.), Proc. 6th Int. Conf. on Wire Chambers, Vienna 1992, *NIM* A **323** (1992) 1

[181] G. Charpak *et al.*, Some Developments in the Operation of Multiwire Proportional Chambers, *NIM* **80** (1970) 13

[182] G.A. Erskine, Electrostatic Problems in Multiwire Proportional Chambers, *NIM* **105** (1972) 565

[183] H. Kapitza, *Bau und Erprobung von Proportionalkammern zur Ortsmessung im Endcap-Schauerzähler des Detektors PLUTO*, Int. Report DESY F14-79/01 (1979)

[184] S. Schmidt, private communication (1992)

[185] R. Venhof, *Drift Chamber Simulation Program Garfield*, CERN/DD Garfield Manual (1984)

[186] G. Charpak, *Filet à Particules*, Découverte 1972

[187] J. Vávra, *Wire Chamber Gases*, SLAC-PUB-5793 (1992)

[188] J. Vávra, Wire Chamber Gases, *NIM* A **323** (1992) 34

[189] Y.H. Chang, Gases for Drift Chambers in SSC/LHC Environments, *NIM* A **315** (1992) 14

[190] T. Trippe, *Minimum Tension Requirement for Charpak-Chamber Wires*, CERN NP Internal Report 69-18 (1969); and P. Schilly *et al.*, Construction and Performance of Large Multiwire Proportional Chambers, *NIM* **91** (1971) 221

[191] M. Chew *et al.*, Gravitational Wire Sag in Non-Rigid Drift Chamber Structures, *NIM* A **323** (1992) 345

[192] H. Netz, *Formeln der Technik*, Hanser, München/Wien (1983)

[193] P. Rennert, H. Schmiedel & C. Weißmantel (eds.), *Kleine Enzyklopädie der Physik*, Harri Deutsch, Zürich/Frankfurt am Main (1987)

[194] R. Roark & W. Young, *Formulas for Stress and Strain*, McGraw-Hill, New York (1975)

[195] W.H. Toki, *Review of Straw Chambers*, SLAC-PUB-5232 (1990)

[196] A. Oed, Position Sensitive Detector with Microstrip Anode for Electron Multiplication with Gases, *NIM* A **263** (1988) 351

[197] F. Angelini *et al.*, The Microstrip Gas Avalanche Chamber: A New Detector for the Next Generation of High Luminosity Machines, *Particle World* **1** (1990) 84

[198] D. Mattern *et al.*, A New Approach for Constructing Sensitive Surfaces: The Gaseous Pixel Chamber, CERN-EF 90-4 (1990); and *NIM* A **300** (1991) 275; and D. Mattern, M.C.S. Williams & A. Zichichi, *New Results on the Development of the Gaseous Pixel Chamber*, CERN-PPE 91-193 (1991)

[199] D. Mattern *et al.*, First Tests of the Gaseous Pixel Chamber Fabricated on a Ceramic Substrate, *NIM* A **310** (1991) 78

[200] H. Stahl *et al.*, First Steps Towards a Microfoil Chamber, *NIM* A **297** (1990) 95

[201] F. Sauli *et al.*, Microstrip Gas Chambers on Thin Plastic Supports, *IEEE Trans. Nucl. Sci.* **39** (1992) 650

[202] F. Angelini *et al.*, A Microstrip Gas Chamber with True Two-dimensional and Pixel Readout, INFN-PI/AE 92/01 (1992); and *NIM* A **323** (1992) 229

[203] T. Nagae *et al.*, Development of Microstrip Gas Chambers with Multichip Technology, *NIM* A **323** (1992) 236

[204] R. Bouclier *et al.*, High Flux Operation of Microstrip Gas Chambers on Glass and Plastic Supports, *NIM* A **323** (1992) 240

[205] R. Bouclier *et al.*, *Development of Microstrip Gas Chambers on Substrata with Electronic Conductivity*, CERN-PPE/93-192 (1993)

[206] S. Schmidt, U. Werthenbach & G. Zech, Study of Thin Substrates for Microstrip Gas Chambers, *NIM* A **337** (1994) 382

[207] P.M. McIntyre *et al.*, Gas Microstrip Chambers, *NIM* A **315** (1992) 170

[208] F. Angelini, A Thin, Large Area Microstrip Gas Chamber with Strip and Pad Readout, *NIM* A **336** (1993) 106

[209] F. Angelini *et al.*, A Microstrip Gas Chamber on a Silicon Substrate, INFN, Pisa PI/AE 91/10 (1991); and *NIM* A **314** (1992) 450

[210] F. Angelini *et al.*, *Results from the First Use of Microstrip Gas Chambers in a High Energy Physics Experiment*, CERN-PPE/91-122 (1991)

[211] J. Schmitz, *The Micro Trench Gas Counter: A Novel Approach to High Luminosity Tracking in HEP*, NIKHEF-H/91-14 (1991)

[212] R. Bouclier *et al.*, *Microstrip Gas Chambers on Thin Plastic Supports*, CERN-PPE/91-227 (1991)

[213] R. Bouclier *et al.*, *Development of Microstrip Gas Chambers on Thin Plastic Supports*, CERN-PPE/91-108 (1991)

[214] R. Bouclier *et al.*, *High Flux Operation of Microstrip Gas Chambers on Glass and Plastic Supports*, CERN-PPE/92-53 (1992)

[215] C.W. Fabjan, *Detectors and Techniques for LHC Experimentation*, CERN-PPE/93-124 (1993)

[216] T. Kondo, *Recent Developments in Detector Technology*, XXVI. Int. Conf. on High Energy Physics, Dallas, USA (1992), and KEK-Preprint 92-163 (1992)

[217] F. Angelini *et al.*, The Micro-Gap Chamber, *NIM* A **335** (1993) 69

[218] W. Bartl & M. Regler, Wire Chambers for Exploring the Elementary Constituents of Matter, *Europhys. News* **23** (1992) 184

[219] L. Cifarelli, R. Wigmans & T. Ypsilantis (eds.), *Perspectives for New Detectors in Future Supercolliders*, World Scientific, Singapore (1991)

[220] E. Iarocci, *Recent Developments in Detectors*, Proc. Int. Europhysics Conference on High Energy Physics; J. Carr & M. Perrottet (eds.), Marseille (1993) p. 725

[221] G. Charpak & F. Sauli, *An Interesting Fall-Out of High Energy Physics Techniques: The Imaging of X-Rays at Various Energies for Biomedical Applications*, Conf. on Computer Assisted Scanning, Padova, Italy (1976), and Topical Meeting on Intermediate Energy Physics, Zuoz, Switzerland (1976)

[222] A.H. Walenta *et al.*, The Multiwire Drift Chamber: A New Type of Proportional Wire Chambers, *NIM* **92** (1971) 373

[223] A. Filatova *et al.*, Study of a Drift Chamber System for a *K-e* Scattering Experiment at the Fermi National Accelerator Lab., *NIM* **143** (1977) 17

[224] G. Marel *et al.*, Large Planar Drift Chambers, *NIM* **141** (1977) 43

[225] U. Becker *et al.*, A Comparison of Drift Chambers, *NIM* **128** (1975) 593

[226] M. Rahman *et al.*, A Multitrack Drift Chamber with 60 cm Drift Space, *NIM* **188** (1981) 159

[227] K. Mathis, *Test einer großflächigen Driftkammer*, Thesis, University of Siegen (1979)

[228] J. Allison, C.K. Bowdery & P.G. Rowe, *An Electrodeless Drift Chamber*, Int. Report, Univ. Manchester MC 81/33 (1981)

[229] J. Allison *et al.*, An Electrodeless Drift Chamber, *NIM* **201** (1982) 341

[230] Yu.A. Budagov *et al.*, How to Use Electrodeless Drift Chambers in Experiments at Accelerators, *NIM* A **255** (1987) 493

[231] A. Franz & C. Grupen, Characteristics of a Circular Electrodeless Drift Chamber, *NIM* **200** (1982) 331

[232] Ch. Becker *et al.*, Wireless Drift Tubes, Electrodeless Drift Chambers and Applications, *NIM* **200** (1982) 335

[233] G. Zech, Electrodeless Drift Chambers, *NIM* **217** (1983) 209

[234] R. Dörr, C. Grupen & A. Noll, Characteristics of a Multiwire Circular Electrodeless Drift Chamber, *NIM* A **238** (1985) 238

[235] A.H. Walenta & J. Paradiso, *The Time Expansion Chamber as High Precision Drift Chamber*, Proc. Int. Conf. on Instrumentation for Colliding Beam Physics; Stanford; SLAC-Report SLAC-250 UC-34d (1982) 34

[236] H. Anderhub *et al.*, Operating Experience with the Mark J Time Expansion Chamber, *NIM* A **265** (1988) 50

[237] E. Roderburg *et al.*, The Induction Drift Chamber, *NIM* A **252** (1986) 285

[238] A.H. Walenta *et al.*, Study of the Induction Drift Chamber as a High Rate Vertex Detector for the ZEUS Experiment, *NIM* A **265** (1988) 69

[239] E. Roderburg *et al.*, Measurement of the Spatial Resolution and Rate Capability of an Induction Drift Chamber, *NIM* A **323** (1992) 140

[240] D.C. Imrie, *Multiwire Proportional and Drift Chambers: From First Principles to Future Prospects*, Lecture delivered at the School for Young High Energy Physicists, Rutherford Lab., Sept. 1979

[241] V.D. Peshekhonov, Wire Chambers for Muon Detectors on Supercolliders, *NIM* A **323** (1992) 12

[242] H. Faissner *et al.*, Modular Wall-less Drift Chambers for Muon Detection at LHC, *NIM* A **330** (1993) 76

[243] W.R. Kuhlmann *et al.*, Ortsempfindliche Zählrohre, *NIM* **40** (1966) 118

[244] H. Foeth, R. Hammarström & C. Rubbia, On the Localization of the Position of the Particle Along the Wire of a Multiwire Proportional Chamber, *NIM* **109** (1973) 521

[245] A. Breskin *et al.*, Two-Dimensional Drift Chambers, *NIM* **119** (1974) 1

[246] E.J. De Graaf *et al.*, Construction and Application of a Delay Line for Position Readout of Wire Chambers, *NIM* **166** (1979) 139

[247] L.G. Atencio *et al.*, Delay-Line Readout Drift Chamber, *NIM* **187** (1981) 381

[248] J.A. Jaros, *Drift and Proportional Tracking Chambers*, SLAC-PUB 2647 (1980)

[249] W. de Boer *et al.*, Behaviour of Large Cylindrical Drift Chambers in a Superconducting Solenoid, Proc. Wire Chamber Conf., Vienna (1980), and *NIM* **176** (1980) 167

[250] PLUTO Collaboration, L. Criegee & G. Knies, e^+e^--Physics with the PLUTO Detector, *Phys. Rep.* **83** (1982) 153

[251] C. Biino *et al.*, A Very Light Proportional Chamber Constructed with Aluminized Mylar Tubes for Drift Time and Charge Division Readouts, *IEEE Trans. Nucl. Sci.* **36** (1989) 98

[252] G.D. Alekseev *et al.*, Operating Properties of Straw Tubes, *JINR-Rapid Communications*, No. 2 [41] (1990) 27

[253] V.N. Bychkov *et al.*, A High Precision Straw Tube Chamber with Cathode Readout, *NIM* A **325** (1993) 158

[254] F. Villa (ed.), *Vertex Detectors*, Plenum Press, New York (1988)

[255] D.H. Saxon, Multicell Drift Chambers, *NIM* A **265** (1988) 20

[256] E. Roderburg & S. Walsh, Mechanism of Wire Breaking Due to Sparks in Proportional or Drift Chambers, *NIM* A **333** (1993) 316

[257] J.A. Kadyk, J. Vávra & J. Wise, Use of Straw Tubes in High Radiation Environments, *NIM* A **300** (1991) 511

[258] U.J. Becker *et al.*, Fast Gaseous Detectors in High Magnetic Fields, *NIM* A **335** (1993) 439

[259] R. Bouclier *et al.*, Fast Tracking Detector Using Multidrift Tubes, *NIM* A **265** (1988) 78

[260] Yu.P. Guz *et al.*, Multi-Drift Module Simulation, *NIM* A **323** (1992) 315

[261] W. Bartel *et al.*, Total Cross-Section for Hadron Production by e^+e^- Annihilation at PETRA Energies, *Phys. Lett.* **88** B (1979) 171

[262] H. Drumm *et al.*, Experience with the JET-Chamber of the JADE Detector at PETRA, *NIM* **176** (1980) 333

[263] A. Wagner, Central Detectors, *Phys. Scripta* **23** (1981) 446

[264] O. Biebel *et al.*, Performance of the OPAL Jet Chamber, CERN-PPE/92-55 (1992), and *NIM* A **323** (1992) 169

[265] F. Sauli, *Experimental Techniques*, CERN-EP/86-143 (1986)

[266] J. Bartelt, *The New Central Drift Chamber for the Mark II Detector at SLC*, Contribution to the 23rd Proc. Int. Conf. on High Energy Physics, Berkeley, Vol. 2 (1986) 1467

[267] S.L. Wu, e^+e^--Physics at PETRA - The First Five Years, *Phys. Rep.* **107** (1984) 59

[268] D.R. Nygren, Future Prospects of the Time Projection Chamber Idea, Phys. Scripta **23** (1981) 584

[269] T. Lohse & W. Witzeling in *Instrumentation in High Energy Physics*, ed. F. Sauli, World Scientific, Singapore (1992)

[270] ALEPH Collaboration, D. Decamp *et al.*, ALEPH: A Detector for Electron-Positron Annihilations at LEP, *NIM* A **294** (1990) 121

[271] W.B. Atwood *et al.*, Performance of the ALEPH Time Projection Chamber, *NIM* A **306** (1991) 446

[272] Y. Sacquin, The DELPHI Time Projection Chamber, *NIM* A **323** (1992) 209

[273] C. Rubbia, *The Liquid Argon Time Projection Chamber: A New Concept for Neutrino Detectors*, CERN-EP 77-08 (1977)

[274] P. Benetti *et al.*, *The ICARUS Liquid Argon Time Projection Chamber: A New Detector for ν_τ-Search*, CERN-PPE/92-004 (1992)

[275] A. Bettini *et al.*, The ICARUS Liquid Argon TPC: A Complete Imaging Device for Particle Physics, *NIM* A **315** (1992) 223

[276] F. Pietropaolo *et al.*, *The ICARUS Liquid Argon Time Projection Chamber: A Full Imaging Device for Low Energy e^+e^- Colliders?*, Frascati INFN-LNF 91-036 (R) (1991)

[277] G. Buehler, *The Liquid Argon Time Projection Chamber*, Proc. Opportunities for Neutrino Physics at BNL, Brookhaven (1987) 161

[278] J. Seguinot *et al.*, Liquid Xenon Ionization and Scintillation: Studies for a Totally Active-Vector Electromagnetic Calorimeter, *NIM* A **323** (1992) 583

[279] P. Benetti *et al.*, A Three Ton Liquid Argon Time Projection Chamber, INFN-Report DFPD 93/EP/05, University of Padua 1993, and *NIM* A **332** (1993) 395

[280] C. Rubbia, *The Renaissance of Experimental Neutrino Physics*, CERN-PPE/93-08 (1993)

[281] G. Carugno *et al.*, A Self Triggered Liquid Xenon Time Projection Chamber, *NIM* A **311** (1992) 628

[282] E. Aprile *et al.*, Test of a Two-Dimensional Liquid Xenon Time Projection Chamber, *NIM* A **316** (1992) 29

[283] G. Charpak, Will Gaseous Detectors Survive the Rapid Progress in the Competing Techniques?, *NIM* A **252** (1986) 131

[284] J. Vávra, Review of Wire Chamber Ageing, *NIM* A **252** (1986) 547 and references therein

[285] E. Roderburg & S. Walsh, *Mechanism of Wire Breaking due to Sparks in Proportional or Drift Chambers*, KfA Forschungszentrum Jülich, Preprint Sept. 1993

[286] J.A. Kadyk, Wire Chamber Ageing, *NIM* A **300** (1991) 436 and references therein

[287] R. Kotthaus, A Laboratory Study of Radiation Damage to Drift Chambers, *NIM* A **252** (1986) 531

[288] J. Wise, *Chemistry of Radiation Damage to Wire Chambers*, Lawrence Berkeley Lab., LBL-32500 (92/08) (1993)

[289] A. Algeri *et al.*, Anode Wire Ageing in Proportional Chambers: The Problem of Analog Response, CERN-PPE/93-76 (1993), and *NIM* A **338** (1994) 348

[290] J. Wise, *Chemistry of Radiation Damage to Wire Chambers*, Thesis, Lawrence Berkeley Lab. Preprint LBL-32500 (92/08) (1992)

[291] M.M. Fraga *et al.*, Fragments and Radicals in Gaseous Detectors, *NIM* A **323** (1992) 284

[292] M. Capéans *et al.*, Ageing Properties of Straw Proportional Tubes with a Xe-CO_2-CF_4 Gas Mixture, CERN-PPE/93-136 (1993), and *NIM* A **337** (1994) 122

[293] L. Malter, Thin Film Field Emission, *Phys. Rev.* **50** (1936) 48

[294] F. Ansorge, *Untersuchungen an einer mit CO_2 gefüllten Driftkammer bei Unterdruck*, Diploma-Thesis, University of Siegen (1993)

[295] D.A. Glaser, Some Effects of Ionizing Radiation on the Formation of Bubbles in Liquids, *Phys. Rev.* **87** (1952) 665

[296] D.A. Glaser, Bubble Chamber Tracks of Penetrating Cosmic Ray Particles, *Phys. Rev.* **91** (1953) 762

[297] D.A. Glaser, Progress Report on the Development of Bubble Chambers, *Nuovo Cim. Suppl.* **2** (1954) 361

[298] D.A. Glaser, The Bubble Chamber, in S. Flügge (ed.), *Handbuch der Physik*, Springer, Berlin, Band **XLV** (1958) 314

[299] L. Betelli *et al.*, *Particle Physics with Bubble Chamber Photographs*, CERN/INFN-Preprint (1993)

[300] P. Galison, *Bubbles, Sparks and the Postwar Laboratory*, Proc. Batavia Conf. 1985, *Pions to Quarks* (1989) 213-251

[301] F. Close *et al.*, *The Particle Explosion*, Oxford University Press (1987)

[302] D.H. Perkins, *Introduction to High Energy Physics*, Addison-Wesley, Menlo Park, Calif. (1987)

[303] V. Barnes *et al.*, Observation of a Hyperon with Strangeness Minus Three, *Phys. Rev. Lett.* **12** (1964) 204; Brookhaven National Laboratory, Public Information Office; private communication 1994

[304] CERN-Courier, *Small Bubble Chambers at CERN*, Vol. 22, No. 1 (1982) 24

[305] CERN-Courier, *CERN: Bubble Chambers Get Smaller*, Vol. 20, No. 2 (1980) 58

[306] H. Bingham *et al.*, Holography of Particle Tracks in the Fermilab 15-Foot Bubble Chamber, E-632 Collaboration, CERN-EF/90-3 (1990), and *NIM* A **297** (1990) 364

[307] W. Kittel, *Bubble Chambers in High Energy Hadron Collisions*, Nijmegen Preprint HEN-365 (1993)

[308] C.T.R. Wilson, Method of Making Visible the Paths of Ionizing Particles, *Proc. R. Soc. Lond.* A **85** (1911) 285; and Expansion Apparatus for Making Visible the Tracks of Ionizing Particles in Gases: Results Obtained, *Proc. R. Soc. Lond.* A **87** (1912) 277

[309] C.T.R. Wilson, Uranium Rays and Condensation of Water Vapor, *Cambridge Phil. Soc. Proc.* 9.7 (1898) 333, and *Phil. Trans. R. Soc. Lond.* **189** (1897) 265

[310] C.M. York, Cloud Chambers, in S. Flügge (ed.), *Handbuch der Physik*, Springer, Berlin, Band **XLV** (1958) 260

[311] G.D. Rochester & J.G. Wilson, *Cloud Chamber Photographs of Cosmic Radiation*, Pergamon Press, London (1952)

[312] W. Wolter, private communication (1969)

[313] U. Wiemken, *Untersuchungen zur Existenz von Quarks in der Nähe der Kerne Großer Luftschauer mit Hilfe einer Nebelkammer*, Ph.D. Thesis, University of Kiel; also U. Wiemken, Diploma-Thesis, University of Kiel (1972); and K. Sauerland, private communication (1993)

[314] A. Langsdorf, Continuously Sensitive Cloud Chamber, *Phys. Rev.* **49** (1936) 422

[315] V.K. Ljapidevski, Die Diffusionsnebelkammer, *Fortschr. der Physik* **7** (1959) 481

[316] E.W. Cowan, Continuously Sensitive Diffusion Cloud Chamber, *Rev. Sci. Instr.* **21** (1950) 991

[317] V. Eckardt, *Die Speicherung von Teilchenspuren in einer Streamerkammer*, Ph.D. Thesis, University of Hamburg (1971)

[318] F. Bulos *et al.*, *Streamer Chamber Development*, SLAC-Technical-Report, SLAC-74, UC-28 (1967)

[319] F. Rohrbach, *Streamer Chambers at CERN During the Past Decade and Visual Techniques of the Future*, CERN-EF/88-17 (1988)

[320] CERN-Courier, *The Collider Marches On*, Vol. 25, No. 4 (1985) 131

[321] CERN-Courier, *Letting Them Sulphur*, Vol. 27 No. 10 (1987) 13 and I (frontpage)

[322] CERN, *Experiments at the Collider*, Annual Report Vol. 1 (1985) 31

[323] CERN-Courier, *Detectors: High Resolution Streamer Chambers*, Vol. 27, No. 6 (1987) 25

[324] G. Charpak, Principes et Essais Préliminaires D'un Nouveau Détecteur Permettant De Photographier la Trajectoire des Particules Ionisantes Dans un Gas, *J. Phys. Rad.* **18** (1957) 539

[325] M. Conversi, *The Development of the Flash and Spark Chambers in the 1950's*, CERN-EP/82-167 (1982)

[326] M. Conversi & A. Gozzini, The 'Hodoscope Chamber': A New Instrument for Nuclear Research, *Nuovo Cim.* **2** (1955) 189

[327] M. Conversi *et al.*, A New Type of Hodoscope of High Spatial Resolution, *Nuovo Cim. Suppl.* **4** (1956) 234

[328] M. Conversi & L. Frederici, Flash Chambers of Plastic Material, *NIM* **151** (1978) 93

[329] C.A. Ayre & M.G. Thompson, Digitization of Neon Flash Tubes, *NIM* **69** (1969) 106

[330] C.G. Dalton & G.J. Krausse, Digital Readout for Flash Chambers, *NIM* **158** (1979) 289

[331] F. Ashton & J. King, The Electric Charge of Interacting Cosmic Ray Particles at Sea-Level, *J. Phys.* A **4**, L31 (1971)

[332] F. Ashton, private communication (1991)

[333] J. Trümper, E. Böhm & M. Samorski, private communication (1969)

[334] J.W. Keuffel, Parallel Plate Counters, *Rev. Sci. Instr.* **20** (1949) 202

[335] S. Fukui & S. Miyamoto, A New Type of Particle Detector: The Discharge Chamber, *Nuovo Cim.* **11** (1959) 113

[336] O.C. Allkofer *et al.*, Die Ortsbestimmung geladener Teilchen mit Hilfe von Funkenzählern und ihre Anwendung auf die Messung der Vielfachstreuung von Mesonen in Blei, *Phys. Verh.* **6** (1955) 166; and P.G. Henning, Die Ortsbestimmung geladener Teilchen mit Hilfe von Funkenzählern, Ph.D. Thesis, University of Hamburg (1955); and *Atomkernenergie* **2** (1957) 81

[337] F. Bella, C. Franzinetti & D.W. Lee, On Spark Counters, *Nuovo Cim.* **10** (1953) 1338; and F. Bella & C. Franzinetti, Spark Counters, *Nuovo Cim.* **10** (1953) 1461

[338] T.E. Cranshaw & J.F. De Beer, A Triggered Spark Counter, *Nuovo Cim.* **5** (1957) 1107

[339] V.S. Kaftanov & V.A. Liubimov, Spark Chamber Use in High Energy Physics, *NIM* **20** (1963) 195

[340] B. Agriniér *et al.*, Variation of Spark Brilliance with Ionization Density Along a Particle Track in a Spark Chamber, *J. Phys.* **24** (1963) 312

[341] S. Attenberger, Spark Chamber with Multi-Track Capability, *NIM* **107** (1973) 605

[342] R. Kajikawa, Direct Measurement of Shower Electrons with 'Glass-Metal' Spark Chambers, *J. Phys. Soc. Japan* **18** (1963) 1365

[343] W. Stamm *et al.*, Electromagnetic Interactions of Cosmic Ray Muons in Iron, *Nuovo Cim.* **51** A (1979) 242

[344] A. Bäcker, *Datenanalyse eines Experiments zur Elektromagnetischen Wechselwirkung von Myonen*, Diploma-Thesis, University of Kiel (1975)

[345] R.C. Uhr, *Das Ansprechvermögen und die Ortsgenauigkeit einer Glasfunkenkammer*, Diploma-Thesis, University of Kiel (1972)

[346] A.S. Gavrilov *et al.*, Spark Chambers with the Recording of Information by Means of Magnetostrictive Lines, *Instr. Exp. Techn.* **6** (1966) 1355

[347] S. Kinoshita, Photographic Action of the α-Particle, *Proc. R. Soc. Lond.* **83** (1910) 432

[348] M.M. Shapiro, Nuclear Emulsions, in S. Flügge (ed.), *Handbuch der Physik*, Springer, Berlin, Band **XLV** (1958) 342

[349] R. Reinganum, Streuung und Photographische Wirkung der α-Strahlen, *Z. f. Phys.* **12** (1911) 1076

[350] D.H. Perkins, *Cosmic Ray Work with Emulsions in the 40's and 50's*, Oxford University Preprint OUNP 36/85 (1985)

[351] C.F. Powell, P.H. Fowler & D.H. Perkins, *The Study of Elementary Particles by the Photographic Method*, Pergamon Press, London (1959)

[352] S. Aoki *et al.*, Fully Automated Emulsion Analysis System, *NIM* B **51** (1990) 466

[353] G.P.S. Occhialini & C.F. Powell, Nuclear Disintegrations Produced by Slow Charged Particles of Small Mass, *Nature* **159** (1947) 186

[354] CERN-Annual Report, *First Sign of the Quark-Gluon Plasma?*, Vol. 1 (1987) 26

[355] M. Simon, Lawrence Berkeley Lab. XBL-829-11834, private communication (1992)

[356] J. Sacton, *The Emulsion Technique and its Continued Use*, University of Brussels, Preprint IISN 0379-301X/IIHF-93.06 (1993)

[357] Th. Wendnagel, University of Frankfurt am Main, private communication (1991)

[358] C. Childs & L. Slifkin, Room Temperature Dislocation Decoration Inside Large Crystals, *Phys. Rev. Lett.* **5**, No. 11 (1960) 502; and A New Technique for Recording Heavy Primary Cosmic Radiation and Nuclear Processes in Silver Chloride Single Crystals, *IRE Trans. Nucl. Sci. Vol.* NS-**9**, No. 3 (1962) 413

[359] Th. Wendnagel *et al.*, *Properties and Technology of Monocrystalline AgCl-Detectors; 1. Aspects of Solid State Physics* and *Properties and Technology of AgCl-Detectors; 2. Experiments and Technological Performance* in S. Francois, *Proc. 10th Int. Conf. on SSNTD, Lyon 1979*, Pergamon Press, London (1980) 47 & 147

[360] A. Noll, *Methoden zur Automatischen Auswertung von Kernwechselwirkungen in Kernemulsionen und AgCl-Kristallen*, Ph.D.-Thesis, University of Siegen (1990)

[361] C.M.G. Lattes, Y, Fujimoto & S. Hasegawa, *Hadronic Interactions of High Energy Cosmic Rays Observed by Emulsion Chambers*, ICR-Report-81-80-3, University of Tokyo (1980)

[362] Mt. Fuji Collaboration (M. Akashi *et al.*), *Energy Spectra of Atmospheric Cosmic Rays Observed with Emulsion Chambers*, ICR-Report-89-81-5, University of Tokyo (1981)

[363] J. Nishimura *et al.*, Emulsion Chamber Observations of Primary Cosmic Ray Electrons in the Energy Range 30 – 1000 GeV, *Astrophys. J.* **238** (1980) 394

[364] I. Ohta *et al.*, *Characteristics of X-Ray Films Used in Emulsion Chambers and Energy Determination of Cascade Showers by Photometric Methods*, 14th Int. Cosmic Ray Conf. München, Vol. 9 (1975) 3154

[365] A.F. McKinley, *Thermoluminescence Dosimetry*, Adam Hilger Ltd., Bristol (1981)

[366] M. Oberhofer & A. Scharmann (eds.), *Applied Thermoluminescence Dosimetry*, Adam Hilger Ltd., Bristol (1981)

[367] Y.S. Horowitz, *Thermoluminescence and Thermoluminescent Dosimetry*, CRC-Press (1984)

[368] Y. Okamoto *et al.*, *Thermoluminescent Sheet to Detect the High Energy Electromagnetic Cascades*, 18th Int. Cosmic Ray Conf., Bangalore, Vol. 8 (1983) 161

[369] R.L. Fleischer, P.B. Price & R.M. Walker, *Nuclear Tracks in Solids; Principles and Application*, University of California Press, Berkeley (1975)

[370] P.H. Fowler & V.M. Clapham (eds.), *Solid State Nuclear Track Detectors*, Pergamon Press, Oxford (1982)

[371] F. Granzer, H. Paretzke & E. Schopper (eds.), *Solid State Nuclear Track Detectors*, Vol. 1 & 2, Pergamon Press, Oxford (1978)

[372] W. Enge, Introduction to Plastic Nuclear Track Detectors, *Nucl. Tracks* **4** (1980) 283

[373] W. Heinrich *et al.*, Application of Plastic Nuclear Track Detectors in Heavy Ion Physics, *Nucl. Tracks Rad. Measurements* Vol. **15**, No. 1-4 (1988) 393

[374] M. Henkel *et al.*, *The Experimental Concept for the ALICE-Instrument and the Measured Elemental Composition*, Proc. 21st. Int. Conf. Cosmic Rays, Adelaide, Vol. 3 (1990) 15

[375] C. Brechtmann & W. Heinrich, Fragmentation Cross Sections of ^{16}O at 60 and 200 GeV/Nucleon, *Z. Phys.* A **330** (1988) 407; and Fragmentation Cross Sections of ^{32}S at 0.7, 1.2, and 200 GeV/Nucleon, *Z. Phys.* A **331** (1988) 463

[376] W. Trakowski *et al.*, An Automatic Measuring System for Particle Tracks in Plastic Detectors, *NIM* **225** (1984) 92

[377] T. Xiaowei *et al.*, A Nuclear Detector with Super-High Spatial Resolution, *NIM* A **320** (1992) 396

[378] T. Hayashi, *Photomultiplier Tubes for Use in High Energy Physics*, Hamamatsu TV Co. Ltd. Application Res.-0791 (1980)

[379] Valvo Datenbuch, *Photovervielfacher, Elektronenvervielfacher, -Einzelkanäle, -Vielkanalplatten*, Philips GmbH, Hamburg (1985)

[380] K.S. Hirata *et al.*, Observation of a Neutrino Burst from the Supernova SN 1987 A, *Phys. Rev. Lett.* **58** (1987) 1490

[381] K.S. Hirata *et al.*, *Observation of ^{8}B Solar Neutrinos in the Kamiokande II Detector*, Inst. f. Cosmic Ray Research, ICR-Report 188-89-5 (1989)

[382] Hamamatsu Photonics K.K., *Measure Weak Light from Indeterminate Sources with New Hemispherical PM*, CERN-Courier Vol. 21, No. 4 (1981) 173; and private communication by Dr. H. Reiner, Hamamatsu Photonics, Germany

[383] K. Oba, *Microchannel Plate Photodetectors*, Hamamatsu TV Co. Ltd., Application Res.-0792 (1980)

[384] Philips, CERN-Courier, *Imaging: From X-Ray to IR*, Vol. 23, No. 1 (1983) 35

[385] J. Chadwick, Observations Concerning Artificial Disintegration of Elements, *Phil. Mag.* **7**, No. 2 (1926) 1056

[386] V. Henri & J. des Bancels, Influences des Diverses Conditions Physiques sur le Rayonnement Ultraviolet des Lampes à Vapeur de Mercure en Quartz, *J. Phys. Path. Gen.* **XIII** (1911) 841

[387] K.W.F. Kohlrausch, Radioaktivität, in W. Wien and F. Harms (eds.), *Handbuch der Experimentalphysik*, Akademische Verlagsanstalt Leipzig, Band **15** (1928)

[388] J.B. Birks, *The Theory and Practice of Scintillation Counting*, Pergamon Press, Oxford (1964,1967) and J.B. Birks, *Scintillation Counters*, Pergamon Press, Oxford (1953)

[389] K.D. Hildenbrand, *Scintillation Detectors*, Darmstadt GSI-Preprint GSI 93-18 (1993)

[390] E.B. Norman, *Scintillation Detectors*, LBL-Report 31371 (1991)

[391] R. Hofstadter, *Twenty-Five Years of Scintillation Counting*, IEEE Scintillation and Semiconductor Counter Symposium, Washington DC, HEPL Report No. 749, Stanford University (1974)

[392] M. Kobayashi *et al.*, A Beam Test on a Fast EM-Calorimeter of Gadolinium Silicate GSO (Ge), *NIM* A **306** (1991) 139

[393] M. Kobayashi & M. Ishii, Excellent Radiation-Resistivity of Cerium-Doped Gadolinium Silicate Scintillators, *NIM* B **61** (1991) 491

[394] G.I. Britvich *et al.*, A Study on the Characteristics of Some Materials for Electromagnetic Calorimeters, *NIM* A **308** (1991) 509

[395] P. Klasen *et al.*, Application of Wavelength-Shifter Techniques to Position Measuring Counters, *NIM* **185** (1981) 67

[396] B.A. Dolgosheim & B.U. Rodionov, The Mechanism of Noble Gas Scintillation, in *Elementary Particles and Cosmic Rays* No. 2 (Atomizdat, Moscow) Sect. 6.3 (1969)

[397] A.J.P.L. Policarpo, The Gas Proportional Scintillation Counter, *Space Sci. Instr.* **3** (1977) 77

[398] A.J.P.L. Policarpo, Light Production and Gaseous Detectors, *Phys. Scripta* **23** (1981) 539

[399] V.A. Monish, Gas Detectors with Detection of Discharge Luminescence (Review), *Prib. Tekh. Éksp.* No. 5 (1980) 7 (English transl. in: *Instr. Exp. Techn.* **23** (1980) 1061)

[400] J. Seguinot *et al.* in *Advances in Cryogenic Engineering*; ed. R.W. Fast, Vol. 37 (1991) 1137

[401] M. Simon & T. Braun, A Scintillation Drift Chamber with 14 cm Drift Path, *NIM* **204** (1983) 371

[402] B.M. Bleichert, *Teilchenidentifikation und Energiemessung mit einem modularen Elektron-Hadron-Kalorimeter*, Ph.D.-Thesis., University of Siegen (1982)

[403] Nuclear Enterprises, *Scintillation Materials*, Edinburgh (1977)

[404] F. Barreiro *et al.*, An Electromagnetic Calorimeter with Scintillator Strips and Wavelength Shifter Read Out, *NIM* A **257** (1987) 145

[405] J. Badier *et al.*, *Shashlik Calorimeter; Beam Test Results*, Paris Ecole Polytechnique Preprint IN2P3 CNRS X-LPNHE 93-04 (1993)

[406] J. Badier, *Radiation Hardness of Shashlik Calorimeters*, Ecole Politechnique, Preprint IN2P3-CNRS, X-LPNHE 93-14 (1993)

[407] D. Acosta *et al.*, *Localizing Particles Showering in a Spaghetti Calorimeter*, CERN-PPE/91-011 (1991)

[408] D. Acosta *et al.*, Lateral Shower Profiles in a Lead Scintillating-Fiber Calorimeter, *NIM* A **316** (1992) 184

[409] M. Livan, RD-Collaboration, *RD1-Scintillating Fiber Calorimetry*, CERN-PPE/93-022 (1993)

[410] D. Acosta *et al.*, Localizing Particles Showering in a Spaghetti Calorimeter, *NIM* A **305** (1991) 55

[411] A. Simon, *Scintillating Fiber Detectors in Particle Physics*, CERN-PPE-92-095 (1992)

[412] M. Adinolfi *et al.*, Application of a Scintillating Fiber Detector for the Study of Short-Lived Particles, CERN-PPE/91-66 (1991), and *NIM* A **310** (1991) 485

[413] D. Autiero *et al.*, Study of a Possible Scintillating Fiber Tracker at the LHC and Tests of Scintillating Fibers, *NIM* A **336** (1993) 521

[414] J. Bähr *et al.*, Liquid Scintillator Filled Capillary Arrays for Particle Tracking, CERN-PPE/91-46 (1991), and *NIM* A **306** (1991) 169

[415] N.I. Bozhko *et al.*, A Tracking Detector Based on Capillaries Filled with Liquid Scintillator, Serpukhov Inst., High Energy Phys. 91-045 (1991), and *NIM* A **317** (1992) 97

[416] CERN-Courier, *CERN: Tracking by Fibers*, Vol. 27, No. 5 (1987) 9

[417] CERN-Courier, *Working with High Collision Rates*, Vol. 29, No. 10 (1989) 9

[418] C. D'Ambrosio *et al.*, Reflection Losses in Polystyrene Fibers, *NIM* A **306** (1991) 549; and private communication by C. D'Ambrosio (1994)

[419] M. Salomon, New Measurements of Scintillating Fibers Coupled to Multianode Photomultipliers, *IEEE Trans. Nucl. Sci.* **39** (1992) 671

[420] CERN-Courier, *Scintillating Fibers*, Vol. 30, No. 8 (1990) 23

[421] D. Acosta *et al.*, Advances in Technology for High Energy Subnuclear Physics. Contribution of the LAA Project, *Riv. del Nuovo Cim.* **13**, No. 10-11 (1990) 1; and G. Anzivino *et al.*, The LAA Project, *Riv. del Nuovo Cim.* **13**, No. 5 (1990) 1

[422] G. Marini *et al.*, *Radiation Damage to Organic Scintillation Materials*, CERN 85-08 (1985)

[423] J. Proudfoot, *Conference Summary: Radiationtolerant Scintillators and Detectors*, Argonne Nat. Lab. -ANL-HEP-CP-92-046 (1992)

[424] G.I. Britvich *et al.*, Investigation of Radiation Resistance of Polystyrene-Based Scintillators, *Instr. Exp. Techn.* **36** (1993) 74

[425] W. Braunschweig, Spark Gaps and Secondary Emission Counters for Time of Flight Measurement, *Phys. Scripta* **23** (1981) 384

[426] M.V. Babykin *et al.*, Plane-Parallel Spark Counters for the Measurement of Small Times; *and* Resolving Time of Spark Counters, *Sov. J. Atomic Energy* **IV** (1956) 627

[427] R. Santonico & R. Cardarelli, Development of Resistive Plate Counters, *NIM* **187** (1981) 377

[428] R. Cardarelli *et al.*, Progress in Resistive Plate Counters, *NIM* A **263** (1988) 20

[429] E. Calligarich *et al.*, The Resistive Plate Counter as a Neutron Detector, *NIM* A **307** (1991) 142

[430] Yu.N. Pestov & G.V. Fedotovich, *A Picosecond Time-of-Flight Spectrometer for the VEPP-2M Based on a Local Discharge Spark Counter*, Preprint IYAF 77-78, SLAC-Translation 184 (1978)

[431] I. Crotty *et al.*, Investigation of Resistive Plate Chambers, *NIM* A **329** (1993) 133

[432] I. Crotty *et al.*, *The Non-Spark Mode and High Rate Operation of Resistive Parallel Plate Chambers*, CERN-PPE/93-180 (1993)

[433] P. Fonte, V. Peskov & F. Sauli, *VUV Emission and Breakdown in Parallel Plate Chambers*, CERN-PPE/91-17 (1991)

[434] P. Astier *et al.*, Development and Applications of the Imaging Chamber, *IEEE Trans. Nucl. Sci.* NS-**36** (1989) 300

[435] V. Peskov *et al.*, Organometallic Photocathodes for Parallel-Plate and Wire Chambers, *NIM* A **283** (1989) 786

[436] R. Bouclier *et al.*, A Very High Light-Yield Imaging Chamber, CERN-PPE/90-140 (1990), and *NIM* A **300** (1991) 286

[437] M. Izycki *et al.*, A Large Multistep Avalanche Chamber: Description and Performance, Proc. 2nd Conf. on Position Sensitive Detectors, London (1990), and *NIM* A **310** (1991) 98

[438] G. Charpak *et al.*, Investigation of Operation of a Parallel Plate Avalanche Chamber with a CsI Photocathode Under High Gain Conditions, CERN-PPE/91-47 (1991), and *NIM* A **307** (1991) 63

[439] G. Bencivenni *et al.*, A Glass Spark Counter for High Rate Environments, *NIM* A **332** (1993) 368

[440] D.F. Anderson, S. Kwan & V. Peskov, *High Counting Rate Resistive Plate Chamber*, Fermilab.-Conf. 93-215 (1993)

[441] W. Schneider, *Neutronenmeßtechnik*, Walter de Gruyter, Berlin, New York (1973)

[442] H. Neuert, *Kernphysikalische Meßverfahren*, G. Braun, Karlsruhe (1966)

[443] M. Banner *et al.*, Observation of Single Isolated Electrons of High Transverse Momentum in Events with Missing Transverse Energy at the CERN p̄p-Collider, (UA2-Collaboration), *Phys. Lett.* **122**B (1983) 476

[444] G. Arnison *et al.*, Experimental Observation of Isolated Large Transverse Energy Electrons with Associated Missing Energy at $\sqrt{s} = 540\,\text{GeV}$ (UA1-Collaboration), *Phys. Lett.* **122**B (1983) 103

[445] P.A. Cherenkov, Visible Radiation Produced by Electrons Moving in a Medium with Velocities Exceeding that of Light, *Phys. Rev.* **52** (1937) 378

[446] P.A. Cherenkov, *Radiation of Particles Moving at a Velocity Exceeding that of Light, and some of the Possibilities for Their Use in Experimental Physics*, and I.M. Frank, *Optics of Light Sources Moving in Refractive Media*, and I.E. Tamm, *General Characteristics of Radiations Emitted by Systems Moving with Super Light Velocities with some Applications to Plasma Physics*, Nobel Lectures Dec. 11, 1958, publ. in *Nobel Lectures in Physics 1942-1962*, Elsevier Publ. Comp., New York, (1964) 426

[447] M. Born & E. Wolf, *Principles of Optics*, Pergamon, New York (1964)

[448] N.W. Ashcroft & N.D. Mermin, *Solid State Physics*, Holt-Saunders, New York (1976)

[449] P. Lecomte *et al.*, Threshold Cherenkov Counters, *Phys. Scripta* **23** (1981) 377

[450] C.W. Fabjan & H.G. Fischer, *Particle Detectors* CERN-EP/80-27 (1980)

[451] J. Seguinot & T. Ypsilantis, Photo-Ionization and Cherenkov Ring Imaging, *NIM* **142** (1977) 377

[452] E. Nappi & T. Ypsilantis (eds.), Experimental Techniques of Cherenkov Light Imaging, Proc. of the First Workshop on Ring Imaging Cherenkov Detectors, Bari, Italy 1993, *NIM* A **343** (1994) 1

[453] T. Ekelöf, *The Use and Development of Ring Imaging Cherenkov Counters*, CERN-PPE/91-23 (1991)

[454] R. Stock, NA35-Kollaboration, private communication 1990

[455] F. Sauli, *Gas Detectors: Recent Developments and Applications*, CERN-EP/89-74 (1989); and Le Camere Proporzionali Multifili: Un Potente Instrumento Per la Ricera Applicata, *Il Nuovo Saggiatore* **2** (1986) 2/26

[456] CERN-Courier, *Fermilab: Striking it Rich*, Vol. 22 (1982) 149

[457] CERN-Courier, *Cherenkov Telescopes for Gamma-Rays*, Vol 28, No. 10 (1988) 18

[458] V.L. Ginzburg & V.N. Tsytovich, *Transition Radiation and Transition Scattering*, Inst. of Physics Publishing, Bristol (1990)

[459] A. Bodek *et al.*, Observation of Light Below Cherenkov Threshold in a 1.5 Meter Long Integrating Cherenkov Counter, *Z. Phys.* C **18** (1983) 289

[460] W.W.M. Allison & J.H. Cobb, Relativistic Charged Particle Identification by Energy Loss, *Ann. Rev. Nucl. Sci.* **30** (1980) 253

[461] W.W.M. Allison & P.R.S. Wright, *The Physics of Charged Particle Identification:* $\mathrm{d}E/\mathrm{d}x$*, Cherenkov and Transition Radiation*, Oxford University Preprint OUNP 83-35 (1983)

[462] B. Dolgosheim, Transition Radiation Detectors, *NIM* A **326** (1993) 434

[463] V.L. Ginzburg & I.M. Frank, Radiation of a Uniformly Moving Electron due to its Transitions from one Medium into Another, *JETP* **16** (1946) 15

[464] G.M. Garibian, *Macroscopic Theory of Transition Radiation*, Proc. 5th Int. Conf. in Instrumentation for High Energy Physics, Frascati (1973) 329

[465] X. Artru *et al.*, Practical Theory of the Multilayered Transition Radiation Detector, *Phys. Rev.* D **12** (1975) 1289

[466] J. Fischer *et al.*, Lithium Transition Radiator and Xenon Detector Systems for Particle Identification at High Energies, JINR-Report D13-9164, Dubna (1975), and *NIM* **127** (1975) 525

[467] C.W. Fabjan *et al.*, Practical Prototype of a Cluster-Counting Transition Radiation Detector, CERN-EP/80-198 (1980), and *NIM* **185** (1981) 119

[468] V. Chernyatin *et al.*, Foam Radiators for Transition Radiation Detectors, CERN-PPE/92-170 (1992), and *NIM* A **325** (1993) 411

[469] K.A. Ispirian *et al.*, X-Ray Transition Radiation Detectors (XTRD) for e/π and $\pi/K/p$ Identification in the TeV Region, *NIM* A **336** (1993) 533

[470] J.N. Marx & D.R. Nygren, The Time Projection Chamber, *Physics Today*, October (1978) 46

[471] T. Miyachi *et al.*, A Thick and Large Active Area Si(Li)-Detector, *Jap. J. Appl. Phys.* **27** (1988) 307

[472] H. Aihara *et al.* (TPC/Two-Gamma Collaboration), *Charged Hadron Production in* e^+e^-*-Annihilation at* $\sqrt{s} = 29\,GeV$, Lawrence Berkeley Laboratory, LBL-23737 (1988)

[473] H. Aihara *et al.* (TPC/Two-Gamma Collaboration), Charged Hadron Inclusive Cross-Sections and Fractions in e^+e^--Annihilation at $\sqrt{s} = 29\,\mathrm{GeV}$, *Phys. Rev. Lett.* **61** (1988) 1263

[474] P.B. Cushman in *Instrumentation in High Energy Physics*, ed. F. Sauli, World Scientific, Singapore (1992)

[475] C.W. Fabjan & R. Wigmans, *Energy Measurement of Elementary Particles*, CERN-EP/89-64 (1989)

[476] J. Straver *et al. One Micron Spatial Resolution with Silicon Strip Detectors* CERN-PPE/94-26 (1994)

[477] G. Hall, Semiconductor Particle Tracking Detectors, Preprint London Imp. Coll. IC-HEP-93-12 (1993), and *Rep. Progr. Phys.* **57** (1994) 481-531

[478] R.M. Baltrusaitis *et al.*, The Utah Fly's Eye Detector, *NIM* A **240** (1985) 410

[479] ORTEC Application Note AN34, *Experiments in Nuclear Science*, (1976)

[480] A.H. Walenta, Principles and New Developments of Semiconductor Radiation Detectors, *NIM* A **253** (1987) 558

[481] R. Horisberger, *Solid State Detectors*, Lectures given at the III ICFA School on Instrumentation in Elementary Particles Physics, Rio de Janeiro, July 1990, and PSI-PR-91-38 (1991)

[482] E. Gatti *et al.* (ed.), Proc. Sixth European Symp. on Semiconductor Detectors, New Developments in Radiation Detectors, *NIM* A **326** (1993) 1

[483] S.P. Beaumont *et al.*, Gallium Arsenide Microstrip Detectors for Charged Particles, *NIM* A **321** (1992) 172

[484] S.P. Beaumont *et al.*, GaAs Solid State Particle Detectors for Particle Physics, *NIM* A **322** (1992) 472

[485] C. del Papa, P.G. Pelfer & K. Smith (eds.), *GaAs Detectors and Electronics for High Energy Physics*, World Scientific, Singapore 1992

[486] S.P. Beaumont *et al.*, GaAs Solid State Detectors for Physics at the LHC, *IEEE Trans. Nucl. Sci.* **40**, No.4 (1993) 1225

[487] Technical Measurement Corporation, *Practical Guide to Semiconductor Detectors* (1965)

[488] B.M. Schmitz, K. Farzine & H. von Buttlar, A 4π-β-Spectrometer Using Si(Li)-Detectors, *NIM* **105** (1972) 427; and K. Farzine & B.M. Schmitz, Fabrication of Si(Li)-Detectors for a 4π-β-Spectrometer, *Kerntechnik* **15**, No. 1 (1973) 27

[489] C. Grupen, *Beta-Spectroscopy with Si(Li)-detectors*, Experiment Description for the Advanced Physics Lab., Siegen University (1989)

[490] H. Ichinose *et al.*, Energy Resolution for Photons and Electrons from ^{207}Bi in LXe Doped with TEA, *NIM* A **322** (1992) 216

[491] A.H. Walenta, Strahlungsdetektoren - Neuere Entwicklungen und Anwendungen, *Phys. Blätter* **45** (1989) 352

[492] D. McCammon *et al.*, *High Resolution X-Ray Spectroscopy Using Microcalorimeters*, NASA-Preprint 88-026 (1988)

[493] F. Cardone & F. Celani, Rivelatori a Bassa Temperatura e Superconduttori per la Fisica delle Particelle di Bassa Energia, *Il Nuovo Saggiatore* **6/3** (1990) 51

[494] A. Matsumura *et al.*, High Resolution Detection of X-Rays with a Large Area $Nb/Al - Al\,O_x/Al/Nb$ Superconducting Tunnel Injection, *NIM* A **309** (1991) 350

[495] R. Turchetta, Spatial Resolution of Silicon Microstrip Detectors, *NIM* A **335** (1993) 44

[496] R. Klanner, *Silicon Detectors*, Max-Planck-Inst. München MPI-PAE/Exp. El. 135 (1984)

[497] P. Delpierre *et al.*, Development of Silicon Micropattern (Pixel) Detectors, *NIM* A **315** (1992) 133

[498] J.P. Egger *et al.*, Progress in Soft X-Ray Detection: The Case of Exotic Hydrogen, *Particle World* **3**, No. 3 (1993) 139

[499] J.L. Culhane, Position Sensitive Detectors in X-Ray Astronomy, *NIM* A **310** (1991) 1

[500] T. Ohsugi *et al.*, Radiation Damage in Silicon Microstrip Detectors, KEK Preprint 87-22 (1987), and *NIM* A **265** (1988) 105

[501] E. Fretwurst *et al.*, Radiation Hardness of Silicon Detectors for Future Colliders, *NIM* A **326** (1993) 357

[502] P. Nieminen, *A Study of the Radiation Tolerance of a Silicon Detector for Space Applications*, University of Helsinki HU-SEFT 1991-11 (1991)

[503] B. Rossi & K. Greisen, Cosmic-Ray Theory, *Rev. Mod. Phys.* **13** (1941) 240

[504] S. Iwata, *Calorimeters*, Nagoya University Preprint DPNU 13-80 (1980)

[505] S. Iwata, *Calorimeters (Total Absorption Detectors) for High Energy Experiments at Accelerators*, Nagoya University Preprint DPNU-3-79 (1979)

[506] W.R. Nelson *et al.*, *The EGS4 Code System*, SLAC-265 (1985)

[507] E. Longo & I. Sestili, Monte Carlo Calculation of Photon-Induced Electromagnetic Showers, in Lead Glass, *NIM* **128** (1975) 283

[508] C.W. Fabjan, Calorimetry in High Energy Physics, in T. Ferbel (ed.), *Proceedings on Techniques and Concepts of High Energy Physics*, Plenum, New York (1985) 281, and CERN-EP/85-54 (1985)

[509] H. Frauenfelder & E.M. Henley, *Teilchen und Kerne*, Oldenbourg, München/Wien (1979)

[510] L.D. Landau, *The Collected Papers of L.D. Landau*, Pergamon Press (1965); and A.B. Migdal, Bremsstrahlung and Pair Production in Condensed Media at High Energies, *Phys. Rev.* **103** (1956) 1811

[511] E. Konishi *et al.*, *Three Dimensional Cascade Showers in Lead Taking Account of the Landau-Pomeranchuk-Migdal Effect*, Inst. for Cosmic Rays, Tokyo, ICR Report 36-76-3 (1976)

[512] CERN-Courier, *Photon Theory Verified after 40 Years*, Vol. 34, No. 1 (1994) 12

[513] R. Becker-Szendy *et al.* SLAC-E-146 Collaboration, *Quantummechanical Suppression of Bremsstrahlung*, SLAC-PUB-6400 (1993)

[514] T. Yuda, Electron-Induced Cascade Showers in Inhomogeneous Media, *NIM* **73** (1969) 301; and Electron-Induced Cascade Showers in Lead, Copper and Aluminium, *Nuovo Cim.* **65**A (1970) 205

[515] O.C. Allkofer & C. Grupen, *Lectures on Space Physics 1*, ed. H. Pilkuhn, Bertelsmann (1973)

[516] J. Nishimura, Theory of Cascade Showers, in S. Flügge (ed.), *Handbuch der Physik*, Springer, Berlin, Band **XLVI/2** (1967) 1

[517] C.W. Fabjan & T. Ludlam, Calorimetry in High Energy Physics, CERN-EP/82-37 (1982), and *Ann. Rev. Nucl. Sci.* Vol. **32** (1982) 335

[518] R. Baumgart, *Messung und Modellierung von Elektron- und Hadron-Kaskaden in Streamerrohrkalorimetern*, Ph.D.-Thesis, University of Siegen (1987)

[519] U. Schäfer, *Untersuchungen zur Magnetfeldabhängigkeit und Pion/Elektron Unterscheidung in Elektron-Hadron Kalorimetern*, Ph.D.-Thesis, University of Siegen (1987)

[520] E. Bernardi, *On the Optimization of the Energy Resolution of Hadron Calorimeters*, Ph.D.-Thesis, University of Hamburg, DESY Int-Rep. F1-87-01 (1987)

[521] A.N. Diddens, A Detector for Neutral-Current Interactions of High Energy Neutrinos, *NIM* **178** (1980) 27

[522] R. Baumgart *et al.*, Performance Characteristics of an Electromagnetic Streamer Tube Calorimeter, *NIM* A **256** (1987) 254

[523] R. Baumgart *et al.*, Test of an Iron/Streamer Tube Calorimeter with Electrons and Pions of Energy between 1 and 100 GeV, *NIM* A **268** (1988) 105

[524] N.V. Rabin, Electron-Photon Calorimeters. Properties of Detector Materials for Calorimeters (Review), *Prib. Tekh. Éksp.* No. 6 (1992) 8 (translated and published by Plenum Publ. Corporation 1992)

[525] N.V. Rabin, Electron-Photon Calorimeters: Main Properties (Review), *Prib. Tekh. Éksp.* No. 1 (1992) 12 (translated and published by Plenum Publ. Corporation 1992)

[526] A. Baranov *et al.*, A Liquid Xenon Calorimeter for the Detection of Electromagnetic Showers, CERN-EP/90-03 (1990), and *NIM* A **294** (1990) 439

[527] V.M. Aulchenko *et al.*, Liquid Krypton Calorimeter for KEDR Detector, *NIM* A **316** (1992) 8

[528] NA48 Collaboration, The NA48 Liquid Krypton Calorimeter, *NIM* A **316** (1992) 1

[529] P. Lecoq, *Homogeneous Calorimeters at LHC/SSC*, CERN-PPE/91-231 (1991)

[530] E.B. Hughes *et al.*, Properties and Applications of Large NaI(Tl) Total Absorption Spectrometers, *IEEE Trans. Nucl. Sci. Vol.* **1-19**, No. 3 (1972) 126; and Properties of a NaI(Tl) Total Absorption Spectrometer for Electrons and γ-Rays at GeV Energies, *IEEE Trans. Nucl. Sci.* NS-**17**, No. 3 (1970) 14, and SLAC-Report Nr. 627 (1972)

[531] C.A. Heusch, *The Use of Cherenkov Techniques for Total Absorption Measurements*, CERN-EP/84-98 (1984)

[532] Y.D. Prokoshkin, *Hodoscope Calorimeters as Basic Coordinate-Energy Detectors of Particles in the Experiments in the 10 TeV-Range*, Proc. of Second ICFA Workshop, Les Diablerets, Oct. 1979

[533] D.F. Anderson *et al.*, *Lead-Fluoride: An Ultra-Compact Cherenkov Radiator for EM Calorimetry*, Fermilab-Pub. 89/189 (1989)

[534] A.A. Aseev *et al.*, $BaYb_2F_8$, a New Radiation Hard Cherenkov Radiator for Electromagnetic Calorimeters, *NIM* A **317** (1992) 143

[535] A. Kusumegi *et al.*, Thallium Formate Heavy Liquid Counter 'Helicon' as a Total Absorption Calorimeter, KEK Preprint 80-11 (1980), and *NIM* **185** (1981) 83

[536] A. Kusumegi *et al.*, *Heavy Liquid Total Absorption Counter: Helicon*, KEK Preprint 81-11 (1981)

[537] P. de Barbaro *et al.*, *Recent R&D Results on Tile/Fiber Calorimetry*, Rochester Univ. Preprint UR-1299 (1993)

[538] J. Badier *et al.*, *Test Results of an Electromagnetic Calorimeter with 0.5 mm Scintillating Fiber Readout*, CERN-PPE/93-20 (1993)

[539] P. Hale & J. Siegrist (eds.), *Calorimetry in High Energy Physics*, Proc. of the 3rd Int. Conf., Corpus Christi, USA (1992)

[540] B.M. Bleichert *et al.*, The Response of a Simple Modular Electron/Hadron Calorimeter to Electrons, *NIM* **199** (1982) 461

[541] D. Bogert *et al.*, *Hadron Showers in a Low-Density Fine-Grained Flash Chamber Calorimeter*, Fermilab-Pub 87-159 (1987); and The Operation of a Large Flash Chamber Neutrino Detector at Fermilab, *IEEE Trans. Nucl. Sci.* NS-**29**, No. 1 (1982) 363

[542] T.C. Weekes, Very High Energy Gamma-Ray Astronomy, *Phys. Rep.* **160** (1988) 1

[543] J. Ranft, Monte Carlo Calculation of Energy Deposition by Nuclear-Cascade (TANC) Counters, *NIM* **81** (1970) 29; and Estimation of Radiation Problems around High Energy Accelerators Using Calculation of the Hadronic Cascade in Matter, *Part. Acc.* **3** (1972) 129

[544] A. Baroncelli, Study of Total Absorption Counters for Very High Energy Particles, *NIM* **118** (1974) 445

[545] T.A. Gabriel & W. Schmidt, Calculated Performance of Iron-Argon and Iron-Plastic Calorimeters for Incident Hadrons with Energies of 5 to 75 GeV, *NIM* **134** (1976) 271

[546] T.A. Gabriel, Uranium Liquid Argon Calorimeters: A Calculational Investigation, *NIM* **150** (1978) 145

[547] R. Wigmans, *Energy Loss of Particles in Dense Matter: Calorimetry*, Lecture Notes, ICFA School on Instrumentation in Elementary Particle Physics Trieste 1987, NIKHEF-H 87-12 (1987)

[548] R. Wigmans, *Advances in Hadron Calorimetry*, CERN-PPE/91-39 (1991)

[549] R. Baumgart *et al.*, Electron-Pion Discrimination in an Iron/Streamer Tube Calorimeter up to 100 GeV, *NIM* **272** (1988) 722

[550] T. Akesson *et al.*, Properties of a Fine-Sampling Uranium-Copper Scintillator Hadron Calorimeter, *NIM* **134** (1985) 17

[551] B. Aubert *et al.* (WALIC-Collaboration), Studies of Compensation of Fe/TMP and Pb/TMP Sampling Calorimeters, *NIM* A **334** (1993) 383

[552] E. Borchi *et al.* (SICAPO Collaboration), Electromagnetic Shower Energy Filtering Effect. A Way to Achieve the Compensation Condition $(e/\pi = 1)$ in Hadronic Calorimetry, *Phys. Lett.* B **222** (1989) 525

[553] E. Borchi *et al.* (SICAPO Collaboration), Systematic Investigation of the Electromagnetic Filtering Effect as a Tool for Achieving the Compensation Condition in Silicon Hadron Calorimetry, *NIM* A **332** (1993) 85

[554] M. Holder *et al.*, A Detector for High Energy Neutrino Interactions, *NIM* **148** (1978) 235

[555] M. Holder *et al.*, Performance of a Magnetized Total Absorption Calorimeter Between 15 GeV and 140 GeV, *NIM* **151** (1978) 69

[556] D.L. Cheshire *et al.*, Measurements on the Development of Cascades in a Tungsten-Scintillator Ionization Spectrometer, *NIM* **126** (1975) 253

[557] D.L. Cheshire *et al.*, Inelastic Interaction Mean Free Path of Negative Pions in Tungsten, *Phys. Rev.* D **12** (1975) 2587

[558] A. Grant, A Monte Carlo Calculation of High Energy Hadronic Cascades in Iron, *NIM* **131** (1975) 167

[559] B. Friend *et al.*, Measurements of Energy Flow Distributions of 10 GeV/c Hadronic Showers in Iron and Aluminium, *NIM* **136** (1976) 505

[560] J.K. Walker, *Neutrino Detector Developments*, Fermilab. Conf. 78/58-Exp. (1978)

[561] F.E. Taylor *et al.*, *A Fine Grain Flash Chamber Calorimeter*, Fermilab-Conf. 77/100-Exp (1977)

[562] M. Aalste *et al.*, Measurement of Hadron Shower Punch-Through in Iron, *Z. f. Phys.* C **60** (1993) 1

[563] O. Botner, New Ideas in Calorimetry, *Phys. Scripta* **23** (1981) 556

[564] S. Denisov *et al.*, A Fine Grain Gas Ionization Calorimeter, *NIM* A **335** (1993) 106

[565] B. Aubert *et al.*, Performance of a Liquid Argon Accordion Calorimeter with Fast Readout, *NIM* A **321** (1992) 467; and *Performance of a Liquid Argon Electromagnetic Calorimeter with a Cylindric Accordion Geometry*, CERN-PPE/92-129 (1992)

[566] B. Aubert *et al.*, Performance of a Liquid Argon Electromagnetic Calorimeter with a Cylindrical Accordion Geometry, *NIM* A **325** (1993) 116

[567] P. Baillon, *Detection of Atmospheric Cascades at Ground Level*, CERN-PPE/91-012 (1991)

[568] S. Barwick *et al.*, Neutrino Astronomy on the 1 km^2 Scale, *J. Phys. G.* **18** (1992) 225

[569] Y. Totsuka, Neutrino Astronomy, *Rep. Progr. Phys.* **55** No. 3 (1992) 377

[570] Chr. Spiering, Neutrinoastronomie mit Unterwasserteleskopen, *Phys. Bl.* **49**, No. 10 (1993) 871

[571] R. Baumgart *et al.*, Interaction of 200 GeV Muons in an Electromagnetic Streamer Tube Calorimeter, *NIM* A **158** (1987) 51

[572] C. Zupancic, *Physical and Statistical Foundations of TeV Muon Spectroscopy*, CERN-EP/85-144 (1985)

[573] M.J. Tannenbaum, *Comparison of Two Formulas for Muon Bremsstrahlung*, CERN-PPE/91-134 (1991)

[574] L. Cifarelli, R. Wigmans & T. Ypsilantis (eds.), *Perspectives for New Detectors in Future Supercolliders*, World Scientific, Singapore (1989)

[575] S. Cooper *et al.*, *Cryogenic Detector Development*, Max-Planck-Inst. München MPI-PhE/91-07 (1991)

[576] G. Gerbier, *Dark Matter: An Overview of Direct Searches*, CEN Saclay, DPhPE 91-13 (1991)

[577] M. Spiro, *Calorimeters for Astroparticle Physics*, Saclay Report DPhPE 91-17 (1991)

[578] P.F. Smith & J.D. Lewin, Dark Matter Detection, *Phys. Rep.* **187**, No. 5 (1990) 203

[579] O. Fackler & J. Tran Thanh Van (eds.), *Proceedings of the 6th Moriond Workshop of the 21st Recontre de Moriond on 'Massive Neutrinos in Astrophysics and in Particle Physics'* (1986)

[580] P. Belli *et al.*, Liquid Xenon Detectors for Dark Matter Experiments, *NIM* A **316** (1992) 55

[581] G. Forster *et al.*, Calorimetric Particle Detectors with Superconducting Absorber Materials, *NIM* A **324** (1993) 491

[582] E. Fiorini, Underground Cryogenic Detectors, *Europhys. News* **23** (1992) 207

[583] W. Seidel, Cryogenic Detectors for Dark Matter Searches, *Ann. N. Y. Acad. Sci.* Vol. **688** (1992) 632

[584]   W. Seidel, Thermal Detectors for Underground Physics, *Nucl. Phys.* B
        (Proc. Suppl.) **32** (1993) 138

[585]   L. Gonzales-Mestres & D. Perret-Gallix (eds.), *Low Temperature
        Detectors for Neutrinos and Dark Matter II*, Edition Frontières 1988

[586]   L. Brogiato, D.V. Camin & E. Fiorini (eds.), *Low Temperature
        Detectors for Neutrinos and Dark Matter III*, Edition Frontières 1990

[587]   N.E. Booth & G.L. Salmon (eds.), *Low Temperature Detectors for
        Neutrinos and Dark Matter IV*, Edition Frontières 1992

[588]   K. Pretzl, N. Schmitz & L. Stodolsky, *Low Temperature Detectors for
        Neutrinos and Dark Matter*, Springer, Berlin, Heidelberg 1987

[589]   J.R. Primack, D. Seckel & B. Sadoulet, Detection of Cosmic Dark
        Matter, *Ann. Rev. Nucl. Part. Sci.* **38** (1988) 751

[590]   K.P. Pretzl, Superconducting Granule Detectors, *Particle World* **1**
        (1990) 153

[591]   V.N. Trofimov, *SQUIDs in Thermal Detectors of Weakly Interacting
        Particles*, Dubna-Preprint E8-91-67 (1991)

[592]   C. Kittel, *Einführung in die Festkörperphysik*, Oldenbourg,
        München/Wien (1980)

[593]   K.H. Hellwege, *Einführung in die Festkörperphysik*, Springer,
        Berlin/Heidelberg/New York (1976)

[594]   A. Allessandrello *et al.*, A Thermal High Resolution Alpha and
        Gamma-Ray Spectrometer, *NIM* A **320** (1992) 388

[595]   D. Yvon *et al.*, *Bolometer Development, with Simultaneous
        Measurement of Heat and Ionization Signals*, Saclay-Preprint
        CEN-DAPNIA-SPP 93-11 (1993)

[596]   M. Frank *et al.*, Study of Single Superconducting Grains for a Neutrino
        and Dark Matter Detector, *NIM* A **287** (1990) 583

[597]   CERN-Courier, *Workshop: Low Temperature Devices*, Vol 27, No. 5
        (1987) 12

[598]   G. Czapek *et al.*, Superheated Superconducting Granule Device:
        Detection of Minimum Ionizing Particles, *NIM* A **306** (1991) 572

[599]   A. Gabutti *et al.*, A Fast, Self-Recovering Superconducting Strip Particle
        Detector Made With Granular Tungsten, *NIM* A **312** (1992) 475

[600]   R.L. Glückstern, Uncertainties in Track Momentum and Direction due
        to Multiple Scattering and Measurement Errors, *NIM* **24** (1963) 381

[601]   G.T. Ewan, J.S. Geiger & R.L. Graham, A One-Meter-Radius Iron-Free
        Double-Focussing $\pi\sqrt{2}$ Spectrometer for β-Ray Spectroscopy with a
        Precision of 1:10^5, *NIM* **9** (1960) 245

[602]   K. Siegbahn & K. Edvarson, β-Ray Spectroscopy in the Precision Range
        of 1:10^5, *Nucl. Phys.* **1** (1956) 137

[603] B. Renk, *Meßdatenerfassung in der Kern- und Teilchenphysik*, B.G. Teubner, Stuttgart (1993)

[604] U. Tietze & Ch. Schenk, *Halbleiterschaltungstechnik*, Springer-Verlag, Berlin (1993)

[605] H1 Collaboration, *The H1 Detector at HERA*, DESY 93-103 (1993)

[606] W.J. Haynes, *The Data Acquisition System for the HERA H1 Experiment*, Rutherford Appleton Laboratory, UK, RAL 90-039 (1990)

[607] J.V. Allaby, *CERN School of Computing*, C. Verkerk (ed.), CERN 88-03 (1988) 240

[608] V. Blobel, *The BOS System, Dynamic Memory Management*, DESY R1-88-01, January 1988.

[609] R. Brun, M. Goossens & J. Zoll, *ZEBRA Dynamic Data Structure and Memory Manager*, CERN Program Library Office, Q100 (1992)

[610] R.K. Böck & J. Zoll, *HYDRA*, CERN, D.PH.II, PROG 74-4(1974)

[611] M. Pimiä, *Track Finding in the UA1 Central Detector at the CERN p$\bar{p}$ Collider*, Univ. of Helsinki, HU-P-D45 (1985); K. Karimaki, *Formulae for the UA1 Track Finding Algorithm*, UA1-TN 84/31 (1984)

[612] W.T. Eadie, D. Drijard, F. James, M. Roos & B. Sadoulet, *Statistical Methods in Experimental Physics*, North-Holland (1971)

[613] J. Hertz, A. Krogh & R.G. Palmer, *Introduction to the Theory of Neural Computation*, Santa Fe Institute, Addison-Wesley (1991)

[614] R. Rojas, *Theorie der neuronalen Netze*, Springer (1993)

[615] C. Peterson, *Neural Networks in High Energy Physics*, LU TP 92-23 (1992); C. Peterson, Track Finding with Neural Networks, *NIM* A **279** (1989) 537; G. Stimpfl-Abele & L. Garrido, Fast Track Finding with Neural Nets, *Comp. Phys. Comm.* **64** (1991) 46

[616] H.F. Teykal, *Elektron- und Pionidentifikation in einem kombinierten Uran-TMP- und Eisen-Szintillator-Kalorimeter*, RWTH Aachen, PITHA 92/28 (1992)

[617] B. Rensch, *Produktion der neutralen seltsamen Teilchen K_s und Λ^0 in hadronischen Z-Zerfällen am LEP-Speicherring*, Univ. Heidelberg, HD-IHEP 92-09 (1992)

[618] B. Adeva *et al.*, (L3-Coll.), The Construction of the L3 Experiment, *NIM* A **289** (1990) 35

[619] G. Altarelli, R. Kleiss & C. Verzegnassi, *Z Physics at LEP 1, Vol. 3: Event Generators and Software*, CERN 89-08 (1989).

[620] S. Bethke, *Hadronic Physics in Electron-Positron Annihilation*, Univ. Heidelberg, HD-PY 93/07 (1993)

[621] Particle Data Group, Review of Particle Properties, *Phys. Rev.* D **50** (1994) 1173

[622] F. James, *Monte Carlo Theory and Practice*, in T. Ferbel, Experimental Techniques in High Energy Physics, Addison-Wesley (1987), p. 627; F. James, A Review of Pseudo Random Number Generators, *Comp. Phys. Comm.* **60** (1990) 329

[623] F. Anselmo, *et al.*, Event Generators for LHC, in G. Jarlskog & D. Rein (eds.), *Large Hadron Collider Workshop Aachen*, CERN 90-10/ECFA 90-133, Vol. 2, (1990) p. 130

[624] W. Buchmüller & G. Ingelmen, *Physics at HERA, Vol. 3: Event generators*, DESY (1991)

[625] H. Plothow-Besch, PDFLIB: A Library of all Available Parton Density Functions of the Nucleon, the Pion and the Photon and the Corresponding α_s Calculations, *Comp. Phys. Comm.* **75** (1993) 396; W. Buchmüller & G. Ingelmen, *Physics at HERA, Vol. 1: Structure Functions - Hadronic Final State - Proton Spin - Photo Production*, DESY (1991)

[626] Application Software Group, CN-Division, *GEANT3 User's Guide*, CERN (1994) W5013

[627] R. Brun & F. Carminati, Detector Simulation and Software Tools, in G. Jarlskog & D. Rein (eds.), *Large Hadron Collider Workshop-Aachen*, CERN 90-10/ECFA 90-133, Vol. 1, (1990) p. 325

[628] W.R. Nelson, H. Hirayama & D.W.O. Rogers, *The EGS Code System*, SLAC-255, UC-32 (1985)

[629] H. Fesefeld, *GHEISHA*, RWTH Aachen, PITHA 86/05 (1986).

[630] K. Hanssgen & J. Ranft, HADRIN, NUCRIN, *Comp. Phys. Comm.* **39** (1986) 53; P. Aarnio *et al.*, *FLUKA: Hadronic Benchmarks and Applications*, CERN TIS 93-08 (1993)

[631] R. Bionta *et al.*, Observation on a Neutrino Burst in Coincidence with Supernova SN 1987 A in the Large Magellanic Cloud, *Phys. Rev. Lett.* **58**, No. 14 (1987) 1494

[632] S. Brandt & H.D. Dahmen, Axes and Scalar Measures of Two-Jet and Three-Jet Events, *Z. Phys.* C **1** (1979) 61

[633] D. Buskulic, *et al.*, Production of K_s and Λ^0 in Hadronic Z^0 Decays, CERN PPE 94-74 (1994), and *Z. Phys.* C **64** (1994) 361

[634] R. Brun, O. Couet, C. Vandoni & P. Zanarini, *PAW Physics Analysis Workstation*, CERN Program Library Office, Q121 (1993)

[635] R. Brun *et al.*, *HBOOK User's Guide*, CERN Program Library Office, Y250 (1994)

[636] F. James & M. Roos, *MINUIT User's Guide*, CERN Program Library Office, D506 (1992)

[637] N.A. Dyson, *Nuclear Physics with Application in Medicine and Biology*, John Wiley & Sons Inc. (Wiley Interscience), New York (1981), and

Radiation Physics with Applications in Medicine and Biology, Ellis Horwood, New York (1993)

[638] F. Sauli, Applications of Gaseous Detectors in Astrophysics, Medicine and Biology, CERN-PPE/92-047, and *NIM* A **323** (1992) 1

[639] K. Kleinknecht & T.D. Lee (eds.), *Particles and Detectors; Festschrift for Jack Steinberger*, Springer Tracts in Modern Physics, Berlin/Heidelberg Vol. 108 (1986)

[640] D.J. Miller, Particle Physics and its Detectors, *NIM* A **310** (1991) 35

[641] G. Hall, Modern Charged Particle Detectors, *Contemp. Phys.* **33** (1992) 1

[642] H.O. Anger, Scintillation Camera with 11-Inch Crystal, *J. Nucl. Med.* **5** (1964) 515

[643] G. Montgomery, *The Mind in Motion*, Discover (1989) 58

[644] V.J. Stenger, *Physics and Psychics*, Prometheus, Buffalo, New York (1990)

[645] S.B. Curtis & M.R. Raju, A Calculation of the Physical Characteristics of Negative Pion Beams Energy-Loss Distribution and Bragg Curves, *Radiation Research* **34** (1968) 239

[646] G.B. Goodman, *Pion Therapy for Cancer — What are the Prospects*, TRIUMF-Preprint TRI-PP-92-134 (1992)

[647] A.J. Lennox, *Hospital-Based Proton Linear Accelerator for Particle Therapy and Radioisotope Production*, Fermilab-Pub. 90/217 (1990)

[648] G. Kraft, *Schwerionenstrahlen in Biophysik und Medizin*, Arbeitsgemeinschaft der Großforschungsanlagen, AGF-Forschungsthemen 6 (1992); G. Kraft, Heavy-Ion Therapy at GSI, *Europhys. News* **25** (1994) 81; Th. Haberer *et al.*, *Magnetic Scanning System for Heavy-Ion Therapy*, GSI-Preprint GSI-93-15 (1993)

[649] U. Braun, *Messung der Radioaktivitätskonzentration in biologischen Objekten nach dem Reaktorunfall in Tschnernobyl und ein Versuch einer Interpretation ihrer Folgen*, Diploma-Thesis, University of Siegen (1988)

[650] C. Grupen *et al.*, *Nuklid-Analyse von Beta-Strahlern mit Halbleiterspektrometern im Fallout*, Symp. *Strahlenmessung und Dosimetrie*, Regensburg (1966) 670

[651] L. Alvarez *et al.* Search for Hidden Chambers in the Pyramids, *Science* **167** (1970) 832

[652] F. El Bedewi *et al.*, Energy Spectrum and Angular Distribution of Cosmic Ray Muons in the Range $50 - 70\,\text{GeV}$, *J. Phys.* A **5** (1972) 292

[653] G. Danby, J.M. Gaillard, K. Goulianos, L.M. Lederman, N. Mistry, M. Schwarz & J. Steinberger, Observation of High Energy Neutrino Reactions and the Existence of Two Kinds of Neutrinos, *Phys. Rev. Lett.* **9** (1962) 36

[654]   CERN-Courier, *Oscillating Neutrinos*, Vol. 20, No. 5 (1980) 189

[655]   H. Faissner, *The Spark Chamber Neutrino Experiment at CERN*, CERN-Report 63-37 (1963) 43

[656]   R. Hillier, *Gamma Ray Astronomy*, Clarendon Press, Oxford (1984)

[657]   P.V. Ramana Murthy & A.W. Wolfendale, *Gamma Ray Astronomy*, Cambridge University Press (1986)

[658]   G.F. Bignami *et al.*, The COS-B Experiment for Gamma-Ray Astronomy, *Space Sci. Instr.* **1** (1975) 245

[659]   P. Léna, *Astrophysique. Méthodes Physiques de L'Observation*, © InterEditions, Paris (1986), and *Observational Astrophysics*, Springer (1988)

[660]   Photo MBB-GmbH, *COS-B, Satellit zur Erforschung der kosmischen Gammastrahlung*, Unternehmensbereich Raumfahrt, München (1975)

[661]   C.E. Fichtel *et al.*, SAS-2 Observations of Diffuse Gamma Radiation in the Galactic Latitude Interval $10° < |b| \leq 90°$, *Astrophys. J. Lett. Ed.* **217**, No. 1, p. L9 (1977), and *Proc. 12th ESLAB Symp.*, Frascati (1977) 95

[662]   J. Linsley, The Highest Energy Cosmic Rays, *Scientific American* **239**, No. 1 (1978) 48

[663]   J. Boone *et al.*, *Observations of the Crab Pulsar near 10^{15}–10^{16} eV*, 18th Int. Cosmic Ray Conf. Bangalore, India, Vol. 9 (1983) 57, and University of Utah, UU-HEP 84/3 (1984)

[664]   G.L. Cassiday *et al.*, Cosmic Rays and Particle Physics, in T.K. Gaisser (ed.), *Am. Inst. Phys.* **49** (1978) 417

[665]   C. Grupen, Kosmische Strahlung, *Physik in unserer Zeit* **16** (1985) 69

[666]   G.L. Cassiday, private communication 1985

[667]   W. Stolz, *Radioaktivität*, Hanser, München/Wien (1990)

[668]   M.A. Geyh & H. Schleicher, *Absolute Age Determination*, Springer; Berlin, Heidelberg (1990)

[669]   ALEPH Collaboration, Alignment of the ALEPH Tracking Devices, *NIM* A **323** (1992) 213

[670]   S.W. Hawking, *Is the End in Sight for Theoretical Physics? — An Inaugural Lecture*, Press Syndicate of the University of Cambridge (1980)

Index